150581464 1

AF606522

Charged Particle Tracks in Solids and Liquids

L. H. Gray

Reading from left to right. **Front row:** Dr T. E. Burlin; Prof. J. R. Greening; Dr S. J. Wyard; Dr A. A. Revina; Prof. I. Adamczewski; Prof. U. Fano; Dr A. K. Pikaev; Prof. N. A. Bach; Dr T. Alper; Prof. P. H. Fowler; Prof. R. L. Platzman; Prof. Kh. S. Bagdasar'yan; Mrs J. Green; Prof. J. W. Boag; Prof. J.F. Fowler; Dr G. E. Adams; Prof. J. Kistemaker; Mr A. Hutton; Dr A. Hummel. **Second row:** Dr J. Bednar; Dr E. J. Kobetich; Prof. R. Katz; Dr M. Ebert; Dr M. Zielczynski; Dr Alma Howard; Dr E. M. Fielden; Dr R. K. Broszkiewicz; Dr G. J. Neary; Dr D. V. Morgan; Dr F. M. Russell; Dr E. W. Emery; Dr D. G. Dearnaley; Dr J. K. Thomas; Dr D. K. Bewley; Dr. M. Marshall. **Third row:** Dr A. P. Sheinker; Dr R. E. Ellis; Dr S. B. Curtis; Dr W. G. Burns; Dr B. Cercek; Dr G. R. Freeman; Dr A. P. Shvedchikov; Dr A. Allisy; Prof. A. G. Tenner; Dr H. Jung; Dr P. R. J. Burch; Dr A. Ore; Dr H. J. Delafield; Dr P. J. Campion; Dr G. J. Clark; Mr J. M. Poate; Prof. H. A. B. Simons; Mr J. A. B. Gibson. **Fourth row:** Mr D. H. Peirson; Dr A. S. Chernyshova; Mr R. M. Fry; Miss R. J. Snelling; Miss S. Piermattei; Dr Yu. D. Tsvetkov; Dr B. G. Ershov; Dr R. Köepp; Dr J. Booz; Dr R. P. O. Hüber; Mr D. E. Watt; ——; Mr M. Säbel; ——; ——; Mr D. A. Sykes; Dr A. C. James. **Back row:** Dr V. A. Zhigunov; Interpreter; Dr D. A. Kritskaya; Dr A. V. Pavlov; Dr A. M. Kellerer; Dr J. S. Bevan; Dr H. B. Maccabee; Dr A. N. Ponomarev; Mr P. D. Holt; Mr J. L. Haybittle; Mrs K. Josefowicz; Mr M. Monnin; Dr W. A. Jennings; ——; Mr E. Rotondi; Dr R. Schiller; Dr V. I. Gusynin; Dr A. V. Vannikov.

Charged Particle Tracks in Solids and Liquids

Proceedings of the Second L. H. Gray Conference organized by the L. H. Gray Memorial Trust held at Trinity College, Cambridge, April 1969

Published by The Institute of Physics and The Physical Society
Conference Series No. 8

The Second L. H. Gray Conference
held at Trinity College, Cambridge
on 8–11th April 1969

Editors
Dr G E Adams, Dr D K Bewley
Professor J W Boag

ISBNO 85498 008 3

Published by The Institute of Physics and The Physical Society, 47 Belgrave Square, London SW1.
Editorial Office 1 Lowther Gardens,
Prince Consort Road, London SW7.

Printed in Great Britain by
Adlard & Son Limited, Bartholomew Press, Dorking

Contents

Preface

The L. H. Gray Memorial Trust

Dr L. H. Gray (see Frontispiece) received his training in research at the Cavendish Laboratory in Cambridge during the exciting period around 1930 when so many of the basic discoveries were being made, on which modern atomic and nuclear theory are based. Gray's contribution was a careful study of the absorption of penetrating gamma radiation and the establishment of an accurate method of measuring the energy absorbed in a material by means of the ionization produced in a gas-filled cavity. On leaving Cambridge he chose to apply these studies to the field of medicine where X- and gamma-radiations were also being used, somewhat empirically, for the treatment of cancer. The rest of his scientific career was spent in attacking problems in this field. To begin with, he showed how his studies on cavity ionization could be applied to the measurement of the radiation beams which were used clinically. Having thus placed radiation dosimetry on a firm basis, he began quantitative studies on the way radiation damages cells and tissues and was one of the small number of pioneers who established a leading place for this country in the new field of radiation biology. His approach to the subject was invariably quantitative and generally simple—he had the vision to see where advances could be made by simple techniques. In spite of the great extension of the subject during and after the second world war and the increasing demands made on him for advice and consultation, his own small laboratory retained a leading place in the field and his personal research retained its characteristic depth and originality.

After Dr Gray's death on 9th July 1965, suggestions were made from many sides that some memorial should be set up to the memory of the man and his work. Three scientific societies with which he had been closely associated—The British Institute of Radiology, The Association for Radiation Research, and The Hospital Physicists' Association—set up a joint committee which recommended that a Memorial Trust be established for the purpose of organizing scientific conferences to be called 'L. H. Gray Conferences'. This was done, and the Trust Deed lays down that these conferences shall provide opportunities for informal discussion on some theme selected for its topical interest from the fields of Radiation Physics, Radiation Chemistry, Radiation Biology, and the application of these fundamental studies to the treatment of disease.

The first L. H. Gray Conference was held at The British Empire Cancer Campaign's Radiobiological Research Unit, situated at Mount Vernon Hospital, Northwood, in September 1967, on the general themes 'radiation sensitizers' and 'dose fractionation'. It brought together some 90 radiation chemists, radiobiologists and radiotherapists in a discussion of the ways in which advances in fundamental knowledge of radiation action could best be applied to clinical treatment. The proceedings of this first meeting were not published.

The 2nd Conference, held at Cambridge in April 1969 is fully reported in the present volume. The 3rd Conference is planned for January 1971 on the topic 'Radiation Effects and the Mitotic Cycle'. It will be held at the Paterson Laboratories, Christie Hospital, Manchester.

Foreword

The distribution of energy along the tracks of charged particles has long been recognised as a fundamental problem in the interpretation of the chemical and biological effects of ionizing radiation. It is well known that this energy is deposited in the first place in a relatively small number of disturbed molecules lying close to the paths of the secondary charged particles. The breakdown of these disturbed molecules and the subsequent reactions of their fragments depend to a greater or lesser extent upon the initial geometry of the 'track'. The word 'track' is used here to denote the nature and initial distribution of the molecular disturbances—it is this *discrete initial* structure of the track which is of such great interest in any attempt to distinguish between the chemical and biological effects of particles of different charge and velocity.

For many years notions of track structure have been based on evidence from (i) the distribution of droplets, clusters of droplets, and short delta tracks seen in cloud-chamber photographs of beta- or alpha-particle tracks in a gas (much of this data is more than 30 years old); and (ii) the Bethe theory of energy loss by a charged particle. Although both these sources of evidence are still extremely relevant and each can be refined and improved today, neither is entirely adequate. On the one hand, formation of droplets is only a very rough indication of the major energy losses (probably ionizations) along a track and the position of the droplets in a photograph is determined not only by the site of formation of the ions but also by their subsequent diffusion. On the other hand, the Bethe theory focuses attention on the energy losses by the *charged particles* averaged over all types of interaction and the final stopping-power formula says nothing about the nature or radial spacing of the separate molecular events along the track.

The object of this conference was to examine all the experimental and theoretical evidence available today which is relevant to track structure in liquids and solids. Much of this information comes from fields of research with little direct contact with radiation chemistry or radiation biology—fields such as solid-state physics, molecular spectroscopy, bubble-chamber studies, electron spin resonance, and so on. For this reason workers from several of these fields were invited to review the relevant information from their own

particular area of research. Those who attended the meeting found these presentations from related fields of the greatest value.
Although this conference did not prepare a new blueprint of track structure, it reviewed the difficulties, both theoretical and experimental, and suggestions for some new approaches arose in the discussion. This collection of papers is likely to remain a useful reference manual for research workers for a considerable time and the method of bringing together a group of people working in a variety of different fields to explore one limited topic of common interest is one which may be more widely adopted as large conferences become increasingly unwieldy.

The L. H. Gray Trust gratefully acknowledges the financial assistance received from the Royal Society, the British Council, and the UK Panel on Gamma and Electron Irradiation towards the expenses of the conference and a contribution from the Royal Society towards the publication. The Trust is also indebted to The Institute of Physics and The Physical Society for undertaking publication.

The formulation of track structure theory[†‡]

U. FANO

Department of Physics, University of Chicago, Chicago, Illinois, USA

Abstract. This brief review concerns the early phases of track structure development. It summarizes current theory and analyses its basic limitations. It also indicates possible lines of development and outlines recent fragmentary progress, particularly with regard to the connection between deflections and impact parameters in collisions along the track.

1. Introduction

The theory of track structure suffers not only from lack of adequate input data but also—and logically foremost—from lack of concepts appropriate to its realistic formulation. Some such concepts can now be discerned but a detailed formulation of track structure theory and the further quantitative evaluation of its parameters are still very remote.

The structure of a track should be described for most purposes in terms of the positions of identifiable events *localized* within an atom or, at least, within a small molecular group. Such an event consists, in essence, of the formation of a new chemical species, e.g., of a radical, of an ion, or of an excitation with a sufficiently long lifetime (say, $>10^{-10}$ s). That is, these events should be of such a nature that one may properly regard them as the identifiable starting points of chemical (or biological) processes. We shall refer to them in this paper as 'localized events', with the understanding that they may be of various kinds (unspecified here but to be specified eventually) and that they mark the end of the physical stage of radiation action and the beginning of a chemical stage.

A long, complex chain of physical processes may—but need not—intervene between the primary action of a fast charged particle and the occurrence of 'localized events'. This point was stressed by Platzman at this conference and reviewed extensively at the Cortina Congress (Silini 1967). In particular, the primary action need not be localized since it may consist of a plasma-type wave distributed over millions of atoms. Also the primary action generally transfers to the electrons of a material, energy quanta much larger than the activation requirements of localized events (amounting, e.g., to ~ 20 eV instead of ~ 2 eV); therefore an energy degradation process often precedes the localized events. The reasonable objective of a track structure theory is to describe and predict the spatial distribution of localized events with a minimum of detailed analysis and of assumptions concerning preliminary physical processes.

This predictive description should be represented by the joint probability of a number of localized events, e.g., of the probability that an event of type A_1 occur at a point P_1, that A_2 occur at P_2, etc. This joint probability constitutes a correlation function, and constitutes the legitimate input data for the study of further stages of radiation action, e.g., for the diffusion and reaction kinetics of chemical action.

† Supported in part by the US Atomic Energy Commission Contract No. COO-1674-18.

‡ This paper consists of material assembled for oral presentation and should not be regarded as a report of completed research. It is submitted here only as a record of the Conference.

Calculation of this correlation function requires, of course, two kinds of input data: on the one hand, the charge, speed and mass of the incident particles, and, on the other hand, parameters representing the *relevant* structural and dynamical properties of the material traversed by them. We have been familiar with parameters such as the oscillator strengths relevant to the production of excitations or other events localized within *isolated* atoms or small molecules. These may be adequate for the description of track structure in low density gases. However, localized events in condensed matter occur within atomic groups *embedded* in a material. Parameters relevant to their production depend on the structure of the whole material through which the radiation action may percolate. A study of track structure should first identify such parameters and then evaluate them.

2. The existing limited theory

To illustrate the need for new concepts and analytical approaches, let us outline first the partial theory of track structure which has existed for a long time (Williams 1933, Neufeld 1953) and point out its limitations. Williams justified the use of the classical concept of trajectory for a fast particle incident on an atom or molecule provided the trajectory passes at a distance b ('impact parameter') from the target much larger than the orbit of the atomic electron to be excited. Under the circumstances the momentum transferred to each atomic electron is (Bohr 1913)

$$q = F\tau = (ze^2/b^2)\tau = 2ze^2/bv \tag{1}$$

where ze and v are the charge and velocity of the incident particle and the effective collision time $\tau = 2b/v$ is utilized. A useful parameter related to this momentum transfer is the corresponding 'recoil energy' of an atomic electron with mass m, assumed to be free and initially at rest,

$$Q = q^2/2m = 2z^2e^4/mv^2b^2. \tag{2}$$

The differential cross section for this 'energy transfer' is

$$\mathrm{d}\sigma_Q = 2\pi b\ \mathrm{d}b = 2\pi b^2\ \mathrm{d}(\ln b) = (4\pi z^2e^4/mv^2Q)\ \mathrm{d}(\ln b). \tag{3}$$

Bohr also pointed out that, irrespective of any details of a future quantum theory, the probability of atomic excitation to a level n would be the product of $\mathrm{d}\sigma_Q$ and of a factor $P_n(Q)$ which is proportional to Q for low Q. Hence

$$\mathrm{d}\sigma_n = [4\pi z^2e^4/mv^2][P_n(Q)/Q]\ \mathrm{d}(\ln b) \tag{4}$$

is distributed uniformly as a function of the distance-from-trajectory parameter ln b, in so far as $P_n(Q)/Q$ is constant and other limitations of the model are fulfilled. The probabilities $P_n(Q)$ constitute the set of relevant parameters of the material for this theory.

The quantum-mechanical theory of collisions (Bethe 1930) does not utilize the concept of incident particle trajectory, but properly regards the momentum transfer $\mathbf{q}$ as a well-defined observable in an isolated collision. Through a different path it yields a formula analytically equivalent to (4), namely,

$$\mathrm{d}\sigma_n = [2\pi ze^4/mv^2][P_n(Q)/Q]\ \mathrm{d}(\ln Q). \tag{5}$$

However, Q as introduced by Bethe is only indirectly related to b so that (5) cannot be interpreted as a track-structure formula. A path for recapturing this interpretation will be indicated in section 4. The Bethe theory goes beyond the earlier one by providing a quantitative expression for $P_n(Q)$, namely,

$$P_n(Q) = |F_n(Q)|^2, \quad F_n(Q) = \left(n\middle|\sum_j \exp(i\boldsymbol{q}.\boldsymbol{r}_j/\hbar)\middle|0\right), \tag{6}$$

where $\boldsymbol{r}_j$ indicates the position of the jth electron of the target and where the value of $F_n(Q)$ as a matrix element can be calculated if the target's wave functions are known.

The following limitations of the track structure formula (4) are apparent:

(i) Restriction to low-Q collisions, which excludes a major part of energy dissipation, though this part is presumably dissipated very near to the trajectory.

(ii) Lack of explicit treatment of δ-ray effects and of other secondaries. Application of the same theory to δ-rays as track branches is limited by the low velocity of secondaries and by their very frequent deflections.

(iii) Assumption that each 'localized event' is produced directly and can be identified as the 'excited state *n*' of a specific atom or molecule.

(iv) Disregard of propagation in space of physical actions prior to the occurrence of a localized event, e.g., propagation of intermediate 'plasmon' excitations; this limitation is, of course, implied by (iii).

Whereas rectifying (i) and (ii) might appear a matter of refinement, overcoming the limitations (iii) and (iv) clearly requires the introduction of new points of view in the theory of charged particle effects. On the other hand, the limitation (ii) might be removed automatically if (iv) is overcome, in so far as the δ-rays themselves may be regarded as a mechanism of propagation of intermediate effects.

3. Possible lines of advance

The preceding analysis suggests the following main developments as prerequisites for a realistic theory of track structure:

(a) A procedure that encompasses complementary models of excitation of matter, i.e., non-localized collective excitations of semimacroscopic aggregates of atoms as well as localized excitations or transformations of small molecules or parts of molecules. The procedure should also include some parametric representation of the influence of degradation of an initial energy-rich excitation upon the location of its eventual effects. A prototype procedure that encompasses two complementary descriptions of a molecule has been developed for the treatment of a quite different phenomenon, namely the microwave absorption of ammonia (Ben-Reuven 1965); its conceptual approach can presumably be applied to a broad class of problems including the track structure.

(b) A procedure that regards the 'final state *n*' of the Bethe collision theory as representing not just the relevant 'localized event' but also side effects of the degradation of energy. This procedure should be developed utilizing not only (a) but also adaptation and extension of existing procedures of the theory of many-body systems, some of which are indicated below.

(c) A procedure for treating the sequence of collision processes along a particle track. This step only involves the application and adaptation of existing methods of quantum mechanics and has already been carried out to a limited extent, as indicated below.

4. Fragmentary extensions

We give here some indication of theoretical steps that have been carried out in the past and should presumably be utilized in a theory of track structure.

(A) Calculations of stopping power involve the probability of definite energy losses $\hbar\omega$ rather than of excitation of specific states n. Accordingly one replaces (4) or (5) by

$$d\sigma_\omega = \sum_n d\sigma_n \delta(\hbar\omega - E_n) = d\sigma_Q \sum_n P_n(Q)\delta(\hbar\omega - E_n) \tag{7}$$

where the Dirac function $\delta(\hbar\omega - E_n)$ enforces the equality of $\hbar\omega$ and E_n. Examination of the $\Sigma_n P_n(Q)\delta(\hbar\omega - E_n)$ with the expression (6) of P_n has shown it to be proportional to $\operatorname{Im}[-1/\epsilon(\omega, q)]$, namely, to a macroscopic dielectric property of the material traversed, which is defined independently of any detailed model of its structure and can be determined in a variety of ways (Fano 1956). For the track problem this treatment should be modified by restricting the Σ_n to final states that include a relevant 'localized event'. The newly defined quantity should partake of the macroscopic character of the dielectric constant.

(B) As noted following equation (5), the momentum transfer $\boldsymbol{q}$ considered in Bethe's quantum-mechanical theory is only indirectly related to the impact parameter b of the semiclassical theory. This relationship consists in essence of a Fourier transformation which can be carried out by a standard procedure of wave mechanics and provides an explicit connection between the semiclassical and quantum-mechanical theories of track structure.

Consider, for purposes of introduction, the following problem of wave mechanics. A particle, in one-dimensional motion along an x axis, is represented by a wave function $\psi(x)$. The probability of finding it between x and $x + dx$ is $|\psi(x)|^2\, dx$. The probability that its momentum lies between p and $p + dp$ is $|\phi(p)|^2\, dp$, where the wave function $\phi(p)$ in the 'momentum representation' is the Fourier transform of $\psi(x)$, namely,

$$\phi(p) = h^{-1/2} \int_{-\infty}^{\infty} dx\, \psi(x) \exp(-ipx/\hbar).$$

The connection between $|\psi(x)|^2$ and $|\phi(p)|^2$ is provided by a Fourier transform (Appendix I) because x and p are conjugate mechanical variables. Similarly, the momentum transfer $\boldsymbol{q}$ experienced by a fast charged particle in a collision is conjugate to the relative position of its trajectory with respect to the isolated target atom (or molecule) considered by the Bethe theory. Accordingly, the Bethe equations (5) and (6) can be reduced to a form analogous to (4) but more accurate than it.

In the reduction procedure one must consider that the momentum transfer $\boldsymbol{q}$ is restricted by energy conservation in the collision. We treat this restriction here, for simplicity, in a small-q approximation analogous to—but less restrictive than—the large-b approximation of the semiclassical theory. Thereby we split $\boldsymbol{q}$ into components $\boldsymbol{u}$ and $\boldsymbol{w}$ perpendicular and parallel to the direction of incidence, respectively,

$$\boldsymbol{q} = \boldsymbol{u} + \boldsymbol{w}, \quad w = E_n/v. \tag{8}$$

The value of w is fixed by the magnitude of the excitation energy E_n, whereas the magnitude and direction of $\boldsymbol{u}$ are unrestricted within a plane perpendicular to the incidence. With this notation, equations (5) and (6) can be rewritten in the form

$$d\sigma_n = 4(ze^2/v)^2 |F_n(Q)|^2 (u^2 + E_n{}^2/v^2)^{-2}\, d\boldsymbol{u}. \tag{9}$$

The quantity conjugate to $\boldsymbol{u}$ is an effective impact parameter vector

$$\boldsymbol{b} = \boldsymbol{\rho} - \boldsymbol{R} \tag{10}$$

perpendicular to the direction of incidence. Here the vector $\boldsymbol{R}$ indicates the position of a reference point (e.g., a nucleus) in the target atom or molecule. The vector $\boldsymbol{\rho}$ indicates, in effect, a point on the incident particle's trajectory, but it can be introduced as a *mathematical variable* in a Fourier transformation of (9) irrespective of any physical interpretation. It is also appropriate to make the dependence of $F(Q)$ on $\boldsymbol{R}$ explicit by rewriting (6) in the form

$$F_n(Q)=\left(n\left|\sum_j \exp\left[\mathrm{i}\boldsymbol{q}.(\boldsymbol{r}_j-\boldsymbol{R})/\hbar\right]\right|0\right)\exp(\mathrm{i}\boldsymbol{q}.\boldsymbol{R}/\hbar)=\bar{F}_n(\boldsymbol{q})\exp(\mathrm{i}\boldsymbol{q}.\boldsymbol{R})/\hbar \qquad (11)$$

thereby $\bar{F}_n(q)$ depends now only on *relative* positions of particles within the target. With these notations (9) can be transformed into

$$\mathrm{d}\sigma_n=4\left|(ze^2/hv)\int \mathrm{d}\boldsymbol{u}\,\bar{F}_n(\boldsymbol{u}+\boldsymbol{w})(u^2+w^2)^{-1}\exp(\mathrm{i}\boldsymbol{u}.\boldsymbol{b}/\hbar)\right|^2\mathrm{d}\boldsymbol{b} \qquad (12)$$

as shown in Appendix I.

This expression is amenable to numerical evaluation given a knowledge of the wave functions in the expression of $\bar{F}_n$. Integration over $\boldsymbol{u}$ before evaluation of the matrix element transforms (12), as shown in Appendix II, into

$$\mathrm{d}\sigma_n=4\left|(ze^2/\hbar v)\left(n\left|\sum_j \mathrm{K}_0(E_n|\boldsymbol{r}_j-\boldsymbol{\rho}|/\hbar v)\exp(\mathrm{i}\boldsymbol{w}.\boldsymbol{r}_j)/\hbar\right|0\right)\right|^2\mathrm{d}\boldsymbol{\rho} \qquad (13)$$

where K_ν is the Bessel function defined on p. 78 of Watson (1944) and $v=0$ in our problem.

Contact between (12) or (13) on the one hand and the results of the semiclassical theory on the other hand is established in the low Q limit where, according to section 2, $P_n(Q)/Q$ becomes independent of Q. Specifically one takes the dipole approximation to $\bar{F}_n(q)$, namely,

$$F_n(\boldsymbol{q})\sim \mathrm{i}\hbar^{-1}\boldsymbol{q}.\left(n\left|\sum_j(\boldsymbol{r}_j-\boldsymbol{R})\right|0\right)=\mathrm{i}\hbar^{-1}\boldsymbol{q}.\boldsymbol{r}_n. \qquad (14)$$

After substitution of this approximation in (12), an analytical calculation of the

$$\int \mathrm{d}\boldsymbol{u}\ldots$$

performed in Appendix II, reduces (12) to a form analogous to (4), namely,

$$\mathrm{d}\sigma_n=(4\pi z^2e^4/mv^2)(f_n/E_n)[\{\mathrm{K}_1{}^2(E_nb/\hbar v)+\mathrm{K}_0{}^2(E_nb/\hbar v)\}(E_nb/\hbar v)^2]\,\mathrm{d}(\ln b). \qquad (15)$$

Here f_n is the oscillator strength for excitation of the state n.

The first factor of (15) is the same as in (4) and the second is the limit of the corresponding factor $P_n(Q)/Q$. The factor in braces, absent in (4) but present in more detailed semiclassical calculations (Fermi 1940, Neufeld 1953), equals unity for $E_nb/\hbar v \ll 1$ and decreases exponentially for $E_nb/\hbar v \gg 1$ in accordance with Bohr's prediction of a maximum effective impact parameter $b\sim E_n/\hbar v$.

(C) The extension of the Bethe theory to provide a cross section for multiple collisions takes a particularly straightforward form in the impact parameter version of the theory. We consider an arbitrary number of target atoms or molecules, $\alpha, \beta, \ldots$, whose reference points lie respectively at $\boldsymbol{R}_\alpha, \boldsymbol{R}_\beta, \ldots$, and ask for the probability of their joint excitation to levels $n\alpha, m\beta, \ldots$ by the passage of a fast particle. The cross section corresponding to (12) is

$$\mathrm{d}\sigma_{n\alpha,\,m\beta,\ldots}=4\prod_{n\alpha,\,m\beta,\ldots}\left|(ze^2/\hbar v)\int \mathrm{d}\boldsymbol{u}\,\bar{F}_{n\alpha}(\boldsymbol{u}+\boldsymbol{w})(u^2+w^2)^{-1}\exp(\mathrm{i}\boldsymbol{u}.(\boldsymbol{\rho}-\boldsymbol{R}_\alpha)/\hbar)\right|^2\mathrm{d}\boldsymbol{\rho}. \qquad (16)$$

This formula involves, of course, one impact parameter vector $\boldsymbol{b}_\alpha = \boldsymbol{\rho} - \boldsymbol{R}_\alpha$ per target, but the differential element of target area is appropriately indicated by $d\boldsymbol{\rho}$. The cross section (16) clearly becomes vanishingly small if any one of the b_α is much larger than $\hbar v/E_{n\alpha}$, i.e., unless all the positions $\boldsymbol{R}_\alpha$ cluster around a track axis identified by $\boldsymbol{\rho}$.

Equation (16) constitutes an example of the correlation formulas we are seeking. Indeed it might serve as a prototype for future track structure formulas. Yet its significance remains subject to the limitations (i) to (iv) listed in section 2, except that the restriction (i) to low Q-collisions has been relaxed here, to an unspecified extent. Moreover (16) possesses now a manifest quantum-mechanical foundation.

The principal task ahead is probably the removal of limitation (iii) by adopting a more flexible definition of the excited states $n\alpha$, $m\beta$, . . . and by performing a summation over the by-products of energy degradation, as suggested in (A) above. A simpler and more immediate task is to generalize (12) and (16) to larger values of $\boldsymbol{q}$ for which the formula (8), i.e., $\boldsymbol{q} = \boldsymbol{u} + \boldsymbol{w}$ with $\boldsymbol{u}$ perpendicular to $\boldsymbol{w}$, is no longer adequate. Another minor task consists of removing from the derivation of the multiple collision formula a tacitly assumed restriction, namely, that the intervals between successive collisions be sufficient to guarantee energy conservation in each single collision, rather than only for the multiple process as a whole. This task should be performed by a more careful derivation of (16) from perturbation theory without discarding relevant corrective terms.

Appendix I

The relationship between the probability distributions $|\psi(x)|^2\,dx$ and $|\phi(p)|^2\,dp$ can be obtained by the following mathematical transformation which utilizes the Fourier representation of the Dirac δ-function

$$\begin{aligned}\int_{-\infty}^{\infty} |\psi(x)|^2\,dx &= \iint \psi(x)\psi^*(x')\delta(x-x')\,dx\,dx' \\ &= \iint \psi(x)\psi^*(x')\,h^{-1}\int dp\,\exp\,(-ip(x-x')/\hbar)\,dx\,dx' \\ &= h^{-1}\int dp \int dx\psi(x)\exp(-ixp/\hbar)\int dx'\psi^*(x')\exp(ipx'/\hbar) = \int_{-\infty}^{\infty} |\phi(p)|^2\,dp.\end{aligned} \tag{A1}$$

Notice that this transformation requires no physical interpretation of p, which is introduced as the parameter of a Fourier representation. Similarly we can express (9) as $d\sigma_n = |A(\boldsymbol{u})|^2\,d\boldsymbol{u}$ and write

$$\begin{aligned}\int d\sigma_n &= \int |A(\boldsymbol{u})|^2\,d\boldsymbol{u} = \iint A(\boldsymbol{u})A^*(\boldsymbol{u}')\delta(\boldsymbol{u}-\boldsymbol{u}')\,d\boldsymbol{u}\,d\boldsymbol{u}' \\ &= \iint A(\boldsymbol{u})A^*(\boldsymbol{u}')h^{-2}\int d\boldsymbol{\rho}\,\exp\,[-i(\boldsymbol{u}-\boldsymbol{u}').\boldsymbol{\rho}/\hbar]d\boldsymbol{u}\,d\boldsymbol{u}' \\ &= \int |h^{-1}\int A(\boldsymbol{u})\,\exp\,(i\boldsymbol{u}.\boldsymbol{\rho}/\hbar)\,d\boldsymbol{u}|^2\,d\boldsymbol{\rho}\end{aligned} \tag{A2}$$

which is equivalent to (12), again without reference to the interpretation of $\boldsymbol{\rho}$.

Appendix II

The basic integral required in section 4 (B) is

$$I(w, \boldsymbol{a}) = \int d\boldsymbol{u}\,(u^2+w^2)^{-1}\exp\,(i\boldsymbol{u}.\boldsymbol{a}) \tag{A3}$$

where $\boldsymbol{u}$ and $\boldsymbol{a}$ are vectors in a plane (perpendicular to $\boldsymbol{w}$) and the integral extends over the whole plane. Utilizing polar coordinates with axis $\boldsymbol{a}$ in this plane, we set $\mathrm{d}\boldsymbol{u}=u\,\mathrm{d}u\,\mathrm{d}\theta$ and $\boldsymbol{u}.\boldsymbol{a}=ua\cos\theta$, whereby

$$I(w, \boldsymbol{a})=\int_0^\infty u\,\mathrm{d}u\int_0^{2\pi}\mathrm{d}\theta\,(u^2+w^2)^{-1}\exp(\mathrm{i}\boldsymbol{u}\boldsymbol{a}\cos\theta)$$

$$=\int_0^\infty u\,\mathrm{d}u\,(u^2+w^2)^{-1}2\pi\mathrm{J}_0(ua)=2\pi\mathrm{K}_0(wa). \quad \text{(A4)}$$

The two steps of integration are found, e.g., in Watson (1944) equation (2) on p. 25 and (5) on p. 425, respectively. Equation (13) is obtained from (12) by utilizing (A4) with

$$\boldsymbol{a}=(\boldsymbol{r}_j-\boldsymbol{R}-\boldsymbol{b})/\hbar=(\boldsymbol{r}_j-\boldsymbol{\rho})/\hbar.$$

To derive (15) from (12) utilizing the form (14) of $\bar{F}_n(\boldsymbol{q})$, one may transform the integrand in (12) by writing

$$\mathrm{i}\hbar^{-1}(\boldsymbol{u}+\boldsymbol{w}).\boldsymbol{r}_n\exp(-\mathrm{i}\boldsymbol{u}.\boldsymbol{b}/\hbar)=(-\boldsymbol{r}_n.\nabla_b+\mathrm{i}\hbar^{-1}\boldsymbol{r}_n.\boldsymbol{w})\exp(\mathrm{i}\boldsymbol{u}.\boldsymbol{b}/\hbar) \quad \text{(A5)}$$

after which the factor in brackets no longer depends explicitly on $\boldsymbol{u}$ and can be factored out of the integral. We have then

$$\int \mathrm{d}\boldsymbol{u}\bar{F}_n(\boldsymbol{q})(u^2+w^2)^{-1}\exp(-\mathrm{i}\boldsymbol{u}.\boldsymbol{b}/\hbar)=(-\boldsymbol{r}_n.\nabla_b+\mathrm{i}\hbar^{-1}\boldsymbol{r}_n.\boldsymbol{w})I(w, -\boldsymbol{b}/\hbar)$$

$$=2\pi\{(\boldsymbol{r}_n)_b\partial/\partial b+\mathrm{i}\hbar^{-1}(\boldsymbol{r}_n)_w w\}K_0(wb/\hbar)$$

$$=2\pi(E_n/\hbar v)\{-(\boldsymbol{r}_n)_b\mathrm{K}_1(E_nb/\hbar v)+\mathrm{i}(\boldsymbol{r}_n)_w\mathrm{K}_0(E_nb/\hbar v)\}. \quad \text{(A6)}$$

Upon taking the squared modulus of (A6), we notice that owing to the—implicitly assumed—random orientation of the target atom the various squared components of its dipole matrix elements, $(\boldsymbol{r}_n)_b{}^2$ and $(\boldsymbol{r}_n)_w{}^2$, are equal to one another and are expressed in terms of the oscillator strength by $\hbar^2 f_n/2mE_n$. We also replace $\mathrm{d}\boldsymbol{b}$ by $2\pi b^2\,\mathrm{d}(\ln b)$, whereupon (15) is obtained.

Note added in proof. Fourier transformation with respect to the transverse momentum transfer $\boldsymbol{u}$ is currently being applied in high energy physics for the same purpose as in this appendix, namely, to connect quantum-mechanical scattering amplitudes to an impact parameter coordinate (Chang and Raman 1969).

References

Ben-Reuven, A., 1965, *Phys. Rev. Lett.*, **14**, 349.
——, 1966, *Phys. Rev.*, **145**, 7.
Bethe, H. A., 1930, *Ann. Phys, Lpz.*, **5**, 325.
Bohr, N., 1913, *Phil. Mag.*, **25**, 10.
Chang, N. P., and Raman, K., 1969, *Phys. Rev.*, **181**, 2048.
Fano, U., 1956, *Phys. Rev.*, **103**, 1202.
Fermi, E., 1940, *Phys. Rev.*, **57**, 485.
Neufeld, J., 1953, *Proc. Phys. Soc.* B **66**, 590.
Radiation Research 1966, 1967, Ed. G. Silini, Amsterdam, North-Holland, pp. 13–95.
Watson, G. N., *Theory of Bessel Functions*, 2nd edn (London: Cambridge University Press).
Williams, E. J., 1933, *Proc. R. Soc.* A **133**, 163.
——, 1945, *Rev. Mod. Phys.*, **17**, 217.

Time dependence of the dielectric constant

R. SCHILLER

Central Research Institute for Physics, Budapest, Hungary

Abstract. Interionic forces and energies are greatly influenced by the dielectric constant of the substance holding the ions. The dielectric constant of dipolar substances varies with the magnitude and rate of change of the electric field, thus one would expect the dielectric constant to depend on the density, velocity and age of the ions distributed in the dipolar material. A simple formula, valid if the net amount of charges in the system is constant, is proposed. The expression shows the overall dielectric constant to depend only on the age of the ions.

The fate of radiation-produced ions and electrons depends largely on the electric forces acting between the charge carriers. Besides the magnitudes and separations of the charges, the forces are determined by the dielectric constant of the irradiated substance. The knowledge of the dielectric constant is apparently vital in the understanding of any ionic process.

Dielectric constants can be easily measured. If one puts a substance between the plates of a capacitor and measures the capacity, this, other things being equal, is proportional to the dielectric constant. It turns out in many cases, that the quantity determined in this way depends on the frequency of the voltage applied to the capacitor. In the case of dipolar substances, i.e. of materials which consist of molecules having a permanent dipole moment, one determines at low frequencies a high value, the so-called static dielectric constant, ϵ_S. Increasing the frequency decreases the dielectric constant converging to its lower limiting value, the infrared dielectric constant, ϵ_∞. The frequency dependence, at least for a large class of dipolar substances can be characterized by a single parameter, the dielectric relaxation time, τ.

Now the question arises which value is to be used in the description of the electric forces brought about by swiftly moving charge carriers. The capacity measurements clearly indicate that the dielectric constant depends on the rate of change of the field strength; hence it must depend on the velocity of charge carriers too. In the case of capacity measurements the applied electric field changes periodically and its rate of change is the same in any point of the system. The charge carriers move randomly, so the electric field varies also at random and its rate of change is different from point to point. The farther a point is from an ion in motion, the slower the field strength changes at this point. Some theoretical considerations are needed to translate the results gained by externally applied, homogeneous, periodic fields to the case of inhomogeneous, randomly changing fields, the sources of which are distributed within the substance.

It can be shown that the overall dielectric constant of a system at time t, $\epsilon(t)$ can be given by the expression

$$\epsilon(t) = \frac{\epsilon_\infty}{1 - \left(1 - \dfrac{\epsilon_\infty}{\epsilon_S}\right)(1 - e^{-t/\theta})}$$

where $\theta = \epsilon_\infty \tau / \epsilon_S$. The time is measured from the instant of charge production. Two graphic representations of this equation are given in figure 1 computed using the

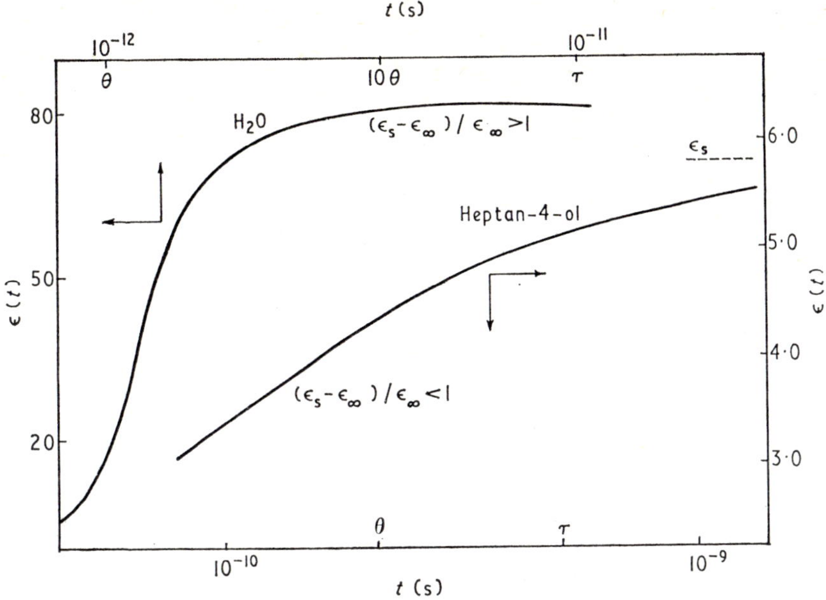

Figure 1. The time-dependent dielectric constant of water and n-heptan-4-ol.

parameters for water ($\epsilon_S = 80$, $\epsilon_\infty = 4{\cdot}9$, $\tau = 1{\cdot}01 \times 10^{-11}$ s) and for *n*-heptan-4-ol ($\epsilon_S = 5{\cdot}8$, $\epsilon_\infty = 3{\cdot}04$, $\tau = 7{\cdot}4 \times 10^{-10}$ s). It is easy to see that θ determines the time dependence of $\epsilon(t)$. For certain substances θ can be one or two orders of magnitude lower than τ. The shape of the curve is influenced by the relative values of ϵ_S and ϵ_∞ too. If $(\epsilon_S - \epsilon_\infty)/\epsilon_\infty > 1$ the curve has a point of inflexion, while for $(\epsilon_S - \epsilon_\infty)/\epsilon_\infty < 1$ it is concave down. If one wants to decide whether a process is fast or slow in comparison with the dielectric relaxation the duration of this process must be compared with θ and not with τ.

The above simple formula is not always true. It holds only for processes during the course of which the net amount of charges in the system is constant. This condition is certainly satisfied in ionization processes where an originally charge-free system is being irradiated, the radiation producing an equal amount of positive and negative charges which disappear again by pairwise neutralization. The net amount of charges is zero throughout the whole process.

If this condition is met, the expression of $\epsilon(t)$ can be applied not only to the description of the field strength but also to the determination of the electric energy of the system. Hence both the total energy of the charge ensemble and the potential energies of the charge carriers depend explicitly on time. This means that the energy of a particular ion or electron depends not only on the actual site of the charge carrier but also on the time which elapses till the charge carrier reaches this site. Physically speaking the time dependence of the dielectric constant makes the electric field non-conservative. This precludes the use of a number of considerations and ideas, which are based explicitly or tacitly on the field being conservative. This is why the celebrated Onsager formula of electron–ion recombination probability, originally deduced for high-pressure rare gases, cannot be applied to dipolar substances like water or alcohols. The time dependence of the dielectric constant is a strong warning against the application of any result based on the assumption of conservative electric field.

Proportional counters

E. W. EMERY, W. GROSS,† H. H. ROSSI† and K. S. J. WILSON

Department of Medical Physics, Royal Postgraduate Medical School, Hammersmith Hospital, London, W.12
†Department of Radiology, Radiological Research Laboratory, College of Physicians and Surgeons, Columbia University, 630 West 168th Street, New York 32, N.Y. 10032, USA.

Abstract. A proportional counter placed in a radiation field may be regarded as simulating a very small volume of condensed matter, such as tissue. The pulse-height spectrum from the counter is closely related to the spectrum of energy deposition in the corresponding small volume. Recent progress in the development of counters for use in this way is reviewed. Artefacts may be introduced by the presence of the counter wall, but these may be avoided by the use of 'wall-less' counters in which the collecting volume is defined by the electrical field arising from a lightweight electrode assembly. Such counters show clearly the effects of delta rays from tracks that do not themselves pass through the collecting volume.

1. Introduction

Interest in track structure arises largely from a desire to account for radiation effects, both chemical and biological, in condensed matter. Our concern is often with systems whose dimensions are at the limit of resolution of the light microscope or smaller; consequently there is an advantage in studying the structure of tracks in gases, and regarding these as large scale models of what occurs in the solid or liquid state. The cloud chamber afforded the first and in many ways the best information available on the structure of tracks, but it is difficult to collect a large sample of measurements from cloud chamber studies. With a proportional counter one can accumulate information rapidly but only on small isolated sections of track. These, however, may well be exactly what one wishes to know about. A proportional counter connected to a multichannel analyser conveniently gives what in effect is an energy deposition spectrum within the volume simulated by the counter. The use of proportional counters in this fashion was initiated by Rossi and Rosenzweig (1955). They used a spherical counter of tissue-equivalent plastic filled with a mixture of methane, carbon dioxide and nitrogen, giving the same atomic composition as the wall material. It is assumed that an ionizing particle track entering the gas continues on a path geometrically similar to the one it would have followed in the solid, but on a linear scale increased in the ratio of the density of the wall material to that of the gas. A linear magnification of 10^5 can thus be obtained, since the counter can be run at a pressure of less than 10 torr. If the counter is properly designed it delivers a pulse proportional to the number of ions formed in the track, and this in turn is proportional, on the average, to the energy deposition. The results of these measurements are usually given in the form of frequency distributions of *event size*, Y. This is defined as the energy deposited divided by the diameter of the unit density sphere simulated (Rossi 1959).

In order that the pulse height should be truly proportional to the number of ionizations in the counter it is necessary that the electric field, and therefore the gas gain in the avalanche, shall be the same at all points on the counter wire. This condition is relatively easily met in a cylindrical counter but special means must be adopted in a sphere. The method originally used (Rossi and Rosenzweig 1955), was to apply a suitable potential to a field compensating electrode at each end of the counter wire and this method has been further developed by Benjamin *et al.* (1963). Rossi and his co-workers, however, have adopted the device of surrounding the centre wire by a helix of fine wire held at a potential intermediate between that of the centre and the counter wall. The centre part of the counter then becomes in effect a cylindrical counter with a transparent wall. A series of counters has been developed following this principle, and their performance has been progressively improved (Annual Reports R.R.L.). A mixture of 55% propane, 39·6% carbon dioxide and 5·4% nitrogen has been found to give a higher available gas gain than did the original mixture of 64·4% methane, 32·4% carbon dioxide and 3·2% nitrogen (Srdoc 1968). The influence of centre wire and helix diameters has been investigated. Increase of either diameter often gives better resolution, but at the expense of gas gain. The size of the counter has been reduced from the original diameter of 10 cm. Much use has been made of a counter of 2·5 cm diameter, and one of 6 mm diameter has been made to

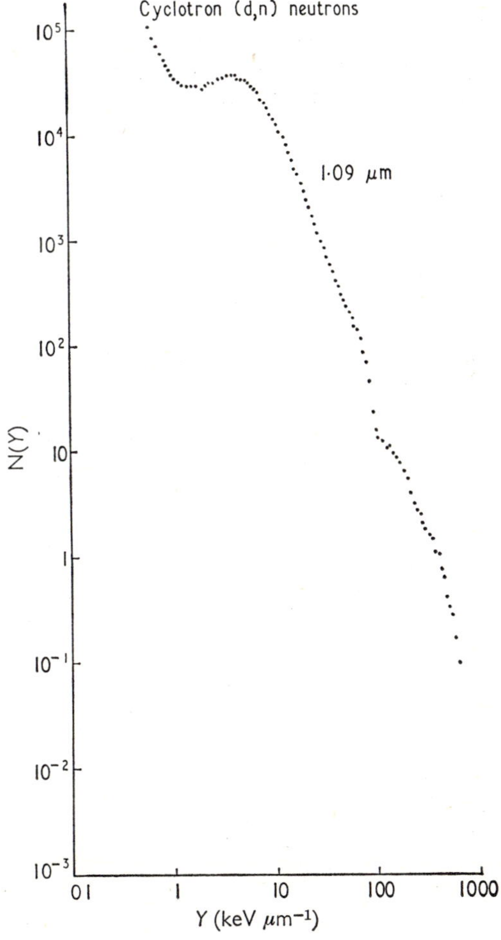

Figure 1. Energy deposition spectrum for (d, n) neutron from the MRC cyclotron in a sphere of equivalent diameter 1·09 μm of tissue. Abscissa scale: Y = energy deposition/equivalent diameter of sphere.

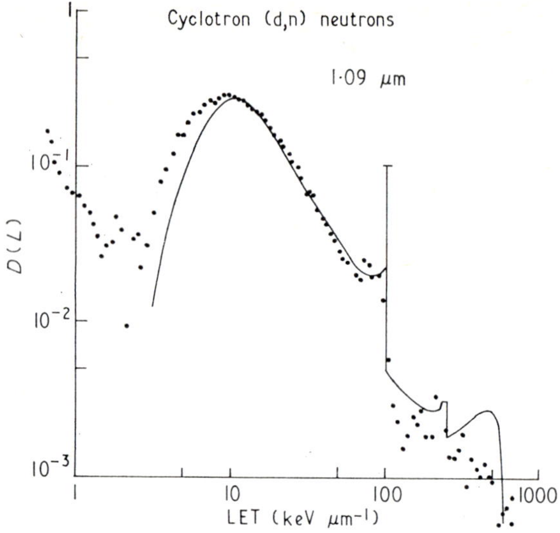

Figure 2. LET distribution for (d, n) neutrons from the MRC cyclotron. Points derived from figure 1, line calculated (Bewley, 1968). The difference at low LET is due largely to γ-ray contamination of the beam, not taken into account in the calculated spectrum.

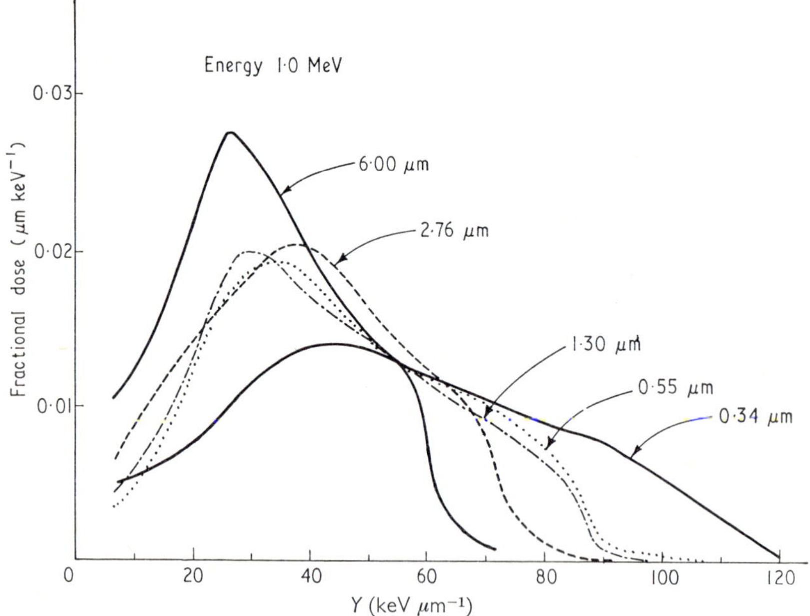

Figure 3. Fractional dose distribution from 1·0 MeV neutrons in spheres of various tissue equivalent diameters. Abscissa scale: Y=energy deposition/diameter of equivalent tissue sphere.

work well. This last counter used a centre wire of 25 μm in diameter. The minimum event size which can be recorded depends of course on the gas gain obtainable which in turn is influenced by the gas pressure. Thus the minimum event size recorded varies with simulated sphere size. By careful attention to the noise level, and using a cooled FET pre-amplifier very close to the wire, pulses corresponding to only one primary ionization have been recorded. The Hammersmith group (K. S. J. Wilson and S. B. Field, to be published) using an electron tube pre-amplifier record pulses down to 5 or 6 primary electrons. The cut-off at which pulses can no longer be distinguished from the background noise is of course not sharply defined.

Spherical counters have the special property that the results are easily convertible into an LET distribution. They are therefore potentially useful in health physics for the assessment of Quality Factors, which are defined in terms of LET. Results of such a conversion are given in figures 1 and 2 which compare the pulse height distribution from neutrons produced by deuteron bombardment of beryllium on the MRC cyclotron with the LET distribution derived from it. The LET distribution is in some ways the more striking representation: the proton and alpha-particle Bragg peaks can be clearly seen, for instance. But the energy deposition spectrum is probably of more fundamental significance. Where LET is used as a parameter in the interpretation of radiation effects it is often necessary to consider a delta-ray cut off. This consideration at once leads towards the concept of a critical volume within which energy depositions are considered to be effective. Figure 2 also compares the experimentally derived LET distribution with one calculated by Bewley (1968). The agreement is good, but it depends on a suitable choice of equivalent counter diameter. The analysis used in deriving the LET spectrum from the event size spectrum (Rossi 1968) assumes that the particle tracks are straight, of uniform LET, and ignores statistical fluctuations in energy loss. It will fail for particles whose tracks are not straight, e.g. electrons, and for counter diameters which are sufficiently large to give an appreciable variation in LET along the track. Figure 3 shows how the fractional dose distribution in a neutron field varies with counter diameter. In the larger counters the Bragg peak is necessarily accompanied by a less densely ionizing track giving a lower average LET for the event within the counter. If one were to go to sufficiently small counter diameters, the statistical variability of energy distribution along the track would become apparent, and again the LET analysis would fail.

2. Wall-less counters

The argument that the cavity simulates an equivalent volume of condensed material is open to objections. These are of two kinds. First, the spectrum of very slow electrons, or of low energy delta-rays, may be appreciably different in the gaseous and condensed states. The second objection arises from the geometrical effects illustrated in figure 4. Where a track branches just outside the solid volume both branches may be included in the equivalent gas volume; similarly, a scattered track may re-enter the gas volume when it would have entirely missed the equivalent solid. Thus, although all the tracks are true analogues of events that would take place in the condensed state, a disproportionate number of them occur in coincidence in a walled counter. An excessively high number of large pulses is therefore to be expected. To investigate this problem 'wall-less' counters have been constructed. These consist of a light electrode assembly to define the collecting volume contained in a large tank of counting gas. Preliminary calculations suggest that a tank of 10 times the diameter of the counter will reduce spurious coincidences to less than 1% of those in a walled counter. Figure 5*a* shows a counter based on the principle of the wall-less ionization chambers designed by Failla (1960). The effective collecting volume is defined by the electric field between the end electrodes and an interposed ring. Figure 6 shows the results of investigating the collecting volume with beams of alpha-particles and soft X-rays and

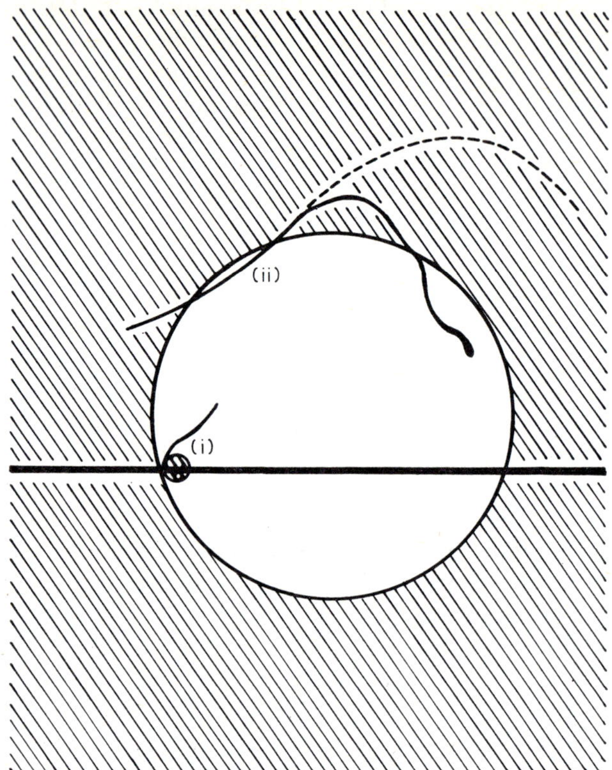

Figure 4. Wall effect. The outer circle encloses the gas volume, the small inner circle encloses the equivalent volume of wall material with respect to track (i). The track forks just before entering the counter. Both branches enter the counter but only one passes through the equivalent solid volume. Track (ii) leaves the counter but is scattered in again from the wall, whereas the equivalent track, if the wall were gaseous (shown dotted) would not re-enter the counting volume.

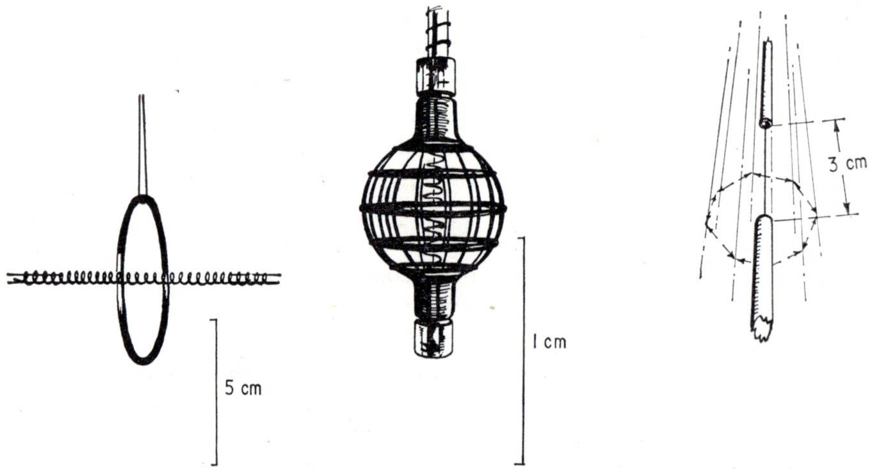

Figure 5. Electrode assemblies for 'wall-less' counters.

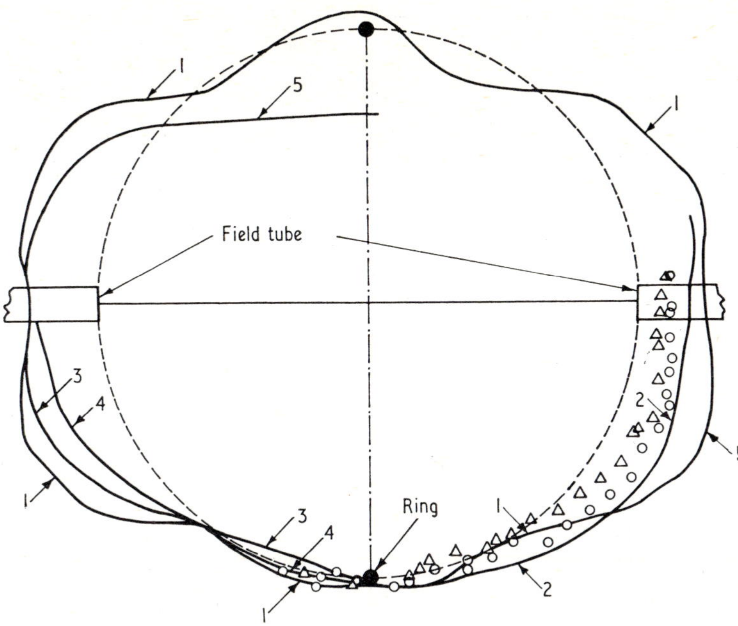

Figure 6. Outline of the sensitive volume of the hoop counter. Solid curves are for alpha-particle tests. Points are for X-ray tests. Curves 1 and 2 are for hoop potentials of -300 and -600 respectively with a helix of 10 turns per inch and a centre wire potential of $+750$ V. Curves 3 and 4 are for same conditions as 1 and 2 respectively but with a helix of 20 turns per inch. Curve 5 depicts the results with the tighter helix but a hoop potential of -150 V. Pressure for all alpha tests was 100 torr. Circles are data for carbon X-rays at 14 torr and triangles for aluminium X-rays at 59 torr. In each case the hoop potential was approximately 60% of the centre wire potential.

figure 7 shows the pulse height distributions from beams of alpha-rays fired through the counter in the indicated positions. It is apparent, as would be expected, that alpha-tracks that pass outside the collecting volume can produce pulses within the counter, by their delta-rays. Irradiation of the whole counter by an uncollimated alpha-source gives a pulse height distribution which differs from the expected triangular distribution by having a predominance of small pulses, due no doubt to those delta-rays penetrating from outside the collecting volume.

As the collecting volume shown in figure 5*a* is not accurately spherical Gross, Biavati and Rossi have constructed a 6 mm diameter counter shown in figure 5*b* defined by a spherical cage of tissue equivalent plastic filaments of 0·1 mm diameter, the transparency of the whole counter being 80%. Wilson and Emery at Hammersmith have constructed an approximately cylindrical wall-less counter in which the collecting volume is defined by an octagonal array of fine wires (figure 5*c*). The collecting volume of this counter is being investigated in the first instance by observing how the collecting volume varies with the voltage on the wire cage when the counter is used simply as an ionization chamber.

In addition to the investigation of the validity of 'cavity theory' as applied to counters, the idea of wall-less counters opens up numerous other possibilities. It would be naive to suppose that all radiobiological effects could be explained in terms of energy depositions within a single volume. More complex models may be needed, Neary (1965) for instance, accounts for chromosome aberrations in terms of events in

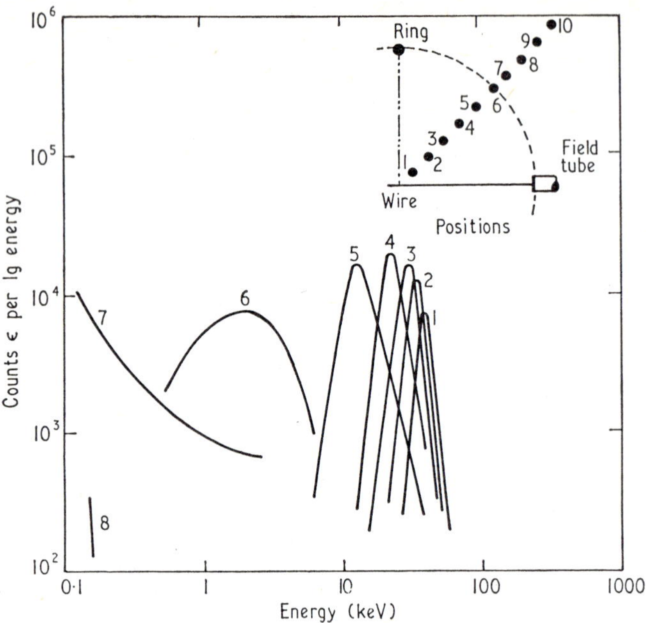

Figure 7. Counts per logarithmic energy interval times the distance, ϵ, of the collimated alpha beam from the centre of the sphere for equal number of alpha-particle traversals as a function of the energy deposited. The various ϵ are indicated on the figure. Simulated size is 0·5 μm.

two associated volumes. Clearly the model could be simulated by two wall-less counters in a single tank, and more complex arrangements could also be considered. More fundamental physical characteristics, such as the lateral spread of delta-rays, could be investigated with 'wall-less' systems. Proportional counters, though not specifically of wall-less design, have of course already been used to investigate energy depositions in regions outside the counter, the ions being attracted by a subsidiary field into the counting volume. In this way Glass (1968) has investigated the energy depositions along short segments of the particle tracks, and Booz (1968) has examined energy depositions in volumes too small to be used directly as a counter.

3. Limitations

There are difficulties in attempting to interpret radiation effects in terms of the number of energy depositions above a certain critical size in a certain critical volume. At the outset there are two parameters to specify: the critical volume and the critical event size. In addition there is the question of shape. Except for its special convenience for deriving LET distributions and its omnidirectional property, there is no special virtue in the sphere, and it may be important to investigate the influence of other shapes on the pulse height distribution. For small volumes, however, the influence of shape may not be very great, because the variance in the primary energy losses is much larger than that due to differences in the length of track segments. Table 1 shows a comparison in the pulse height variance and the variance in cross section of the cylinder as viewed by isotropic radiation. (The cross section is inversely proportional to mean path length, and its variation is a conveniently calculated index of the variation of path length distribution with change of aspect). It is clear that with energy deposition of less than a few hundred eV, the statistical fluctuation exceeds the

Table I. Pulse height variance in a cylindrical proportional counter

Height/ diameter	Variance of cross section with aspect*	No. of ionizations† to give pulse height variance equal to cross-section variance	Approx. energy deposition (eV)
0·2	0·115	11	300
0·5	0·028	44	1300
1	0·0058	224	6000
2	0·0193	67	2000
5	0·0435	30	900

* Variance of cross section about its mean value calculated for the formula:

$$V = \frac{\displaystyle\int_0^{\pi/2} \left[(\pi \cos \theta + 4x \sin \theta)^2 - \frac{\pi^2}{4} (1+2x)^2 \right] \sin \theta \, d\theta}{\left[\displaystyle\int_0^{\pi/2} \frac{\pi}{2} (1+2x) \sin \theta \, d\theta \right]^2}$$

where x = height/diameter.

† Variance of pulse height due to statistical fluctuations from the formula $V = 1{\cdot}3/N$ (Emery 1966).

geometrical fluctuation, even in a volume which is far from spherical. The variability of counter pulses arises about equally from statistical fluctuations in the gas multiplication and from fluctuations in the number of ions formed for a given energy deposition (Kellerer 1968). The former is a disadvantage introduced by the counter and it is conceivable that a more refined instrument could be devised that would avoid it. The second cause, however, is an inherent limitation imposed by the concept of energy deposition. It is not to be supposed that energy deposition is of itself the direct cause of radiation effects, but rather that it is a parameter closely related to whatever chemical or physical change is the direct cause. In just the same way that the number of ions formed imperfectly represents the energy deposition, so it is to be expected that the energy deposition imperfectly represents the early chemical or physical events from which the radiation effects arise. If these effects arose from just such ionizations as take place in the gas, then the response of the counter would be a thoroughly satisfactory index of radiation effect. But, although primary ionizations with large energy transfer may be expected to be the same in tissue and in the gas, secondary ionizations and lower energy loss processes can be expected to be different because of the different molecular arrangements in these media.

Finally the concept of a gas volume as an analogue of condensed matter must break down if the volume is made sufficiently small, although it is not easy to assess at what size the breakdown occurs. There are at least two types of process involved. First, the effective volume for energy absorption will be determined in part by the propagation of energy and the diffusion of chemical products away from the site of the initial interaction of the radiation. The counter will not behave as a scale model of tissue with respect to those processes. Secondly, the initial interactions themselves are not precisely localized, and this is particularly true of collective excitations. Again, the counter will not be a scale model of tissue in respect of such excitations. The insensitivity of the stopping power formula to the individual processes of energy transfer ensures that the integral energy transfer will be correct, but does not guarantee the validity of the event size spectrum in small regions.

References

Benjamin, P. W., and Nicholls, G. S., 1963, *A.W.R.E. Rep.* No. 5/63.

Bewley, D. K., 1968, *Radiat. Res.*, **34**, 437.

Booz, J. and Smit, Th., 1968, *Proc. Symp. Microdosimetry* (Brussels: EUR 3747, Euratom), p. 125.

Emery, E. W., 1966, *Radiation Dosimetry*, Vol. II, Eds. F. H. Attix and W. C. Roesch (New York: Academic Press), p. 90.

Failla, P. M., and Failla, G., 1960, *Radiat. Res.*, **13**, 61.

Glass, W. A., 1968, *Proc. Symp. Microdosimetry* (Brussels: EUR 3747, Euratom), p. 189.

Kellerer, A. M., 1968, *Annual Report on Research Project*, Radiological Research Laboratory, Columbia University NYO-2740-5.

Neary, G. J., 1965, *Int. J. radiation Biol.*, **9**, 477.

Rossi, H. H., 1959, *Radiat. Res.*, **10**, 522.

Rossi, H. H., and Rosenzweig, W., 1955, *Radiology*, **64**, 404.

——, 1968, *Radiation Dosimetry*, Vol. I, Eds F. H. Attix and W. C. Roesch (New York: Academic Press), p. 43.

Srdoc, D., 1968, *Annual Report on Research Project*, Radiological Research Laboratory, Columbia University NYO-2740-5.

Annual Reports, Radiological Research Laboratories, Columbia University, New York, NYO-2740-4 U.S. AEC (1967), NYO-2740-5 U.S. AEC (1968).

The influence of track structure on initial recombination in gases

J. BOOZ*

Institut für Medizinische Physik und Biophysik der Universität Göttingen

Abstract. A survey is given on the theory of the recombination coefficient for the recombination of ions in gases. Cluster recombination, columnar recombination, volume recombination and the boundary regions are discussed. The formulae are applied to measurements of ion recombination after well-defined recombination times in order to evaluate the time dependence of the variance of linear ion density. For recombination times t greater than 1 ms this variance decreases as $t^{-1/2}$.

1. Introduction

Recombination of ions in gases can be described by the formula

$$\frac{\mathrm{d}\overline{n(t)}}{\mathrm{d}t} = -\alpha\overline{n(t)}^2. \tag{1}$$

Here it is presumed that the mean concentrations $\overline{n(t)}$ and the mobilities of the positive and negative ions are equal. The recombination coefficient α in this equation is not a constant, but depends on the local ion density distribution n and the so-called recombination time t, the time elapsed after ionization. The reason for this is that immediately after ionization the distribution of ions is not homogeneous. Within the tracks are regions of high local ion density which expand and equilibrate by diffusion. Because of this initial high density the recombination coefficient at the beginning is large in comparison with the coefficient for volume recombination, which is the recombination in a homogeneous ion distribution. Therefore this effect is called initial recombination.

The measurements I will describe here have been carried out in the period between 1960 and 1965. During that time we were mainly interested in the absolute value of the volume recombination coefficient of electrons and ions in gases, and therefore we did some measurements on initial recombination in order to know how much the recombination coefficient depends on the inhomogeneity of the charge distribution after ionization. These measurements have already been described in detail (Booz and Ebert 1963, Ebert 1964, Ebert *et al.* 1964, Jütting *et al.* 1965a and b, Gottsmann 1966, Koepp and Gottsmann 1966). However, the results on track structure which emerge from the measurements of initial recombination have not yet been published and will be presented in this paper.

2. Method and apparatus

The so-called decay method has been used for these measurements. This means that the gas is ionized briefly in a field free chamber, and after the recombination time t all remaining charges are collected and measured by applying a strong electrical field.

* Present address: Biology Service, Joint Research Center Ispra, Euratom.

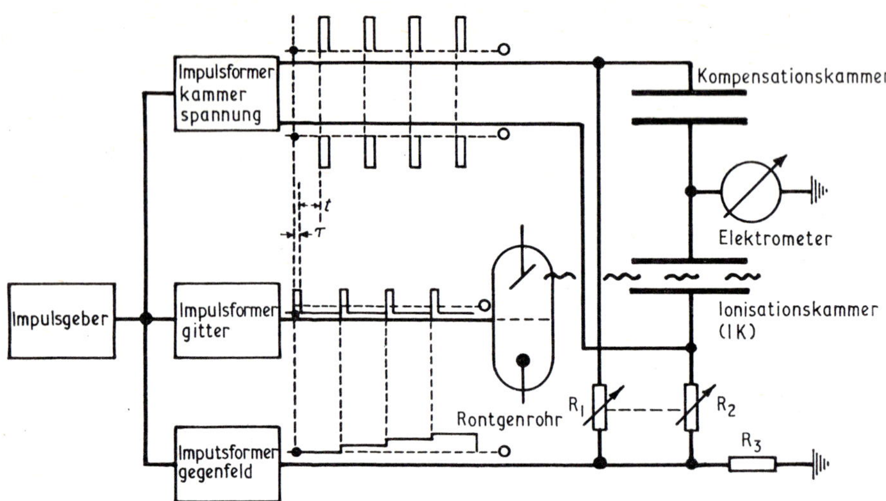

Figure 1. Block diagram of the apparatus. The X-ray tube (Röntgenrohr) was opened for the small time τ by a pulse former (Impulsformer Gitter) which was triggered by the master pulser (Impulsgeber). After the recombination time t another pulse generator (Impulsformer Kammerspannung) opens the gate for the high tension to the ionization chamber (Ionisationskammer) and the compensation chamber (Kompensationskammer).

Figure 1 shows a block diagram of the apparatus. By using a grid controlled X-ray tube there was no need for mechanical switch elements. This allowed the ionization time τ to be in the range 30–300 μs, which was small compared to the recombination time t. The switching time of the X-ray tube was 0·1 μs. The recombination time t was varied from 1 to 500 ms. By changing the intensity of the radiation it was possible to vary the average initial ion density between $7{\cdot}5\times10^3$ and $1{\cdot}5\times10^6$ cm^{-3} independently of the ionization time. The charge induced on the vibrating reed electrometer due to the pulse of high tension on the ionization chamber was compensated by an equal and opposite pulse applied to a second chamber. For measuring the collected charge with sufficient accuracy at the lower ion densities the total charge was collected over several identical pulses. In order to maintain a zero field in the ionization chamber during the whole recombination time the charge already collected was compensated by a staircase generator which was connected to the high tension electrode of the ionization chamber (Ebert *et al.* 1964). X-rays of 50 and 80 kV have been used with an effective filtration of 1 mm Al.

3. Summary of the theory of initial recombination

The recombination coefficient of initial recombination can be described as a function of any local and temporal distribution $n(\boldsymbol{r}, t)$ of ion density (Ebert 1964):

$$\alpha=\alpha_\infty V\frac{\int_V n^2(\boldsymbol{r},t)\,\mathrm{d}V}{\left(\int_V n(\boldsymbol{r},t)\,\mathrm{d}V\right)^2}=\alpha_\infty\frac{\overline{n^2}}{\overline{n}^2} \tag{2}$$

This equation is based only on the assumption that the density distributions and the mobilities of the two charge carriers are equal. V is the volume over which the integration is carried out. If the ion density is independent of the local vector $\boldsymbol{r}$, then we have

$$\alpha=\alpha_\infty \tag{3}$$

This means that α_∞ is the recombination coefficient of volume recombination. The following formulae for the coefficient of initial recombination in clusters, columns, and in the intermediate region between columnar and volume recombination have been calculated with equation (2) and with the assumption that the local distribution of ions in clusters and columns is a Gaussian one. The details of the calculation can be found in Appendix I.

3.1. *Cluster recombination*

Let $\nu(t)$ be the number of ions per cluster at time t, $\nu(0)=\nu_0$, D the diffusion coefficient, and $r(t)$ the average radius of the cluster.

$$r(t)=b/\sqrt{\pi} \quad \text{with} \quad b^2=4Dt+b_0^2 \tag{4}$$

where $b_0/\sqrt{\pi}$ is the average cluster radius immediately after ionization. With these definitions we get for the recombination coefficient:

$$\alpha=\alpha_\infty \frac{\overline{\nu^2}}{\bar{\nu}^2} \frac{\nu_0}{\pi^{3/2}b^3n_0\sqrt{8}} \tag{5}$$

Under conditions of pure cluster recombination the recombination coefficient is therefore inversely proportional to the average ion density n_0 and proportional to $t^{-3/2}$. Equation (5) is valid as long as $N_0\sqrt{(2Dt)}\ll 1$ and $\nu_0/\pi^{3/2}b^3n_0\sqrt{8}\gg 1$; that means as long as the recombination is neither transformed into columnar recombination nor into volume recombination.

3.2. *Columnar recombination*

Let $N(t)$ be the linear ion density of the track at time t, $N(0)=N_0$, and δ the average cross section of the column:

$$\delta(t)=4Dt+\delta_0 \tag{6}$$

where δ_0 is the cross section immediately after ionization. With this definition the columnar recombination coefficient is

$$\alpha=\alpha_\infty \frac{\overline{N^2(t)}}{\overline{N(t)}^2} \frac{N_0}{2\pi\delta(t)n_0} \tag{7}$$

Under conditions of pure columnar recombination the recombination coefficient is therefore inversely proportional to the average ion density n_0 and to the recombination time t. Equation (7) is valid when $N_0\sqrt{(2Dt)}\gg 1$ and as long as $N_0/2\pi\delta n_0\gg 1$.

3.3. *Columnar and volume recombination*

For $N_0/2\pi\delta n_0\leqslant 1$ we have only volume recombination. For $N_0/2\pi\delta n_0\sim 1$, i.e. for the intermediate region between columnar and volume recombination, a theoretical treatment is very difficult (Ebert 1964). The measurements that are being reported here have been made in the region $10^{-3}\leqslant(N_0/2\pi\delta n_0)\leqslant 10^3$. It has been shown (Ebert *et al.* 1964) that the experimental results in the whole of this region can be described very well by

$$\alpha = \alpha_\infty \frac{\overline{N^2(t)}}{\overline{N(t)}^2} \left(1 + \frac{N_0}{2\pi\delta(t)n_0}\right) \tag{8}$$

Cluster recombination has had little effect on the experimental results and has caused only slight deviations from equation (8). This is due to the relatively high ion density (Jütting *et al.* 1965b) which varies between 10^3 and 10^4 cm^{-1}.

3.4. *Influence of track structure on cluster and columnar recombination*

The equations (5), (7), and (8) contain the expressions $\overline{\nu^2}/\overline{\nu}^2$ and $\overline{N^2(t)}/\overline{N(t)}^2$ which give information on the variance of the number of ions per cluster and of the linear ion density respectively. The results in this paper have been obtained in the following way. The experimental value α/α_∞ which has been measured as a function of n_0 at a constant recombination time t has been plotted against $1/n_0$. This plot gives a straight line which crosses the ordinate at the value $\overline{N^2}/\overline{N}^2$ (equation (8)). By repeating such series of measurements for several recombination times t one obtains $\overline{N^2(t)}/\overline{N(t)}^2$ as a function of t.

4. Results

Figure 2 shows the measured recombination coefficients in air (1 atm) as a function of the recombination time. The figure demonstrates especially that α/α_∞ becomes very large at small recombination times and small ion densities.

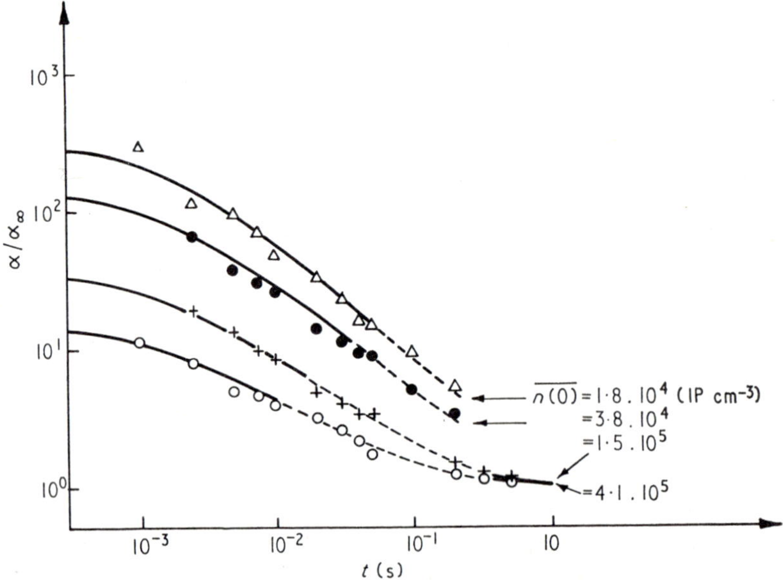

Figure 2. Relative recombination coefficient α/α_∞ in air at 1 atm as a function of the recombination time t for several average ion densities $n(0) = n_0$. Ionization time $\tau = 300$ μs. 50 kV X-rays.

In figure 3 the recombination coefficient measured in air at 1 atm at a recombination time $t = 10$ ms has been plotted against $1/n_0$. This example demonstrates that α is proportional to $1/n_0$ in accordance with equation (8).

Figure 4 shows the results for measurements in O_2 at different pressures plotted in

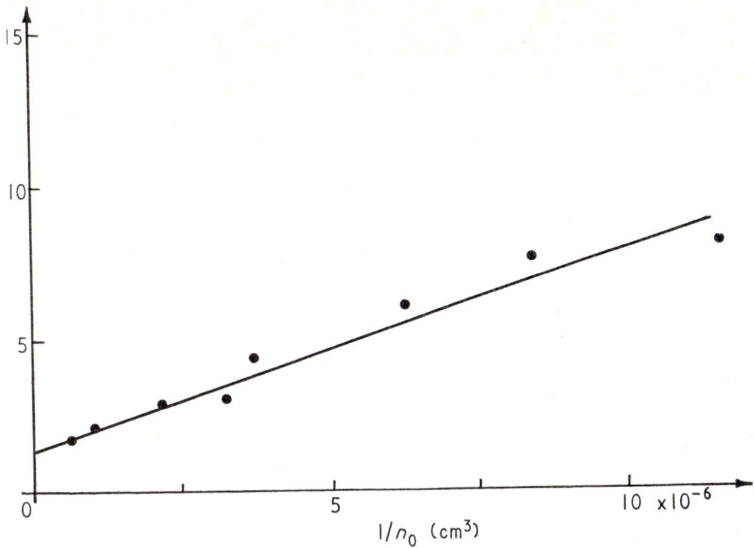

Figure 3. Recombination coefficient α in air at 1 atm as a function of $1/n_0$. Recombination time $t=10$ ms. Ionization time $\tau=300$ μs.

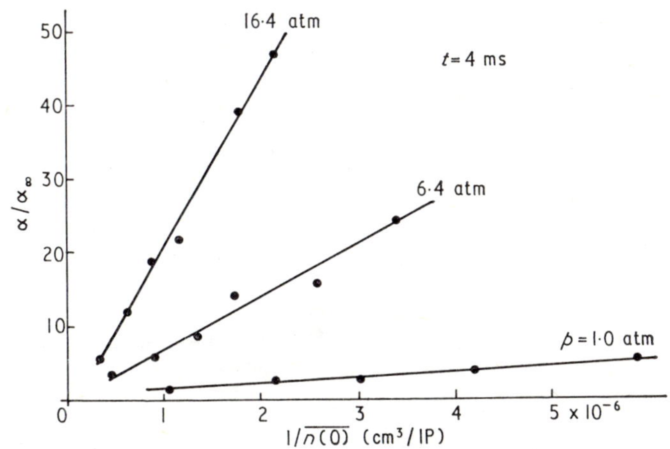

Figure 4. Relative recombination coefficient α/α_∞ in O_2 at 1·0 atm, 6·4 atm, and 16·4 atm as a function of $1/n_0$. Recombination time $t=4$ ms.

the same way. The higher the pressure, the higher the initial recombination because of the increase of N_0/n_0 with rising pressure.

In figure 5 α/α_∞ has been plotted for different recombination times as a function of $1/n_0$. The data are taken from the same series of measurement as was used for figure 3. However, the experimental points have been omitted for the sake of clarity. The intercepts on the ordinate are equal to $\overline{N^2(t)}/\overline{N(t)}^2$ (equation (8)).

The variation of this value with recombination time is shown in figure 6. There is no significant difference between X-rays of 50 and 80 kV. This is the reason for plotting the means of these values in this figure. The points at $t<2{\cdot}5$ ms are based on measurements with $\tau=30$ μs only.

Figure 7 shows the relative variance of the linear ion density which has been calculated with the results of figure 6. Within the limits of experimental accuracy the

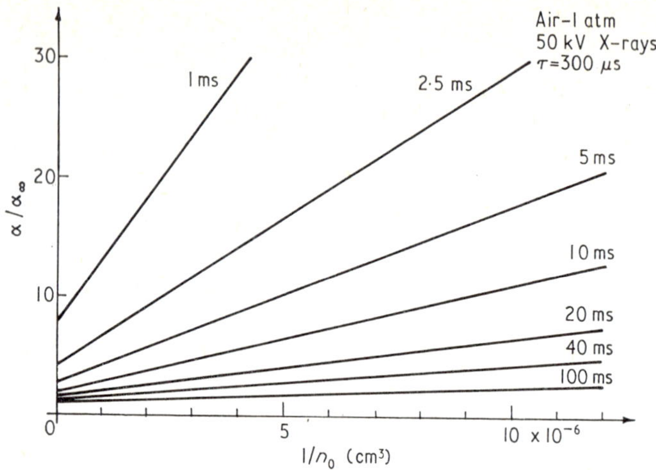

Figure 5. Relative recombination coefficient α/α_∞ in air at 1 atm for different recombination times t as a function of $1/n_0$. Ionization time $\tau=300$ μs. The experimental points have been omitted for the sake of clarity.

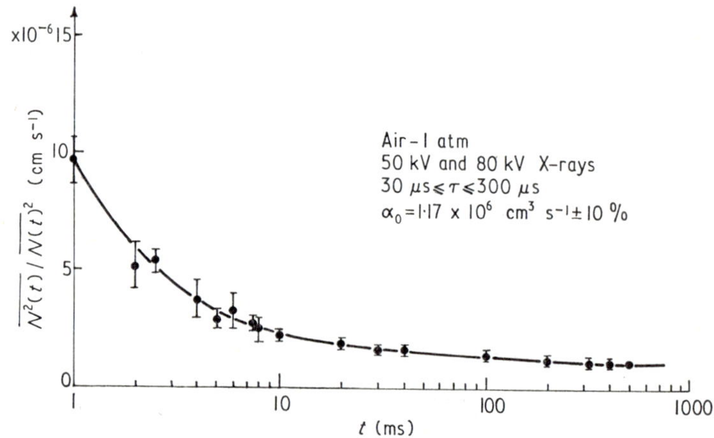

Figure 6. $\overline{N^2(t)}/\overline{N(t)}^2$ as a function of t for 50 kV and 80 kV X-rays in air at 1 atm. Ionization time τ between 30 and 300 μs for $t \geqslant 2{\cdot}5$ ms. $\tau=30$ μs for smaller recombination times t. The points have been calculated with $\alpha_\infty=1{\cdot}17\times10^{-6}$ cm^3 s^{-1}.

points in figure 7 form a straight line with a slope of $-0{\cdot}41$. This slope depends strongly on the value chosen for α_∞. The results in figures 6 and 7 have been calculated with $\alpha_\infty=1{\cdot}17\times10^{-6}$ cm^3 s^{-1} (see Appendix II). By a slight increase of α_∞ to $1{\cdot}25\times10^{-6}$ cm^3 s^{-1}, which is still within the limits of experimental error, the straight line in figure 7 would have a slope of $-0{\cdot}5$, i.e. $\sigma_N/N \sim 1/\sqrt{t}$.

Figure 8 shows an example of the measurement of the time dependence of the recombination coefficient. Equation (8) gives

$$\alpha=\alpha_\infty \frac{\overline{N^2(t)}}{\overline{N(t)}^2}\left(1+\frac{N_0}{2\pi\delta(t)n_0}\right)=\alpha_\infty\frac{\overline{N^2(t)}}{\overline{N(t)}^2}+\frac{\beta(t)}{n_0} \tag{9}$$

In this equation $\beta(t)$ is the experimentally evaluated slope of the line in figure 3.

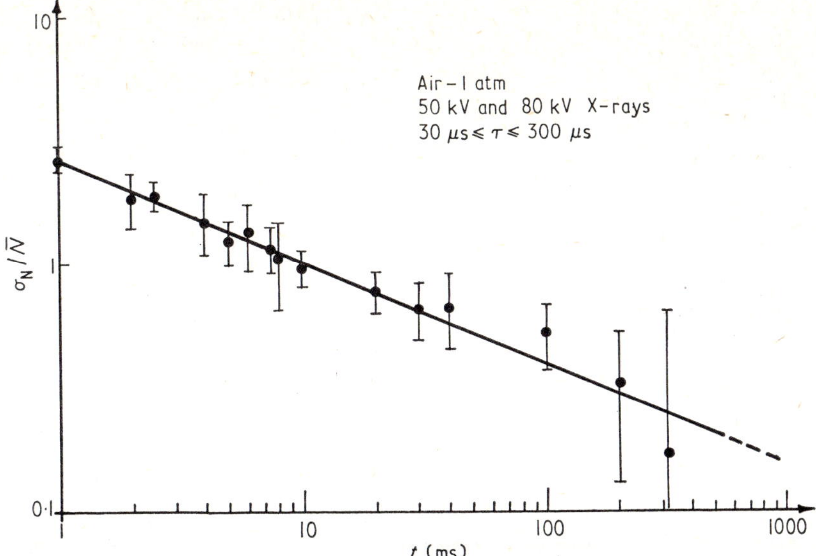

Figure 7. Relative variance of the linear ion density as a function of t. The other parameters are the same as in the preceeding figure.

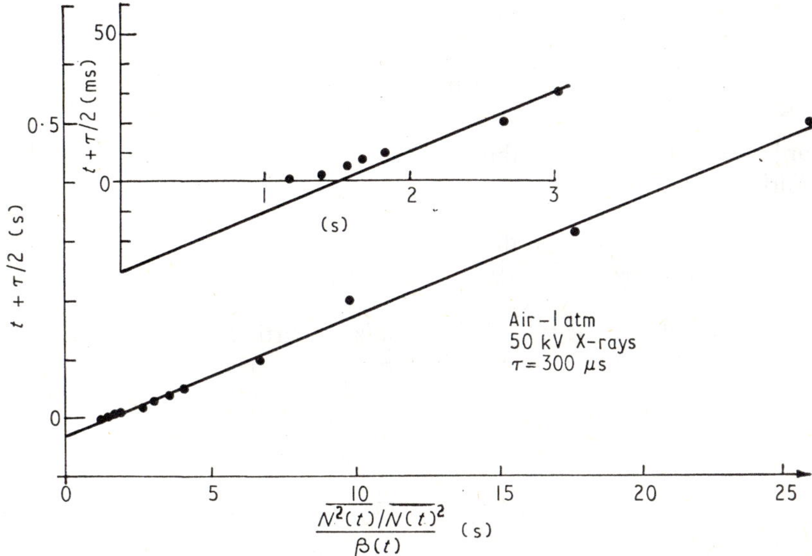

Figure 8. $t+\tau/2$ as a function of $(\overline{N^2}/\overline{N}^2)/\beta(t)$. Air at 1 atm, 50 kV X-rays, ionization time $\tau=300$ μs.

From equation (9)

$$t=\frac{N_0\alpha_\infty}{8\pi D}\,\frac{\overline{N^2(t)}/\overline{N(t)}^2}{\beta(t)}-\frac{\delta_0}{4D} \tag{10}$$

If the ionization time τ be no longer negligible in relation to the recombination time t, one has to use $t+\tau/2$ instead of t on the left side of equation (10). Figure 8 demonstrates that a straight line is obtained in accordance with equation (10), when $t+\tau/2$ is plotted against $(\overline{N^2}/\overline{N}^2)/\beta(t)$. Therefore columnar recombination seems to be the dominating process. Only for $t+\tau/2 \leqslant 10$ ms do the experimental points deviate from the straight line drawn in figure 8. Information on cluster recombination cannot be expected as long as $N_0\sqrt{2Dt} > 1$. This means for X-rays that t must be smaller than 10 μs.

5. Summary

It has been shown how the variance of the linear ion density as a function of the time after ionization can be obtained by measuring the recombination coefficient of initial recombination. The shortest recombination time which has been studied was 1 ms. This means that the measurements have been done in the regions of columnar and volume recombination, the boundary region between these two types of recombination being dependent on the average ion density n_0. The results have shown that in the observed region the relative variance of the linear ion density decreases as $t^{-1/2}$.

Appendix I

Columnar recombination and volume recombination have been treated extensively by Jaffé (1913, 1929) and by Ebert *et al.* (1964). In the following the recombination coefficient for cluster recombination will be studied in a similar way.

Lea (1934) has calculated the distribution of ion density $n(x, y, z, t)$ in an isolated cluster, by neglecting at first the recombination term in the general differential equation. His solution was

$$n(x, y, z, t) = \frac{\nu}{\pi^{3/2} b^3} \exp\left(-\frac{x^2+y^2+z^2}{b^2}\right); \quad b^2 = 4Dt + b_0^2 \tag{11}$$

Afterwards he corrected this distribution function like Jaffé by making ν time dependent and requiring

$$\frac{\mathrm{d}\nu}{\mathrm{d}t} = -\alpha \frac{\nu^2(t)}{\pi^{3/2} b^3 \sqrt{8}} \tag{12}$$

Let it be assumed that the tracks of the ionizing particles are orientated only in the z-direction and are infinitely long. They may be distributed regularly in space so that they form a regular lattice in cross section. The distance from each track to the nearest neighbours is a, and l is the constant separation of the clusters. By overlapping the ion densities of the different clusters one obtains with equation (11)

$$n(x, y, z, t) = \frac{\nu(t)}{\pi^{3/2} b^3} \sum_{\mathrm{i,j,k}=-\infty}^{+\infty} \exp\left\{-\frac{(x-\mathrm{i}a)^2+(y-\mathrm{j}a)^2+(z-\mathrm{k}l)^2}{b^2}\right\}$$

$$\mathrm{i, j, k} = 0, \begin{matrix} +1, +2, +3, \dots \\ -1, -2, -3, \dots \end{matrix} \tag{13}$$

For calculating equation (2) with the distribution (13) one must choose the volume V over which the integration is made in such a way, that the number of ions in V does not change by diffusion. If this rule is obeyed, equation (12) is fulfilled too. Let V equal $a^2 l$. From this follows

$$\int_V n(x, y, z, t)\, \mathrm{d}V = \overline{\nu(t)} \tag{14}$$

and

$$\int_V n^2(x, y, z, t)\, dV = \overline{\nu^2(t)}\, \frac{1}{\pi^{3/2} b^3 \sqrt{8}} \sum_{i,j,k} \exp\left\{-\frac{1}{2b^2}(a^2 i^2 + a^2 j^2 + l^2 k^2)\right\} \tag{15}$$

For $a^2/b^2 \geqslant 10$ and $l^2/b^2 \geqslant 10$ the sum converges very quickly and one obtains

$$\int_V n^2(x, y, z, t)\, dV = \frac{\overline{\nu^2(t)}}{\pi^{3/2} b^3 \sqrt{8}} \tag{16}$$

As $V = a^2 l = \bar{\nu}_0 / n_0$, the equations (2), (13), and (15) give

$$\alpha = \alpha_\infty \frac{\overline{\nu^2(t)}}{\overline{\nu(t)}^2} \frac{\bar{\nu}_0}{\pi^{3/2} b^3 n_0 \sqrt{8}} \tag{17}$$

where n_0 is the average ion density at $t = 0$.

Appendix II

Table 1. Measured coefficients of volume recombination in different gases

Gas	99·9% O_2	Air	99·99% N_2 + 5 ppm O_2
Type of recombination	ion–ion $O_2^+ + O_3^-$ (6)	ion–ion $O_2^+ + O_3^-$ (6)	ion–ion + electron–ion
Authors:			
Booz and Ebert		$1 \cdot 35 \times 10^{-6}$ cm³ s⁻¹	
Ebert *et al.*	($1 \cdot 10 \times 10^{-6}$ cm³ s⁻¹)	($1 \cdot 17 \times 10^{-6}$ cm³ s⁻¹)	
Koepp	$1 \cdot 25 \times 10^{-6}$ cm³ s⁻¹ ±6%	$1 \cdot 41 \times 10^{-6}$ cm³ s⁻¹ ±7%	
Gottsmann	$1 \cdot 32 \times 10^{-6}$ cm³ s⁻¹ ±6%	$1 \cdot 37 \times 10^{-6}$ cm³ s⁻¹ ±7%	$1 \cdot 42 \times 10^{-6}$ cm³ s⁻¹
Most probable value	$1 \cdot 29 \times 10^{-6}$ cm³ s⁻¹	$1 \cdot 39 \times 10^{-6}$ cm³ s⁻¹	

All measurements have been made with carefully dried gases (dry ice) and have been corrected for macroscopic diffusion following the method of Jaffé (1941).

References

Booz, J., and Ebert, H. G., 1963, *Z. angew. Phys.*, **16**, 180.
Ebert, H. G., 1964, *Z. Physik*, **181**, 181.
Ebert, H. G., Booz, J., and Koepp, R., 1964, *Z. Physik*, **181**, 187.
Gottsmann, H., 1966, *Bundesministerium für wissensch Forschung-forschungsbericht-FG* **1**, 307.
Jaffé, G., 1913, *Ann. Phys.*, IV, **42**, 303.
—— 1929, *Ann. Phys.*, V, **1**, 977.
—— 1941, *Phys. Rev.*, **59**, 652.
Jütting, H., Koepp, R., Booz, J., Ebert, H. G., 1965a, *Z. Naturf.*, **20a**, 213.
—— 1965b, *Z. Naturf.*, **20a**J, 217.
Koepp, R., 1963, *Diplomarbeit Göttingen.*
Koepp, R. and Gottsmann, H., 1966, *Proc. 3rd Int. Congr. of Radiation Research*, Cortina D'Ampezzo.
Lea, D. E., 1934, *Proc. Camb. Phil. Soc.*, **30**, 80.

Preferential recombination of ions in gases at high pressures

R. KOEPP

Institut für Medizinische Physik und Biophysik, Universität Göttingen, Germany

Abstract. Preferential recombination is studied in gases (N_2, O_2, Air) at high pressures (p_{max} = 200 atm), irradiated by 50 kV X-rays. After definition of the terms preferential recombination of the first and second kind there is described a method to differentiate between initial (columnar) and preferential recombination. The result is represented as a plot of escape probability versus relative density. The track parameters are determined by Jaffé's theory. They can only give an order of magnitude. Nevertheless they are in agreement with parameters used to fit the curve of escape probability by Onsager's theory.

1. Introduction

By preferential* recombination we mean that process where the electron originated by ionization cannot escape from the sphere of Coulombic attraction of its parent ion or enter the sphere of attraction of another positive ion. The radius of this sphere is defined as the distance from the positive ion at which the Coulombic energy is equal to the thermal energy:

$$R_k = \frac{e^2}{AkT}$$

with e = electronic charge, A = numerical factor ~ 1, k = Boltzmann constant and T = absolute temperature.

This is demonstrated in figure 1. In the upper part let us assume high density and in line (a) ionization by low LET radiation (for instance ^{60}Co γ-rays), resulting in a low linear ion density, N_0. The small black circles represent the positive ions and the circles R_k, the spheres of active attraction. If the mean distance the electrons can travel before they are thermalized is R and $R \ll R_k$ then they recombine preferentially (preferential recombination of the first kind). In figure 1*b* a high LET particle has passed, resulting in an overlapping of the Coulombic spheres; electrons escaping their parent ions frequently get into the sphere of attraction of another ion. This we call preferential recombination of the second kind, the condition for this being $RN_0 \gg 1$.

Table 1. Examples for RN_0-values for various radiations

Radiation	RN_0
Co$-\gamma$	$\sim 0{\cdot}1$
soft X-rays	~ 2
α-particles	~ 30

At low density the preferential recombination of the first kind is of little importance

* There are also in use nomenclatures such as 'geminate recombination' and 'primary recombination'.

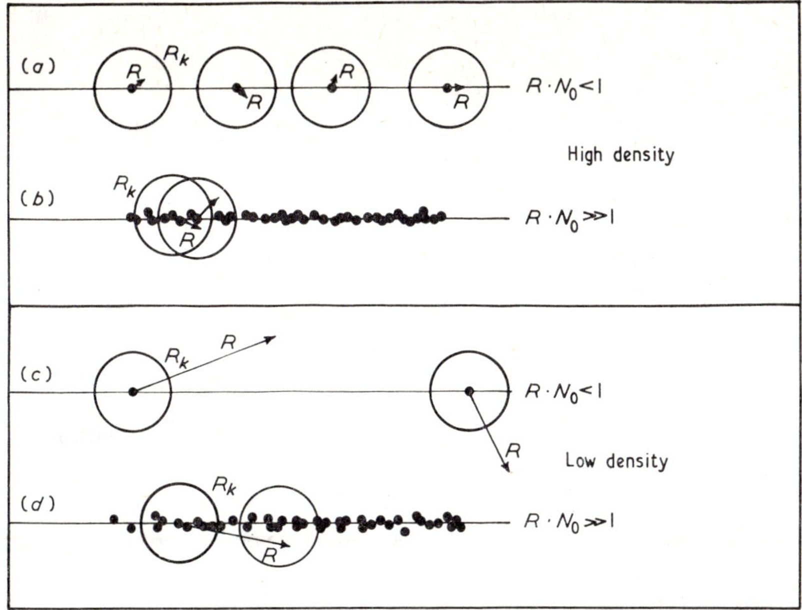

Figure 1. Schematic representation of the possibilities for thermalization of electrons in proximity of positive ions. (a) High density, low LET particle. $RN_0 < 1$: Preferential recombination of the first kind. (b) High density, high LET particle. $RN_0 \gg 1$: Preferential recombination of first and second kind. (c) Low density, low LET particle. $RN_0 < 1$: Very small probability for recombination of the first kind. (d) Low density, high LET particle. $RN_0 \gg 1$: A certain amount of preferential recombination of second kind.

(figure 1*c*) (Ebert, Booz and Koepp 1964). However one can have a certain percentage of preferential recombination of the second kind (figure 1*d*).

Most of the experimental approach to the preferential recombination has been done in liquids with high energy X- and γ-rays (Freeman 1963, Hummel and Allen 1967, Jacobi 1968, Schmidt and Allen 1968). Then the Onsager theory which describes the preferential recombination of isolated ion pairs may be valid (Onsager 1938). To study preferential recombination in gases as a function of pressure, wall effects of the ionization chamber have to be absent. This is the case with low energy X-rays. However there is still the explained overlapping of the spheres of attraction and also a great amount of initial recombination. We differentiated between initial and preferential recombination by irradiating the gas in an ionization chamber with X-rays of constant energy flux and varied the gas pressure from 1 atm to 200 atm. The experimental saturation current determined in a suitable manner corresponds to initial recombination only. If we compare this current with the theoretically expected ionization current which increases proportional to gas density, we obtain information on the loss by preferential recombination.

There are three theories concerned with the influence of initial recombination on the current voltage curve. The theory of Kramers (1952) was found to be inapplicable as it deals with the case where recombination losses are large compared with diffusion losses out of the track (it is valid for liquids at very low temperature). The extrapolation methods of Lea and Jaffé for high electrical field strength lead to the same extrapolation current (Kara-Michailova and Lea 1940). Therefore the method of Jaffé–Zanstra (Zanstra 1935) was used. This describes the dependence of ionization current as a function of field strength E according to the well-known formula

$$\frac{1}{i}=\frac{1}{i_J}+C\,f(x)$$

where i_J = saturation current corresponding to initial recombination, C = constant containing ionic constants and N_0, and $f(x)$ = complicated function of field strength E and Jaffé parameter b.

If in addition to initial recombination, preferential recombination is present then

$$i_J=i_s\phi$$

with i_s = theoretical current proportional to gas density and ϕ = escape probability from preferential recombination.

This corresponds to Onsager

$$\phi=\phi_0\phi_E$$

containing a constant part ϕ_0 which describes the escape probability without electrical field E

$$\phi_0=\exp\,(-R_k/R)$$

and a field dependent part ϕ_E which will be shown to be negligible in our case.

2. Apparatus and gases

In figure 2 a construction diagram of the ionization chamber is shown. A guard ring condenser is built up isolated in a commercial high pressure vessel. 50 kV X-rays

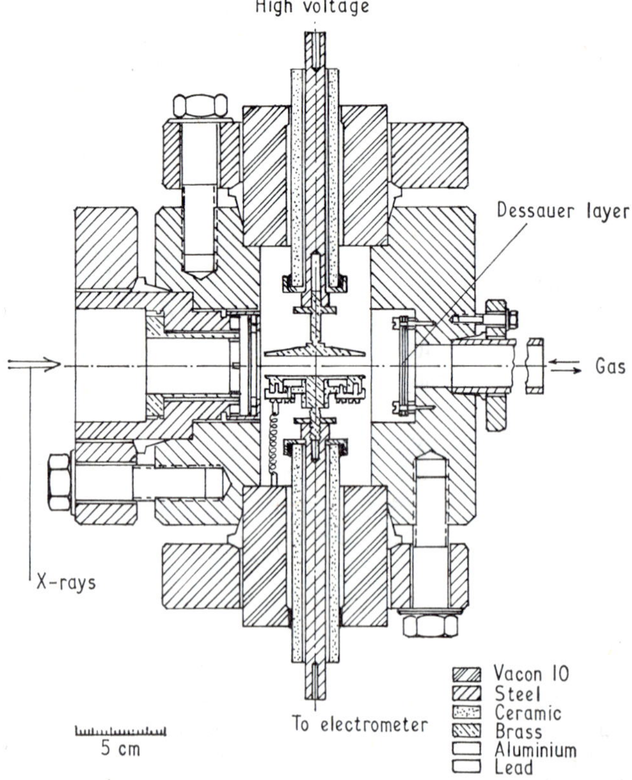

Figure 2. Construction diagram of the high-pressure ionization chamber.

enter from the left side by a 6 mm Al-window into the chamber and are absorbed behind the condenser by a Dessauer layer. The X-ray beam has a width of 1 mm in a plate spacing of 1 cm. So wall effects which take place in the former works are mostly prevented.

The gases used were oxygen, nitrogen, and synthetic air (21% O_2; 79% N_2)
Purity: O_2: 99·9%; $H_2O < 50$ ppm, $N_2 + Ar < 1000$ ppm
N_2: 99·99% incl. rare gases; $O_2 < 5$ ppm, $H_2 < 1{\cdot}5$ ppm.
The highest gas pressure in the chamber was determined by the pressure of 200 atm available in commercial tanks.

3. Measurements and evaluation

In figure 3 an example for curves measured with oxygen is shown. Ionization

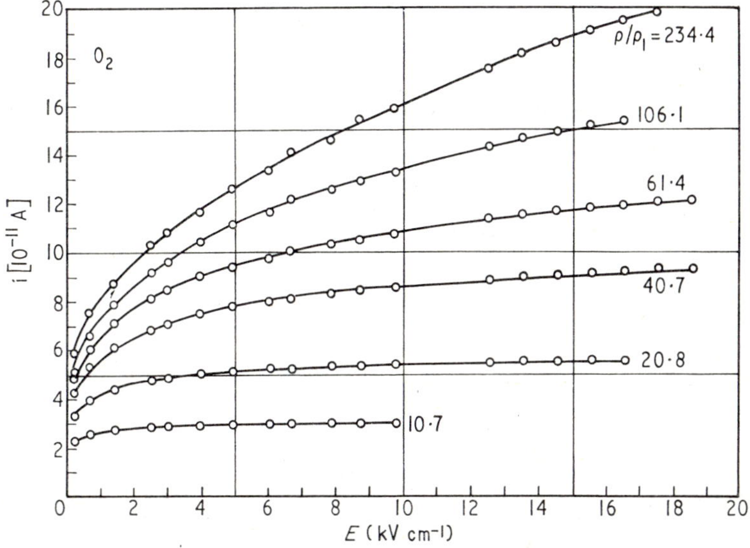

Figure 3. Current–voltage characteristic for oxygen; parameter is relative density. Corrected for attenuation of X-rays ($U_A = 50$ kV).

current is plotted against field strength; parameter is the relative density. Currents are corrected with respect to attenuation of X-rays by increasing the density of the gas layer before the condenser. It can be seen that at low pressures saturation is approximated. At higher pressures we are far from saturation even with a field strength of about 20 kV cm^{-1}.

Figure 4 represents a plot of $1/i$ versus $f(x)$ according to the theory of Jaffé–Zanstra for the same oxygen measurements. We see in the diagram a series of parallel straight lines which are extended in broken lines to the intersections $1/i_J$ with the ordinate. I have to emphasize that the column parameter b, used to calculate $f(x)$, is only once fitted and then reduced according to pressure as the theory demands (Jaffé 1913).

As the straight lines are parallel it is evident that the influence of the electrical field on preferential recombination is small, in other words $\phi \sim \phi_0$.

The saturation currents determined in this manner are now compared as normalized values i_J/i_1 (i_1 = current at 1 atm) with the theoretical expected current i_s/i_1 (figure 5).

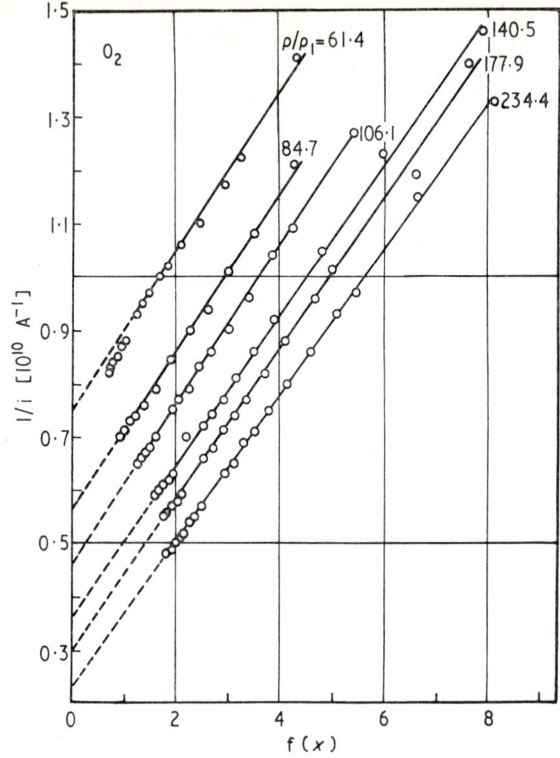

Figure 4. Jaffé–Zanstra plot of the values out of Figure 3.

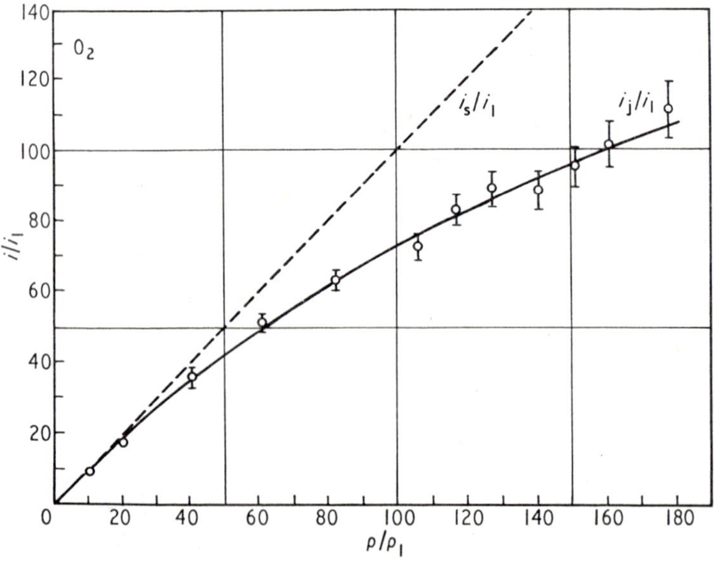

Figure 5. Comparison of the expected current i_s (dashed line) with the experimentally determined Jaffé-saturation currents i_J (solid curve) as a function of relative density. All currents are normalized (i_1 = current at 1 atm).

The theoretical current normalized as above and plotted against ρ/ρ_1 gives a straight line with a slope of 1. The deviation of the measured values from the expected due to preferential recombination is remarkable, about 40% for oxygen at 200 atm.

The ratio $\phi = i_J/i_s$ therefore represents the escape probability from preferential recombination. In figure 6 this quantity is plotted against the relative density for all

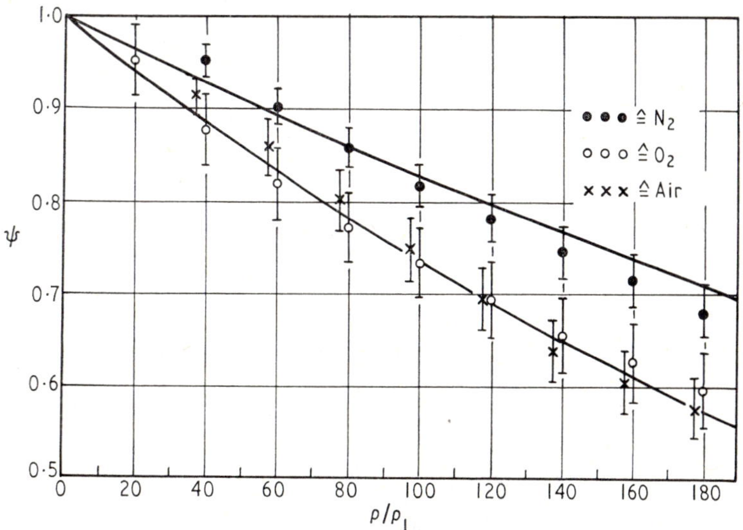

Figure 6. Escape probability $\phi = i_J/i_s$ versus relative density for all measurements with nitrogen, oxygen and air. The curves corresponding to the Onsager expression $\phi_0 = \exp(-R_k/R(\rho))$.

measurements I have made. Oxygen and air show rather similar escape probabilities, while nitrogen at all pressures yields a higher percentage of escaping ions. This result with nitrogen is surprising because until now the attachment behaviour of gases was considered essential for preferential recombination and nitrogen should have shown no preferential recombination. The curves in the diagram are theoretical ones according to the theory of Onsager for zero field escape probability.

Substituting into the equation

$$\phi_0 = \exp(-R_k/R)$$

the following values: $R_k = 350$ Å, $R\,(1\text{ atm}) = 1{\cdot}1 \times 10^{-3}$ cm (oxygen and air), $R\,(1\text{ atm}) = 1{\cdot}8 \times 10^{-3}$ cm (nitrogen) we obtain the curves drawn in the diagram.

4. Results

As the measurements are in good agreement with the Jaffé theory, we may conclude that the ionization originated by low energy X-rays seems to exist in the form of columns. Nevertheless the resulting track parameters give only an order of magnitude on account of the simplifying assumptions of Jaffé's theory.

These parameters are determined as follows

(i) Column parameter $b\,(1\text{ atm}) = 3{\cdot}4 \times 10^{-4}$ cm for air, oxygen,

$b\,(1\text{ atm}) = 9{\cdot}0 \times 10^{-4}$ cm for nitrogen.

(ii) The slope of Jaffé–Zanstra straight lines delivers the mean linear ion density

$$N_0\,(1\text{ atm}) : 2300\,\frac{JP}{\text{cm}}$$

corresponding to LET ~ 6(keV μm^{-1}).

(iii) Hence follows the mean distance of ion pairs in the column

$$\frac{1}{N_0}\,(1\text{ atm}) \sim 4\times 10^{-4}\text{ cm.}$$

(iv) From Onsager theory we obtain an average distance for the electrons from the parent ion:

$$R\,(1\text{ atm}) = 1{\cdot}1\times 10^{-3}\text{ cm for oxygen and air.}$$

$$R\,(1\text{ atm}) = 1{\cdot}8\times 10^{-3}\text{ cm for nitrogen.}$$

which is in the order of magnitude of the column parameter b.

(v) The product $RN_0 \sim 2$ shows that there must have been a certain amount of preferential recombination of the second kind.

References

Ebert, H. G., Booz, J., and Koepp, R., 1964, *Z. phys. Chem.*, **43**, 304–10.
Freeman, G. R., 1963, *J. chem. Phys.*, **39**, 1580–6.
Hummel, A., and Allen, A. O., 1967, *J. chem. Phys.*, **46**, 1602–6.
Jacobi, W., 1968, *Studia biophys.*, **7**, 195–202.
Jaffé, G., 1913, *Ann. Phys., Lpz.*, **42**, 303–44.
Kara-Michailova, E., and Lea, D. E., 1940, *Proc. Camb. Phil. Soc.*, **36**, 101–26.
Kramers, H. A., 1952, *Physica*, **18**, 665–75.
Onsager, L., 1938, *Phys. Rev.*, **54**, 554–7.
Schmidt, W. F., and Allen, A. O., 1968, *Science*, **160**, 301–2.
Zanstra, H., 1935, *Physica*, **2**, 817–24.

An assessment of a low pressure cloud chamber for measuring the spatial distribution of ionization produced by high LET particles

H. J. DELAFIELD and P. D. HOLT

Health Physics and Medical Division, AERE, Harwell, Didcot, Berkshire.

Abstract. The potentialities of a low pressure cloud chamber for measuring the spatial distribution of energy deposition by ionizing particles of low energy have been investigated. Experimental studies, using pulsed 1·5 keV characteristic Al X-rays, have shown that at low pressures (10 cm Hg or less) a cloud chamber must be operated with considerable precision to avoid the diffusion of ions during the early stages of droplet growth. Chord ranges of electrons down to about 1 keV can be measured but individual droplets cannot be readily resolved.

Theoretical considerations have shown that the ultimate performance of such a chamber is dependent upon the lowest operating pressure which can be utilized before diffusion becomes significant, and upon the resolution of the optical system. Using an improved expansion mechanism it should be possible to resolve the details of the tracks of particles whose LET does not exceed 10 keV μm^{-1} in unit density material.

The observed chord ranges of the electrons produced in the chamber, of energy $E = 1{\cdot}0$–$1{\cdot}5$ keV, were consistent with the range–energy relationship $R = 4{\cdot}55\ E^{1{\cdot}5}\ \mu g\ cm^{-2}$.

1. Introduction

This paper describes an experimental investigation with a pulsed source of characteristic 1·49 keV X-rays, into the potentialities and limitations of a low pressure cloud chamber for studying the initial structure of the tracks produced by particles with high specific ionization. The factors involved in the interpretation of the experimental data are assessed.

In a Wilson-type cloud chamber the path of an ionizing particle is made visible by the controlled condensation of vapour upon the individual ions formed in the trajectory of the particle. To obtain precise data from photographs of the droplets it is essential that the ions should not have diffused very far between the time of their formation and the time the droplets are photographed. This requirement determined the operating sequence of the present chamber, which has been fully described elsewhere (Delafield 1968):

1. At zero time the cloud chamber was expanded adiabatically.
2. At 108 ms the electrostatic clearing field was shorted out.
3. At about 110 ms the X-ray tube was pulsed for a few milliseconds to produce characteristic X-rays, and hence photoelectron and Auger electron tracks in the chamber during its very short sensitive time.
4. At 150 ms the droplets formed were photographed using a xenon flash tube.

The chamber was operated typically at a filling pressure of about 60 mm hydrogen, 18 mm of ethyl alcohol vapour and 11 mm of water vapour which gave a density on expansion of about 6×10^{-5} gm cm^{-3}. Characteristic aluminium X-rays (1·49 keV) were obtained from a simple tube in which 3 kV electrons were accelerated from an electron gun on to a thin aluminium target which also served as the window. Spectral measurements with a proportional counter showed that characteristic X-rays were produced without any appreciable continuous background radiation.

2. Performance of cloud chamber at low pressures

Experimental studies showed that, to obtain reproducible results as the operating pressure in the chamber was reduced, it was necessary to increase the precision of the timing sequence. The sensitive time of the chamber, considered here to be that time during which condensation giving rise to a droplet will occur immediately on any ion formed, was estimated to be only about 1 ms at an operating pressure of 90 mm of mercury. By contrast the expansion of the cloud chamber caused by the rapid collapse of a neoprene diaphragm took about 110 ms, and was not reproducible to probably much better than $\pm 5\%$.

Since the expansion time was not reproducible, it was difficult to make the X-ray pulse coincide with the short sensitive time of the chamber. Hence a large number of photographs were taken and a selection made for analysis. Tracks (two shown in figure 1 and one in figure 2) or diffuse clusters (figure 3) were observed according to whether the X-ray pulse had occurred during or before the sensitive time.

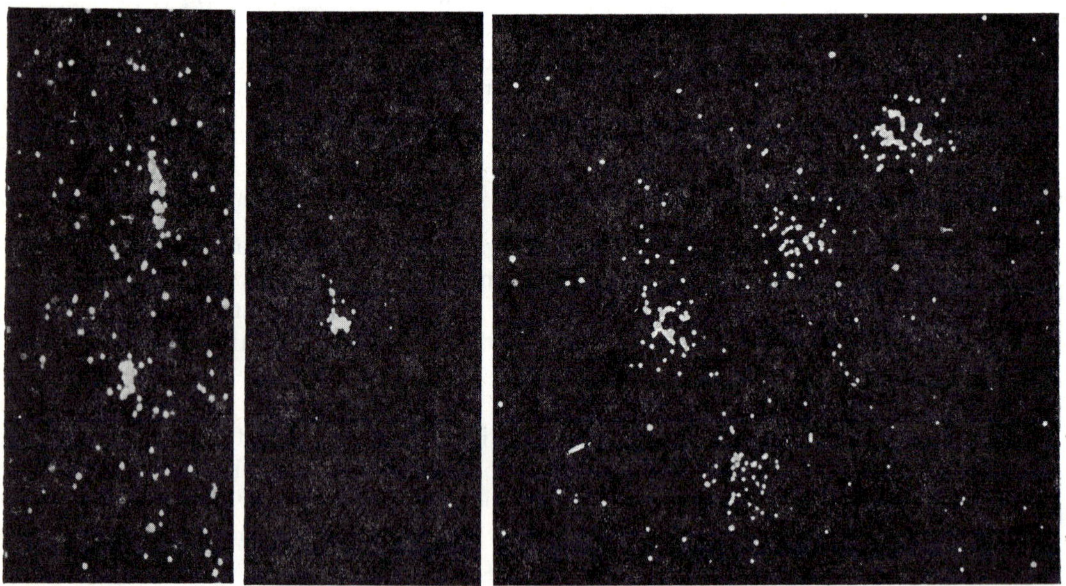

Figures 1–3. Tracks and diffuse clusters produced by photoelectric absorption of 1·5 keV X-rays (1 cm on photographs $= 1{\cdot}03 \times 10^{-2}$ mg cm^{-2}).

2.1. *Track magnification and resolution*

In order to resolve the individual droplets in tracks of high specific ionization, it is clearly desirable to operate with a high optical magnification (m), and at as low a chamber-filling pressure as possible.

Optical considerations have shown (Delafield 1968), that there is little advantage in operating at a magnification of $m > 1$ because the object depth of field decreases as $1/m^2$. If the resolution is not limited by the optics, coalescence of adjacent droplets in the chamber may be the limiting factor. In the practical case using $m < 1$, the ability to resolve adjacent droplets is limited by the resolution of the photographic film.

Photographs taken with Ilford FP3 film at a magnification of $m = 0{\cdot}26$ enabled droplets separated by about 0·0085 cm in the chamber to be just resolved. Therefore a track of 39 ions cm^{-1} specific ionization could be resolved in the cloud chamber, allowing a factor of 3 for the random nature of ionization. This may be compared with the expected theoretical resolution of 44 ions cm^{-1} at this magnification for a good fine grain emulsion film, which according to Welford (1962) has a resolution half-width of about 20 μm.

Because the optical magnification is limited by the considerations outlined, it becomes essential to operate a cloud chamber at as low a pressure as possible to obtain the maximum overall track magnification. With the present chamber the operating mechanism (maximum expansion ratio 1·62) limited the lowest pressure achieved to 10 mm hydrogen and 18 mm of water vapour. Although by using a larger expansion ratio, and perhaps a more suitable vapour, operating pressures of a few mm mercury may be achieved, as has been demonstrated by Nielson (1956), Mills (1952) and Vladimir *et al.* (1959), the problems of operating a chamber as a precision instrument become much greater.

2.2. *Diffusion*

From selected photographs of the least diffused tracks (figures 1 and 2), it was deduced that the diffusion of newly formed ions had been controlled to within certain limits. Some of the tracks, which were 'tadpole' shaped (figure 2) were believed to be from the absorption of 1·47 keV photoelectrons. It is not known whether the apparent width of the 'tail of the tadpole', corresponding to the beginning of the electron track, is due to diffusion or whether it represents the genuine scattered path of the electron. Hence it is only possible from the half-width of the tail of these tracks to conclude that the ions had not diffused more than about 0·01 cm (less if scattering had occurred) in the chamber at a filling of density 6×10^{-5} gm cm^{-3}.

Difficulty was experienced at this filling density in operating the chamber with sufficient precision to control the degree of diffusion adequately. At lower filling pressures the problem of operating the chamber becomes more acute, as the sensitive time is expected to decrease with pressure (Wilson 1951) and the rate of diffusion to increase.

In a theoretical study Holt (private communication) has shown that the average distance a droplet diffuses is proportional to the mean free path of molecules in the vapour and hence inversely proportional to the pressure. Because of this increased diffusion it may not be possible to utilize the higher resolution obtainable by lowering the operating pressure to increase the mean spacing between the ions. Holt showed that the average diffusion distance is critically dependent upon the sticking probability of the first few vapour molecules colliding with the newly formed ion and would be expected to be less than the 0·01 cm mentioned above unless the initial sticking probability is very low. Unfortunately little experimental data is available on the sticking probability of various vapour molecules on charged ions.

Meanwhile assuming that the diffusion could be adequately controlled using a chamber with an improved design of expansion mechanism, it is probable that the present resolution could be improved by a factor of two by using a higher optical magnification ($m = 0{\cdot}5$), and a further factor of two by reducing the chamber pressure to give a filling density of 3×10^{-5} gm cm^{-3}. This would enable a track of about 85 ions cm^{-1} in the chamber to be resolved, corresponding to a track of 10 keV μm^{-1} LET in unit density material.

3. Interpretation of cloud chamber photographs

Characteristic aluminium X-rays (1·49 keV) were absorbed by the cloud chamber filling gas to produce a mixture of photoelectrons and Auger electrons. The relative frequency of the interactions together with the energies of the electrons produced (table 1) were calculated using mass absorption coefficients given by Henke, White and

Table 1. Production of electrons in the cloud chamber filling gas

Partial pressures 59·3 mm hydrogen, 18·1 mm ethyl alcohol, 10·6 mm water

Interacting atom	Shell	Relative frequency of interaction (%)	Photoelectron energy (keV)	Auger electron energy (keV)
O	K	58·4	0·955	0·518
O	L	14·6	1·480	—
C	K	21·4	1·203	0·270
C	L	5·4	1·480	—
H	K	0·2	1·473	—

Lundberg (1957), assuming that 20% of the interactions for oxygen and carbon were with L-shell electrons (Evans, 1955). The energies of the electrons produced were derived from the energy levels given by Lederer, Hollander and Perlman (1967). Since the K-fluorescence yield is very small (about 10^{-3}), for both oxygen and carbon the photoelectrons will practically always be accompanied by Auger electrons, so that for each photon absorbed about 1·47 keV will be imparted to the electron(s) formed.

3.1. *The average energy per ion pair, W*

From photographs of clusters which had diffused sufficiently for the droplet overlap correction to be small the mean number of droplets per cluster was observed to be 41·3 with a standard error of $\pm 0{\cdot}8$. The results presented as a histogram in figure 4 are compared with a curve for a Poisson distribution with the same mean and enclosing the same area. If it is now assumed that the observed droplets were all formed by 100% condensation of vapour upon positive ions, the much more mobile electrons having rapidly diffused away from their point of origin, the average energy to produce an ion pair (W) in the cloud chamber gas mixture becomes $35{\cdot}6 \pm 0{\cdot}7$ eV ion pair^{-1}.

Based on the formula of Bortner and Horst (1954), a value of W for the cloud chamber gas mixture was calculated to be 28·3 eV ion pair^{-1}, assuming that the energy loss to the component gases was a simple function of their partial pressures and molecular stopping powers. Energy-averaged values of stopping powers for a 1 keV electron slowing down were derived from M. Marshall, P. D. Holt and J. A. B. Gibson's (private communication) modified form of the Bethe–Bloch equation. Values of W for water and alcohol vapours were taken from Cooper and Mooring (1968), and for hydrogen from Whyte (1963). The difference between the experimental and calculated values of W reflect the uncertainty that was involved in deciding to what size the clusters had diffused, and also the error introduced in correcting for droplet overlap.

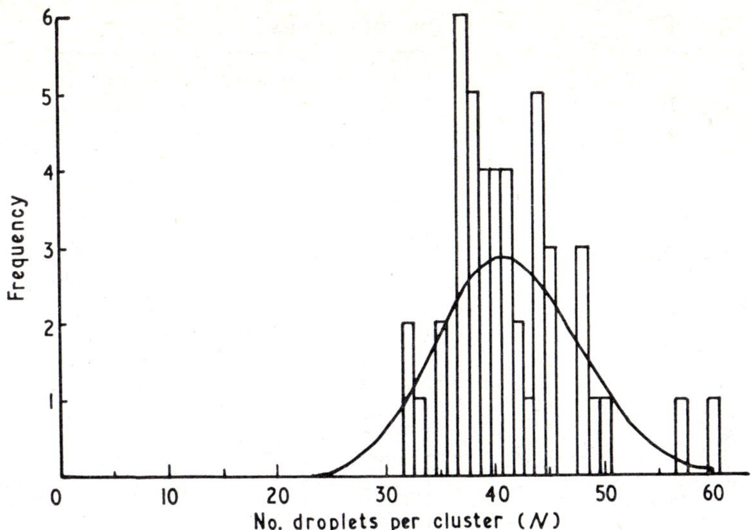

Figure 4. Relative frequency of clusters containing N droplets per cluster.

3.2. *The chord range of low energy electrons*

Since individual ions could only be resolved at the beginning of the tracks, it was more meaningful to derive information on the chord range of low energy electrons. This is defined as the greatest straight line dimension of the track. It may differ from the penetration range and from the straight line distance between the start and end of the track. The expected distribution of electron ranges projected into the plane of the field of view of a camera viewing at 60° to the direction of the incident X-rays has been calculated.

At first projected range distributions were calculated for each energy of photoelectron, assuming that the distribution of chord ranges in space was the same as the distribution of directions of generation of photoelectrons. Corrections were made for the presence of accompanying Auger electrons where appropriate. The photoelectrons were assumed to be isotropically distributed in space, after it had been shown that this assumption gave a very similar range-probability distribution to that obtained using data given by Evans (1955). Corrections for the Auger electrons, which are isotropically distributed in space, were made by folding together the photoelectric and Auger probability distributions.

Then taking into account the frequency of each type of event as given in table 1, the results for the different energy photoelectrons were added together using as the chord range–energy relationship

$$R = KE^x$$

where R = range of electrons in cloud chamber filling gas ($\mu g\ cm^{-2}$), E = electron energy (keV), $x = 1{\cdot}5$, chosen to be the most suitable value from Holt (1969) for the energy of electrons being considered and K = a constant of normalization.

The calculated distribution of projected ranges as they would be observed by the camera is compared directly with a histogram of the actual observations in figure 5. The two distributions were superimposed so that the mean ranges of the distributions are equal. This gave a value of $K = 4{\cdot}55\ \mu g\ cm^{-2}\ keV^{-1{\cdot}5}$ for the cloud chamber gas filling (partial pressures 59·3 mm H_2, 10·6 mm H_2O, 18·1 mm C_2H_5OH). The mean projected range in the chamber was $(7{\cdot}32 \pm 0{\cdot}49) \times 10^{-2}$ cm based on photographs

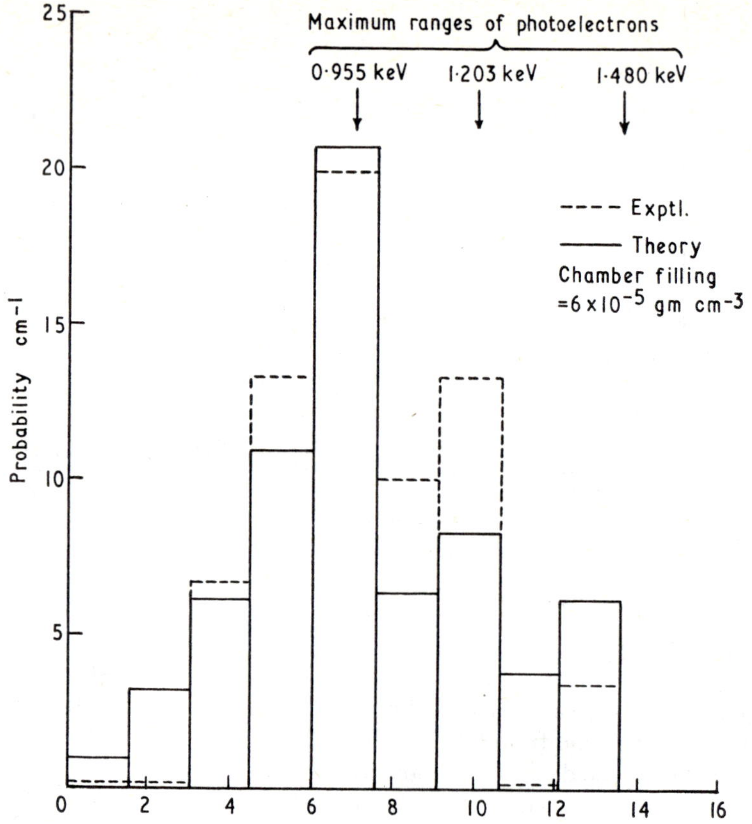

Figure 5. Comparison of experimental and theoretical probability distributions for ranges of electrons in cloud chamber. The abscissa gives the electron range in the cloud chamber as observed by camera viewing at 60° to direction of incident X-rays.

of 20 tracks showing minimum diffusion, the quoted error being the standard error of the mean. For each observation the range was taken as the overall length of the track minus the minimum width of the track, the latter being taken as an approximate correction for diffusion. Both the expected and experimental histograms show that one is most likely to see the track of the photoelectron lying in the field of view of the camera. The effect of the Auger electrons is to increase the maximum observed ranges for the 0·955 and 1·203 keV photoelectrons.

The derived value of K is subject to the mean error of the experimental results of $\pm 7\%$ quoted above, as well as the uncertainties introduced in correcting for diffusion, in over-simplifying the calculation, and in ignoring the effect of range-straggling.

Chord ranges based on the relationship $R = 4{\cdot}55\ E^{1\cdot 5}\ \mu\text{g cm}^{-2}$ obtained from the present measurements, together with Tikvah Alper's (1932) range measurements of low energy δ-rays in air using a cloud chamber, are compared in figure 6 with calculated slowing down range–energy curves given by the modified Bethe–Bloch equation of Marshall *et al.* (private communication). One would expect that the chord range given by this experiment (and essentially by Alper's work as well, since maximum range δ-rays were observed), would be less than the calculated range along the path. However both cloud chamber experiments give ranges which are some 10–20% greater than the calculated values. We are unable to decide whether this discrepancy

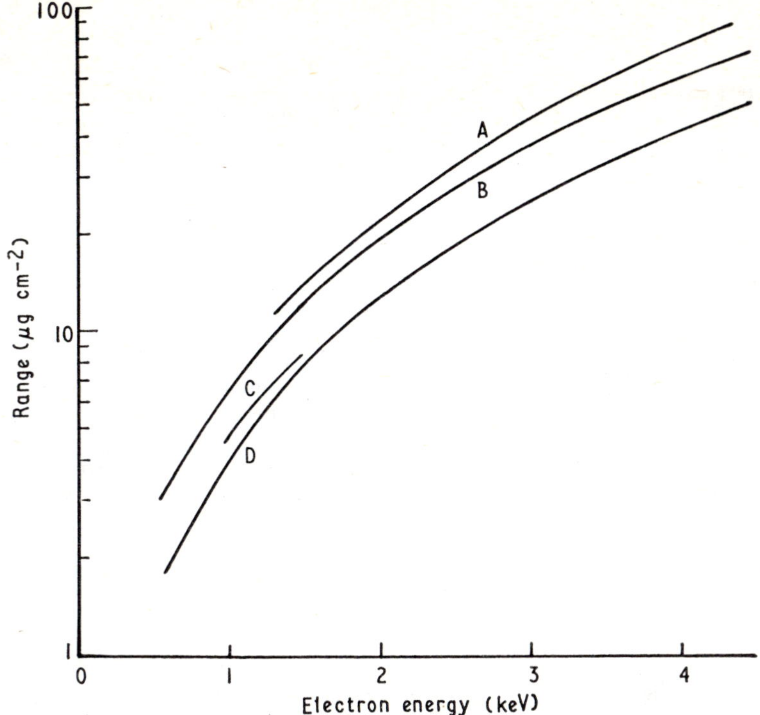

Figure 6. Comparison of measured and calculated electron ranges. (A) Air; experimental curve by Alper. (B) Air; calculated curve by Marshall. (C) Hydrogen/water/alcohol mixture; experimental curve by Delafield. (D) Hydrogen/water/alcohol mixture; calculated curve by Marshall.

is due to experimental errors in the present and earlier cloud chamber measurements or to the assumptions implicit in the calculations.

Acknowledgments

The authors wish to thank J. A. Dennis for his encouragement and much helpful discussion, G. Verney for his experimental assistance and S. J. Boot for computing assistance.

References

Alper, T., 1932, *Zeit. für Physik*, **76**, 172.

Bortner, T. E., and Hurst, G. S., 1954, *Phys. Rev.*, **93**, 1236.

Cooper, R., and Mooring, R. 1968, *Aust. J. Chem.*, **21**, 2417.

Delafield, H. J., 1968, *AERE Rep.*, No. 5991.

Evans, R. D., 1955, *The Atomic Nucleus* (New York: McGraw-Hill Book Co., Inc.), p. 695.

Henke, R. L., White, R., and Lundberg, B., 1957, *Brit. J. appl Phys.*, **28**, 98.

Holt, P. D., 1969, this volume.

Lederer, C. M., Hollander, J. M., and Perlman, I., 1967, *Table of Isotopes*, Appendix III, 6th Edn. (Chichester: John Wiley and Sons).

Mills, R. G., 1952, *UCRL*-1815.

Nielson, C. E., 1956, *Conf. on recent development in cloud chamber and associated technique*, Eds N. Norris and M. J. B. Duff, University College, London, p. 200.

Vladimir, A. S., Miho, C. A., Živan, D. DJ., and Aleksandar, M. B., 1959, *Bull. Inst. Nucl. Sci., Boris Kidrich*, **10**, 33.

Welford, W. T., 1962, *Journ. Photogr. Sci.*, **10**, 242.

Whyte, G. N., 1963, *Radiat. Res.*, **18**, 265.

Wilson, J. G., 1951, *The Principles of Cloud-chamber Technique* (London: C.U.P.), p. 39.

The range of electrons below 1 keV

P. D. HOLT

Atomic Energy Research Establishment, Harwell, Didcot, Berkshire

Abstract. The quantum-mechanical formula for the energy loss of electrons by ionization and excitation fails at low energies (below about 200 eV) and this energy region is of critical importance for target theory. An attempt has been made to calculate the stopping power using a formula based on classical principles, proposed by Gryzinski. Cross section of atoms for ionization and excitation by electron impact, calculated from related formulae, are compared with experimental data.

Ranges calculated using Gryzinski's formula are compared with those proposed by Marshall *et al.* (1969). An examination of the discrepancies between previously proposed range–energy relationships is given.

1. Introduction

Recent work in this laboratory (Marshall *et al.* 1969) has shown that a knowledge of the range–energy relation of low energy electrons in dry protein is of crucial importance for the application of target theory to the radiation inactivation of enzymes. Published figures for the energy loss and range of electrons below 1 keV differ widely (see section 3 and table 1), particularly in the important energy

Table 1. Estimate of the range of a 100 eV electron

Author	Material	Range (μg cm^{-2})	Type of range
Lea (1955)	Water	0·30	Total path length
Davis (1955)	Dry protein	0·2	Penetration
Mozumder and Magee (1966)	Water	0·15	Total path length
Butts and Katz (1967)	Water, Aluminium	1·0	Penetration
Present calculation	Water	0·16	Total path length
Present calculation	Dry protein	0·17	Total path length
Present calculation	Aluminium	0·29	Total path length
Marshall *et al.* (1969)	Water	0·22	Total path length
Marshall *et al.* (1969)	Dry protein	0·24	Total path length
Marshall *et al.* (1969)	Aluminium	0·48	Total path length

region below 100 eV. The usual quantum-mechanical formula for energy loss of electrons (Berger and Seltzer 1964) is not applicable in this energy region since it uses an average excitation–ionization potential which involves all the electron shells, and the velocity of the incident electron is not high compared with that of the K-shell electrons. In fact, the stopping power calculated from this formula goes negative at some energy in the range under consideration.

Gryzinski (1965) has derived formulae, on a classical basis, for the cross section of an atom for excitation, ionization, and total stopping power for charged particles. The effects of the different electron shells are calculated separately, and the contribution of the K-shell to the stopping power is very small below 1 keV, so that the stopping power formula can be used down to energies comparable with the binding energy of the valency electrons. The purpose of this paper is to present calculations of excitation, ionization and stopping cross sections, using Gryzinski's formulae, to compare the excitation and ionization cross sections with published values in order to test the validity of the method, and to compare electron ranges calculated by this method with other published values. A discussion of the reason for the discrepancies between the published ranges is also given.

2. Calculations of atomic cross sections for excitation, ionization and total stopping power

Gryzinski's formulae for incident electrons are as follows:

Excitation:

$$\text{Cross section per atomic electron} = \frac{\sigma_0}{U_n{}^2} g_Q\left(\frac{U_i}{U_n}; \frac{E}{U_n}\right) - \frac{\sigma_0}{U_{n+1}{}^2} g_Q\left(\frac{U_i}{U_{n+1}}; \frac{E}{U_{n+1}}\right) \quad (1)$$

where

$$g_Q\left(\frac{U_i}{U_n}; \frac{E}{U_n}\right)$$
$$= \frac{U_i}{E}\left(\frac{E}{U_i+E}\right)^{3/2}\left[\frac{U_n}{U_i}+\frac{2}{3}\left(1-\frac{U_n}{2E}\right)\ln\left\{2\cdot 7+\left(\frac{E-U_n}{U_i}\right)^{1/2}\right\}\right]\left(1-\frac{U_n}{E}\right)^{1+U_i/(U_i+U_n)}$$

Ionization:

$$\text{Cross section per atomic electron} = \frac{\sigma_0}{U_i{}^2} g_i\left(\frac{E}{U_i}\right) \quad (2)$$

where

$$g_i\left(\frac{E}{U_i}\right) = \frac{U_i}{E}\left(\frac{E-U_i}{E+U_i}\right)^{3/2}\left[1+\frac{2}{3}\left(1-\frac{U_i}{2E}\right)\ln\left\{2\cdot 7+\left(\frac{E-U_i}{U_i}\right)^{1/2}\right\}\right]$$

Stopping power:

$$\text{Stopping cross section per atomic electron} = \frac{\sigma_0}{U_{\text{exc}}} g_s\left(\frac{U_i}{U_{\text{exc}}}; \frac{E}{U_{\text{exc}}}\right) \quad (3)$$

where

$$g_s\left(\frac{U_i}{U_{\text{exc}}}; \frac{E}{U_{\text{exc}}}\right) = \frac{U_i}{E}\left(\frac{E}{U_i+E}\right)^{3/2}\left[\frac{U_{\text{exc}}}{U_i}\left(1-\frac{U_{\text{exc}}}{E}\right)\ln\frac{E}{U_{\text{exc}}}\right.$$
$$\left.+\frac{4}{3}\left(1-\frac{U_{\text{exc}}}{E}\right)\ln\left\{2\cdot 7+\left(\frac{E-U_{\text{exc}}}{U_i}\right)^{1/2}\right\}\right]\left(1-\frac{U_{\text{exc}}}{E}\right)^{U_i/(U_i+U_{\text{exc}})}$$

In these formulae $\sigma_0 = \pi e^4 = 6\cdot 56\ 10^{-14}\ \text{eV}^2\ \text{cm}^2$, U_i = ionization potential of the electron shell considered (eV), U_n = excitation energy of the shell to which an electron is being excited (eV), U_{n+1} = excitation energy of next highest level (eV), U_{exc} = lowest excitation energy (eV), E = energy of incident electron (eV).

Calculations were carried out on several materials using these formulae. Energy levels of electron shells were obtained mainly from the American Institute of Physics Handbook (1963). In figure 1 the calculated cross section for excitation of the $2p$ level in atomic hydrogen by electron impact is plotted, with an experimental curve taken from the review of Moiseiwitsch and Smith (1968). The theoretical curve reproduces the general shape of the experimental curve, and in view of the general difficulties, both experimental and theoretical, in this energy region (see Moiseiwitsch and Smith) the quantitative agreement may be considered reasonable.

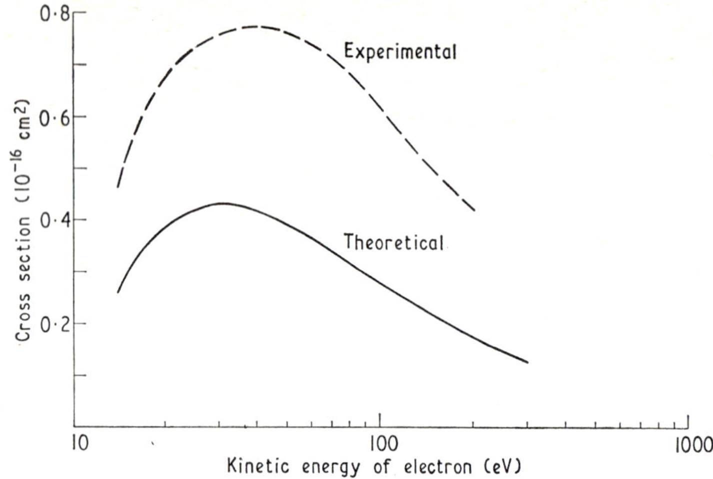

Figure 1. Comparison of experimental and theoretical cross sections for the $1s-2p$ excitation in atomic hydrogen by electron impact.

In figure 2 the calculated cross sections for single ionization of atomic hydrogen and atomic oxygen by electron impact are plotted with the experimental results of Fite and Brackmann (1958, 1959) and of Boksenberg, as quoted by Kieffer and Dunn (1966). The agreement between calculation and experiment is good for atomic hydrogen, but not so good for atomic oxygen, though here there is a considerable disagreement between the two sets of experimental results, and the general shape of the calculated curve, with a maximum at about 100 eV, is again similar to those of the experimental curves.

In using Gryzinski's formula to calculate total stopping power, a discrepancy of 30–40% (different for different materials) was found between the values calculated at 1 keV using Gryzinski's formula and using the quantum-mechanical formula (Berger and Seltzer 1964), except for hydrogen. The reason for the discrepancy is not known, and so all the stopping powers calculated at all lower energies by Gryzinski's formula have been multiplied by a constant factor to make the results from the two formulae agree at 1 keV.

3. Discussion

The calculated stopping power of electrons in water is shown in figure 3, and the range (total path length) in figure 4. The stopping power calculated by straightforward application of the quantum-mechanical formula is shown in figure 3, together with a modification proposed by Marshall *et al.* (1969) for the low energy region where the formula fails. Because of the uncertainty at low energies, the path length calculated from the quantum-mechanical stopping power is subject to an additive constant, but assuming that the constant is equal to that derived from Marshall's modification it is very close to the path length we have calculated ($5\cdot15$ μg cm^{-2} at 1 keV as compared with our value of $5\cdot0$ μg cm^{-2}).

Also shown in figure 4 is the path length in water given by Mozumder and Magee (1966), which is considerably lower than our value at energies above 100 eV. However they used the same stopping power formula as for protons of equal velocity, and this is incorrect.

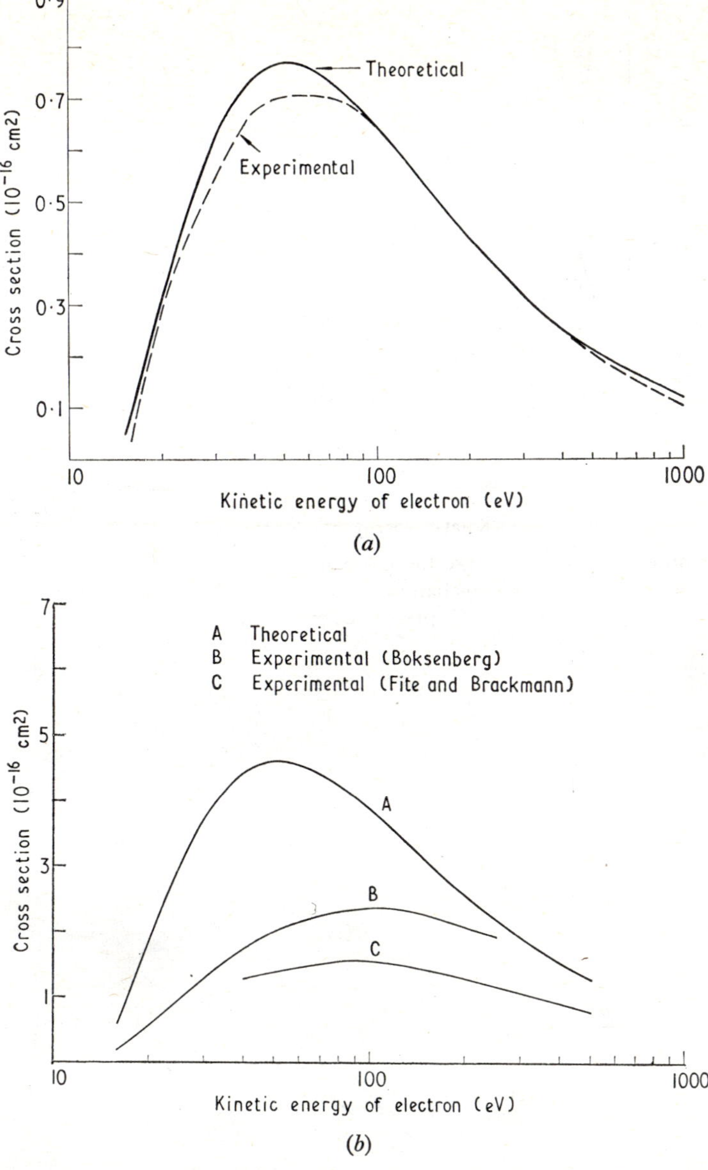

Figure 2. Comparison of experimental and theoretical cross sections for single ionization in (*a*) atomic hydrogen and (*b*) atomic oxygen by electron impact.

Measurements of the penetration ranges of low energy electrons in dry protein were made by Davis (1955), and these are compared with our calculated path length in dry protein in figure 5. Also in figure 5 the range–energy relation of Butts and Katz (1967) is shown. This is a penetration range and is assumed to apply for water though it is derived from measurements of Kanter and Sternglass (1962) in aluminium.

Our calculated path length in aluminium is also shown. It is $9{\cdot}0\ \mu g\ cm^{-2}$ at 1 keV compared with $9{\cdot}9\ \mu g\ cm^{-2}$ for the quantum-mechanical path length, with the modification of Marshall *et al.* (1969). It is surprising that the experimental

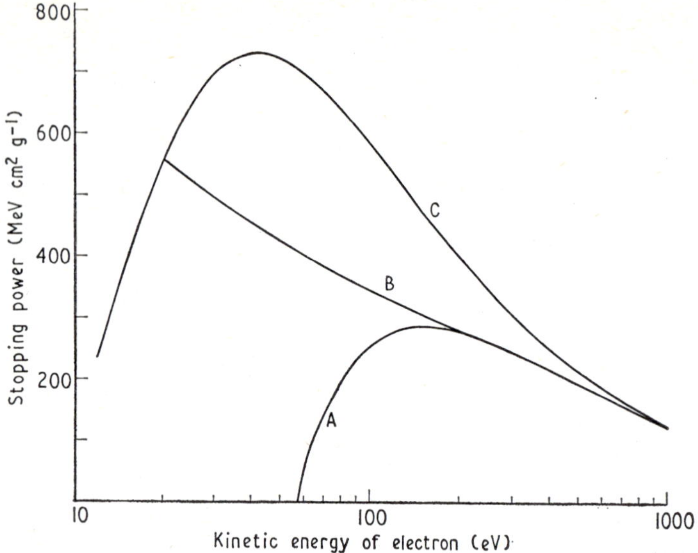

Figure 3. Stopping power of water for electrons according to: A, uncorrected quantum-mechanical formula; B, quantum-mechanical formula with correction of Marshall *et al.* (1969); C, present calculation.

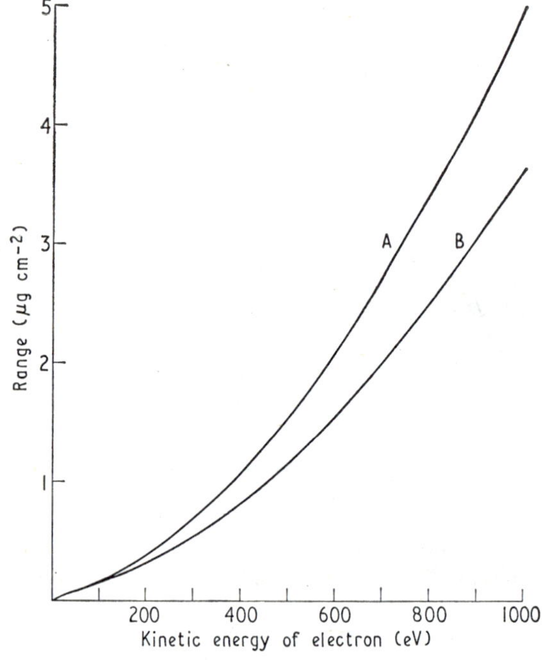

Figure 4. Range of low energy electrons in water according to: A, present calculation; B, Mozumder and Magee (1966).

penetration ranges in aluminium are greater than the calculated path length, while in dry protein the experimental ranges are lower. However, the aluminium ranges were obtained from a plot of electron penetration through an aluminium foil of fixed thickness, as a function of electron energy, extrapolating the straight line portion of the graph back on to the energy scale, while the protein results were

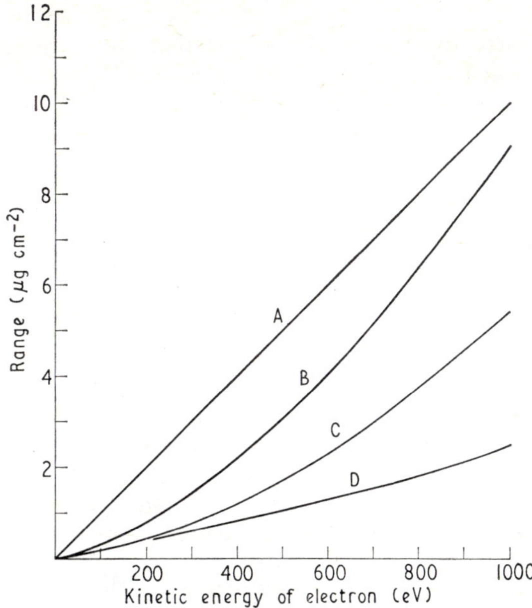

Figure 5. Range of low energy electrons in aluminium, according to: A, Butts and Katz (1967) and B, present calculation; and in dried protein, according to: C, present calculation and D, Davis (1955).

obtained essentially from dose survival curves for thin layers of protein, extrapolating the plateau region on to the survival scale. It is by no means certain that these two methods of averaging give the same result.

It should also be borne in mind that in both cases the 'penetration range' is not an average or median penetration range; in fact only a small percentage of the electrons penetrate further than this. Hence there is less difference than might be expected between the penetration range and the total path length.

Furthermore, a 100 eV electron will undergo on average only two or three energy loss events, and hence there will be a large amount of straggling in the path length. This will be incorporated in the measured curve of electron penetration versus foil thickness, and the maximum range, or path length derived from such a curve will in fact be the maximum of the straggling distribution. Theoretical formulae on the other hand give the average energy loss and hence an average path length.

4. Conclusion

The chief difficulty to date in calculating the ranges of electrons of energy below about 200 eV has been that the quantum-mechanical stopping power formula fails—it goes negative at some energy in this region. The uncertainty about the range–energy relationship in this low energy region leads to an uncertain additive constant at higher energies. The experimental excitation and ionization cross sections which have been quoted indicate that the stopping power will increase with decreasing energy down to some energy around 50–100 eV, and will then fall as the ionization potential of the valency electrons is approached. The classical formula discussed here gives this type of behaviour but the excitation and ionization cross sections calculated do not agree too well with experiment, and the formula itself cannot be

more than an approximation, if only because of the normalization required at 1 keV. Taking all the evidence together it is thought that the average path length for a 100 eV electron derived from this calculation is probably correct to ±50%.

References

American Institute of Physics Handbook, 2nd edn, 1963 (New York: McGraw-Hill).

Berger, M. J., and Seltzer, S. M., 1964, In *Nuclear Science Series, Rep. No.* 39 (NAS–NRC Publication 1133), p. 205.

Butts, J. J., and Katz, R., 1967, *Radiat. Res.*, **30**, 855.

Davis, M., 1955, *Nature, Lond.*, **175**, 427.

Fite, W. L., and Brackmann, R. T., 1958, *Phys. Rev.*, **112**, 1141.

—— 1959, *Phys. Rev.*, **113**, 815.

Gryzinski, M., 1965, *Phys. Rev.*, **138**, A336.

Kanter, H., and Sternglass, E. J., 1962, *Phys. Rev.*, **126**, 620.

Kieffer, L. J., and Dunn, G. H., 1966, *Rev. mod. Phys.*, **38**, 1.

Lea, D. E., 1955, *Actions of Radiations on Living Cells*, 2nd edn (London: Cambridge University Press), p. 24.

Marshall, M., Gibson, J. A. B., and Holt, P. D., 1969, *Proc. Second Symp. on Microdosimetry*. Euratom.

Moiseiwitsch, B. L., and Smith, S. J., 1968, *Rev. mod. Phys.*, **40**, 238.

Mozumder, A., and Magee, J. L., 1966, *Radiat. Res.*, **28**, 203.

Bubble chambers

A. G. TENNER

Zeeman-Laboratorium der Universiteit van Amsterdam

Abstract. A survey is presented of recent measurements on bubble chamber liquids. A model calculation has been made to explain the nucleation in hydrogen. The behaviour of heavier liquids is discussed.

In a bubble chamber the tracks of charged particles become visible as strings of vapour bubbles. The liquid is made sensitive by a fast expansion which reduces the pressure to below the saturated vapour pressure at the liquid's temperature. Boiling then starts around nuclei that have been created by the charged particles. We are concerned with the mechanism by which part of the charged particle's energy is made available for this nucleation.

For producing bubbles in the liquid, vapour nuclei must be created with a radius

$$R=\frac{2\sigma}{\Delta p} \tag{1}$$

where σ is the surface tension and Δp is the difference between the saturated vapour pressure and the pressure of the liquid after the expansion. Bubbles larger than the nucleus will in general grow further and reach macroscopic size; bubbles smaller than the nucleus have a high probability for disappearing again. In order to create the bubble of critical size in an adiabatic process, the following energy must be available:

$$E=4\pi R^2\left(\sigma-T\frac{\mathrm{d}\sigma}{\mathrm{d}T}\right)+\tfrac{4}{3}\pi R^3H+\tfrac{4}{3}\pi(R^3-R'^3)p \tag{2}$$

The first term stands for the surface energy of the nucleus, the second for the vapourization energy, whereas the last is necessary for expanding the bubble against the pressure of the liquid. Theprenu clear volume $\tfrac{4}{3}\pi R'^3$ contains the liquid that will constitute the bubble. Further terms of (2) which take account of fluid motion will be neglected here, since their existence is not established in a satisfactory way.

In figure 1 some values for R and E are shown, as derived from recent measurements by Horlitz *et al.* (1968)*. These data deviate considerably from older, less accurate data, so that a new discussion of the nucleation in hydrogen should be made. In the figure the nuclear radius and the nucleation energy are plotted at constant temperature against the number of bubbles produced per cm tracklength by 2 GeV c^{-1} electrons. The bubble density is a monotonic increasing function of the pressure drop in the chamber. It is seen from the figure, that the nucleation energy for hydrogen lies around 140 eV under normal bubble chamber operation conditions (20 bubbles per cm for a relativistic track) and ranges until 300 eV at the limit of sensitivity.

The measurements of Horlitz *et al.* (1968) allow for a comparison between hydrogen and deuterium which have been investigated under the same experimental conditions. In the plots the comparison is made at the same reduced temperature. The deuterium curves are of a somewhat smaller accuracy than the hydrogen curves, since no measured

* Similar measurements on hydrogen have been carried out by Biswas *et al.* (1966).

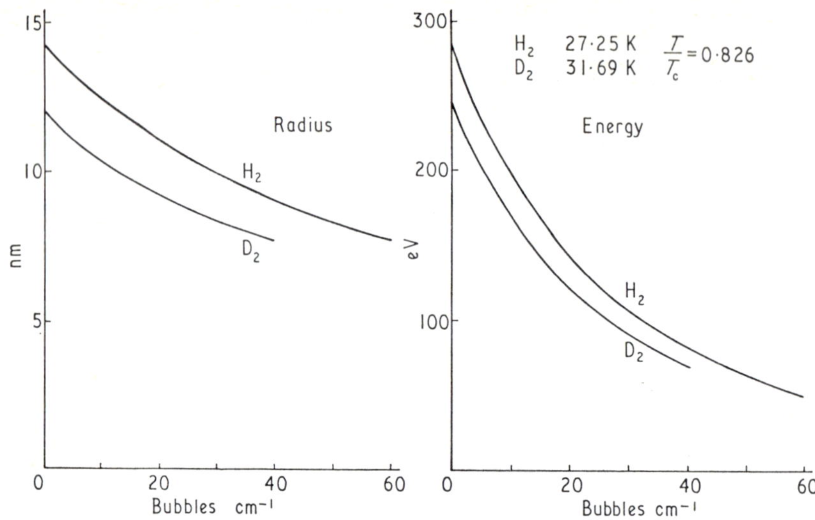

Figure 1. Nuclear radius and nucleation energy for hydrogen and deuterium at equal reduced temperature.

values of the deuterium surface tension are available at the bubble chamber temperature. A value has been obtained by interpolating between values at lower temperature (Grigor'ev and Rudenko 1965) and the critical point where the surface tension disappears. An error of 5% in this interpolation would produce a 15% shift in the energy curve which is the actual separation between the deuterium and hydrogen curves.

If this uncertainty is neglected, one may conclude that the nucleation energy is somewhat smaller in deuterium than in hydrogen. When comparing nucleation energies in different liquids, it must be noted that these energies are very sensitive functions of temperature and pressure drop and that a great variety of values may be obtained for the same liquid. Therefore, the quoted values are by no means characteristic of the thermodynamics of the liquid. They are obtained under such conditions as to produce a desired bubble density along the track of a charged particle. If we introduce a bubble formation efficiency, some quantity indicating at what rate a particle makes available portions of energy that serve for nucleation, we see that this efficiency is larger in hydrogen than in deuterium. Larger amounts of energy exist in hydrogen at a frequency of 20 per cm than in deuterium; for obtaining the same bubble density in both liquids, somewhat less favourable thermodynamic conditions must be chosen for hydrogen which results in an increase of the nucleation energy.

New and accurate data about propane as a bubble chamber liquid are presented by Kunkel (1967). Nuclear radius and nucleation energy as a function of the bubble density of 20 MeV c^{-1} electrons are plotted in figure 2. Whereas the nuclei are somewhat smaller than in hydrogen, the nucleation energy is larger and lies around 400 eV under normal bubble chamber conditions. It seems that the nucleation efficiency increases with increasing complexity of the liquid. Comparisons with helium show, that the nucleation energy is very low in that liquid. Unfortunately, the measurements on helium have not the precision of the above mentioned investigations, so that an accurate comparison is not possible (Tenner 1963).

There is little doubt that the nucleation in bubble chambers goes through a stage in which the whole nucleation energy is available as heat (Seitz 1958). If the nucleation

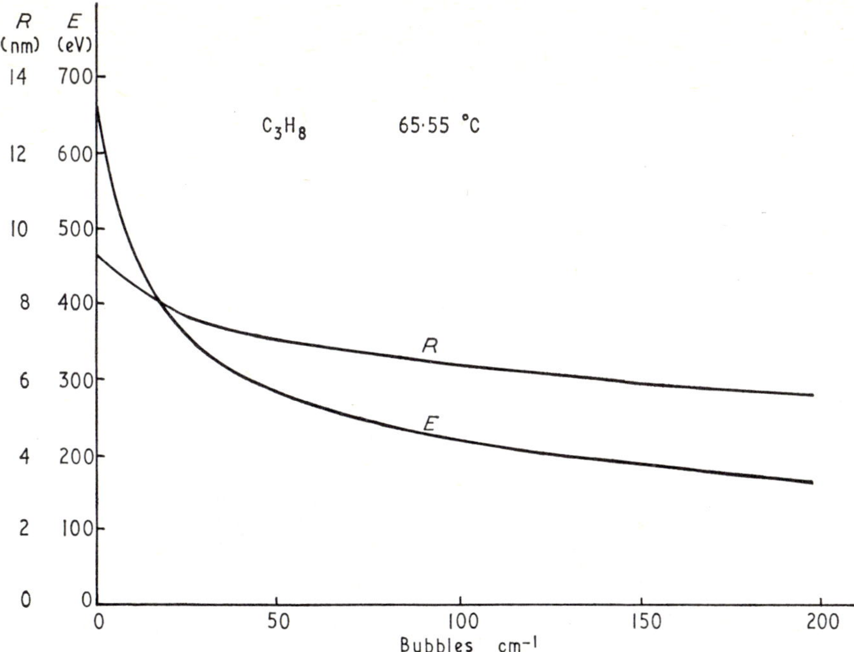

Figure 2. Nuclear radius and nucleation energy for propane.

process is completely adiabatic, this heat should be present in the prenuclear volume before the nucleus can be created. It is conceivable that nucleation actually occurs under somewhat less stringent conditions, e.g. that the temperature distribution may be somewhat wider. If in the densest part of this distribution the liquid explodes into a subnuclear bubble, it may cool down below the temperature of the surroundings. Then—in a non adiabatic stage of the nucleation—heat from the surrounding liquid will contribute to the growth of the bubble. In that case, the total amount of energy in the distribution will have to be higher than the nucleation energy E. Nevertheless when trying to explain the nucleation we must look for processes by which the charged particle dissipates heat of the amount given by equation (2) within volumes of the order of the prenuclear volume.

It is immediately clear that along a charged particle's track the areas of highest interaction density lie at the end of secondary electron tracks. The electron will be scattered and produces a cascade that becomes denser towards the end of its track. Since major overlaps are rare, only the final area will be effective in general. Therefore, secondary electrons will have equal probabilities for making a bubble if their energy lies above a certain value.

When trying to explain the nucleation in hydrogen, we assume this value to be 500 eV, since the number of δ-electrons of 500 eV and more is just equal to the number of observed bubbles. In order to investigate the process in more detail, a Monte-Carlo calculation has been carried out for δ-electrons of various energies. In this calculation the liquid is considered as a gas of high density. For the elastic and inelastic cross sections, values have been employed as obtained from measurements in low pressure atomic hydrogen. Where not enough measurements were available, as for the energy distribution of secondary electrons in the cascade, Born approximation data have been employed. Under these crude assumptions the behaviour of the

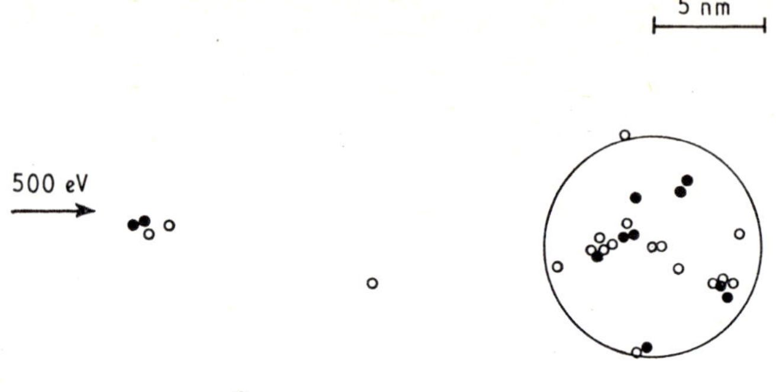

Figure 3. Example of a 500 eV electron cascade as generated by a Monte-Carlo calculation. Open circles are excitations, black circles ionizations. The cascade is shown in projection. At the densest part the size is shown of the pre-nuclear volume (5 nm radius).

electron and its secondaries has been calculated through and an estimate of the distribution of excitations and ionizations in the liquid has been made. A typical distribution, as generated by the computer, is shown in figure 3. The size of the cascade is of the right order of magnitude, however, it exceeds the prenuclear volume. In the densest area of the electron cascade, the density of inelastic events has been adapted to a Gaussian distribution

$$A = A_0 \exp\left(-4r^2/\pi\bar{r}^2\right). \tag{3}$$

The average radius $\bar{r}$ of this distribution is shown in table 1. As expected, the density in the densest part of the distribution is not much dependent on the energy of the δ-electron.

Table 1. Results of the Monte-Carlo calculation for liquid hydrogen

Electron energy (eV)	400	500	600	800
Number of inelastic events	25	31	38	50
Number of excitations	16	20	24	32
Average radius $\bar{r}$ as defined by (3) (nm)	5·15	5·21	5·25	5·45
Number of excitations within 5 nm from the centre	5·3	5·5	5·6	6·0

We now must investigate how the liquid will be heated by the activity of ionized and excited molecules and by all the electrons in the cascade. The life-time of a heat distribution of the mentioned size may be estimated from the heat diffusion coefficient and the specific heat of the liquid. We arrive at the result, that the temperature at the centre of the distribution drops to half its value in a time of $0{\cdot}8 \times 10^{-10}$ s. A somewhat larger value would be obtained, if proper account were taken of the fact that the liquid is near the critical temperature at the centre of the distribution. Nevertheless, we must conclude that all heat that contributes to the nucleation, is available within 0·1 ns after the arrival of the particle; all heat being converted at a later time will not participate in the nucleation.

In classifying the different secondary processes according to the time in which they degrade energy of the excited systems into heat, we may consider both the predissociation of excited and ionized molecules and the deexcitation of excited molecules as fast enough for our purpose. Energy is made available as heat by a chain of reactions, in which unstable molecules fall into pieces and excited fragments attach themselves to unexcited molecules or groups of molecules. Of course, only part of the excitation energy will be degraded into heat, the rest being stored as chemical energy. It is extremely difficult to make an estimate of the ratio between heat and chemical energy in the liquid state.

Next we must discuss the behaviour of the subexcitation electrons of the cascade. These electrons are in a diffusive motion when being cooled down to the liquid temperature. Again making use of experimental data for hydrogen gas, we assume that the mean free path of an electron between 0·5 and 5 eV in liquid hydrogen for change of its direction is 0·55 nm (Engelhardt and Phelps 1963). The mean free path for a loss of 0·5 eV to a molecular vibration is 20 times larger (Phelps 1968). As a result, an electron starting with 5 eV and having reached the energy limit of 0·5 eV, at which no vibrational excitation is possible any more, has a Gaussian distribution around the positive ion and an average distance of 10 nm to it. We see, that only part of the subexcitation electron's energy falls in the prenuclear region. The distance would become even larger than 10 nm, if coherent scattering in the liquid were to be taken into account.

The behaviour of the subexcitation electrons is of crucial importance for the recombination. An electron of 0·5 eV at a distance of 10 nm from an ion has a kinetic energy which is still three times larger than the potential energy in the Coulomb field of the ion. It is not to be expected that this electron will immediately return to the ion; it first has to lose more energy to rotation and translational motion of molecules (Tenner 1963, Thomson 1924). When doing so, it will reach still larger distances from the ion. In recent publications a fast recombination mechanism has been proposed (Freeman 1967). Due to this mechanism electrons could indeed recombine with ions within the time of 0·1 ns. The initial separations between ions and electrons assumed in this theory, are smaller than in our case. Since the recombination velocity goes with the third power of this separation, it becomes questionable whether fast recombination really will occur within 0·1 ns in hydrogen. On the other hand, the recombination in the electron cascade is certainly of the 'cluster' type, rather than of the 'preferential' type. An electron at 0·5 eV may approach any ion of the cascade within 3 nm and then may give rise to a fast recombination. As a conclusion we may say, that probably only a fraction of the electrons will recombine within 0·1 ns and that most of the ionization energy will be lost for the nucleation process.

As a matter of fact, the discussion of recombination is only meaningful, if the ratio between numbers of excitations and ionizations is the same for liquid and gas. The possibility exists that in the liquid state most inelastic events result in excitations, no electron being separated from the excited system.

Making the balance we find that from a 500 eV electron 320 eV may contribute to the nucleation, part of this energy being stored as chemical energy. Nevertheless, it seems to be possible to explain the nucleation phenomenon in this way, since the nucleation heat only amounts to 140 eV.

If we may attach significance to the difference in nucleation energy between hydrogen and deuterium, we are able to understand it according to the previous discussion. The behaviour of the subexcitation electrons will be different in both liquids. The cross sections for vibrational excitation are the same for hydrogen and deuterium, but the vibrational level lies lower in deuterium. Therefore, the liquid is less intensively heated by the subexcitation electrons in the deuterium case. The energy release to rotational and translational motion of deuterium molecules is also smaller, so that the diffusive motion goes over larger areas in deuterium than in hydrogen. Recombination will become less probable for deuterium.

We will now discuss the dependence of the bubble density on the velocity of the charged particles. According to the sketched model, the bubble density should follow the $1/\beta^2$ dependence of the δ-electron production. In hydrogen this behaviour has been established by various investigations. Since the electron binding energy in hydrogen is not more than 15 eV and the energies under discussion are far above this value, no significant conclusion may be drawn from this fact. Significant information is expected to come from heavier bubble chamber liquids. Figure 4 shows the bubble density measurements of Hofmann and Hugentobler (1965) in freon ($CBrF_3$). The curve shows no $1/\beta^2$-dependence for high values of γ, but demonstrates a relativistic rise of more than 30%. A relativistic rise of 16% has been measured for both propane and SF_6, although these measurements have not the precision of the freon data.

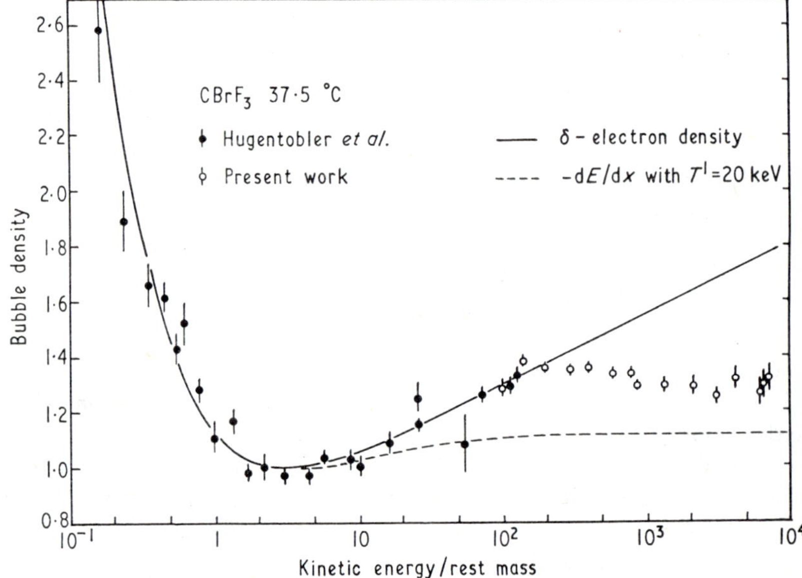

Figure 4. Number of bubbles per mm against $\gamma-1$ of particles in a freon ($CBrF_3$) bubble chamber. Taken from Hofmann and Hugentobler.

We must conclude that a simple δ-electron theory does not apply to the heavier liquids. Hofmann and Hugentobler, nevertheless, present a computation upon the δ-electron concept. They press down the lower limit of the δ-electron spectrum to a value of about 100 eV, in which case the whole electron spectrum should be able to produce the observed relativistic rise. The probability for creating a nucleus should be approximately the same for all electrons out of this spectrum. The disadvantage of the theory lies in the fact, that the limit of 100 eV is below the nucleation energy in the freon chamber which under normal conditions amounts to 170 eV. Of course, uncertainties in the thermodynamic description of the liquid and in the particle interaction mechanism may be due to this discrepancy, but a completely different approach seems to be more attractive.

It has already been anticipated that nucleation in freon may occur through Auger cascades (Tenner 1963, Darup and Platzman 1961). An electron from the K- or L-shell of a bromine atom is removed by the charged particle, the ionization energy being

13 keV for the K-shell and 1·6–1·8 keV for the L-shell. This energy is shared by a number of relatively slow electrons, all emerging from the same point. This situation would be extremely convenient to explain the occurrence of narrow heat distributions in the liquid. The production rate of these ionizations must yield the relativistic rise. A calculation shows that the number of Auger cascades in $CBrF_3$ is sufficient to explain bubble density. It is attractive to assume, that the relativistic rise in propane also occurs through this mechanism. Here the electron binding energy in the K-shell is of the same order of magnitude as the nucleation energy (284 eV). In order to achieve sufficient energy for the nucleation, we must add the kinetic energy of the electron ejected initially to the ionization energy. If the first electron is not too fast, it will dissipate heat in the vicinity of its origin and therefore contribute to the local heat dissipation.

References

Biswas, N. N., Cason, N. M., Derado, I., Kenney, V. P., Poirier, J. A., Shephard, W. D., and Clinton, Sr. E. M., 1966, *Proc. Int. Conf. on Instrumentation for High Energy Physics*, Stanford.

Durup, J., and Platzman, R. L., 1961, *Discuss. Faraday Soc.*, **31**.

Engelhardt, A. G., and Phelps, A. V., 1963, *Phys. Rev.*, **131**, 2115.

Freeman, G. R., 1967, *J. chem. Phys.*, **46**, 2822.

Grigor'ev, V. N., and Rudenko, N. G., 1965, *Sov. Phys.–JETP*, **20**, 63.

Hofmann, J., and Hugentobler, E., 1965, *Helv. Phys. Acta*, **38**, 783.

Horlitz, G., Wolff, S., and Harigel, G., 1968, *Deutsches Elektronen Synchrotron Rep.* 68/45, Hamburg.

Kunkel, P., 1967, *Thesis*, Würzburg.

Phelps, A. V., 1968, *Rev. mod. Phys.*, **40**, 399.

Seitz, F., 1958, *Phys. Fluids*, **1**, 2.

Tenner, A. G., 1963, *Nucl. Instrum. Meth.*, **22**, 1.

Thomson, J. J., 1924, *Phil. Mag.*, **47**, 337.

Energy deposition in microscopic volumes*

W. A. GLASS, W. C. ROESCH and L. A. BRABY

Radiological Sciences Department, Battelle Memorial Institute, Pacific Northwest Laboratory, Richland, Washington.

Abstract. The initial energy deposited by a proton is distributed radially, in a random fashion, about the proton path. One way of investigating this distribution of energy is to probe the proton track with a small cylindrical absorber (a few hundred ångströms equivalent radius simulated by a wall-less proportional counter in rarefied gas). When the probe is traversed by monoenergetic protons along various known chords, the spectra of absorbed energies show large statistical fluctuations as predicted by straggling theories. However, the observed mean and variance of these spectra can be explained only by consideration of the energetic δ-rays produced in primary interaction of the proton. Many of the δ-rays have ranges greater than the dimensions of the probe. Significant energies are deposited in the probe by protons passing up to 500 Å from its edge.

The initial energy deposited in a medium by an energetic proton is distributed about the path of the particle in a manner determined by the statistical nature of the energy transfers in primary collisions and in subsequent interactions of the secondary electrons (δ-rays). Thus the energy deposited in any small bounded volume in or near the path of an incident proton must be treated as the statistical result of combining the many energy transfer processes with the geometry of the absorber. A physical quantity of basic interest in radiochemistry and radiobiology is the distribution of energy initially deposited in sites of microscopic dimensions as they are traversed by ionizing particles. We assume that the initial transfer of energy from the charged particle to the medium is essentially independent of the phase of the medium so long as the atomic composition is equivalent.

The energy initially deposited in a gas appears in many forms of excitation and ionization; however, only that part resulting in ionization is amenable to direct experimental measurement. It is possible by using proportional counter techniques to measure a single ionization in volumes having dimensions of a few micrograms per cm^2; i.e. equivalent to absorbers of a few hundred ångströms in unit density material. The correlation of energy with ionization, however, improves as the number of ionizations increase because the various stochastic processes involved in energy loss tend to average out. Recent measurements (D. Srdoc 1969, private communication) have shown ionization to be a reliable measure of deposited energy for electron energies as small as 250 eV. That is, W, the quotient of absorbed energy by number of ions produced has the same value at low energy as at higher energies. Ionizations corresponding to a few hundred electron volts absorbed energy are easy to determine experimentally.

We have, therefore, begun a systematic investigation of energy deposition spectra in cylindrical sites of a few hundred ångströms radius placed in and near the path of monoenergetic protons. The data to be presented in this paper are for protons of 1·7 MeV and an absorber of unit elongation with a radius equivalent to 1250 Å of soft

* This paper is based on work performed under United States Atomic Energy Commission Contract AT(45-1)-1830.

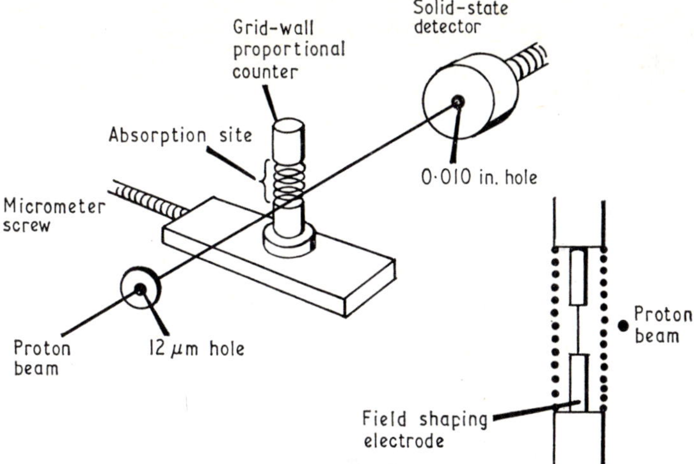

Figure 1. Schematic diagram of the apparatus used to determine energy absorption spectra.

tissue. This size is representative of volumes of biological interest and is convenient because no special experimental techniques are involved in measuring the ionization. A somewhat smaller probe would perhaps be desirable if the experimental goal were to investigate the stochastic structure of the proton track. We believe that proportional counter methods can be used to represent absorbing volumes with dimensions as small as 500 Å.

Figure 1 is a schematic diagram of the experimental apparatus. Protons enter through a 12 μm diameter hole shown at the lower left of the figure and pass through (or near) a grid-walled proportional counter and then through a 0·025 cm aperture covering a solid-state detector. By requiring coincidence between signals from the proportional counter with those from the solid-state detector a high degree of collimation is attained. Only those events in the proportional counter are accepted which are produced by a proton that has traversed the apparatus along a line determined by the entrance and exit apertures. This procedure reduces the uncertainty in the position of the proton path to about $\pm$0·005 cm. The diameter of the grid-walled proportional counter is 1·2 cm. The 12 μm hole also served as a differential pumping aperture so no foil window was necessary between the accelerator and the experimental apparatus. Count rates of a few hundred proton events per second were common.

The proportional counter was mounted on a track driven by a micrometer screw so that it could be located at known positions relative to the proton path. The protons passed normal to the axis of the counter and midway between the ends of the collecting volume. An FET preamplifier was built inside the apparatus to reduce noise due to additional capacitance which would have been present with external connectors. The resulting electronic noise level was about 300 r.m.s. electrons; thus, when the proportional counter gas gain was greater than 10^3 the electronic noise was negligible.

The density of the gas flowing through the system was maintained at 20·8 μg cm^{-3} by a feedback circuit which compared the gas density in the chamber with a captive reference volume at the desired pressure. Equilibrium flow was established with the gas input by adjusting the exhaust rate through an electrically operated needle-valve connected in the feedback loop. The flow system maintained the clean gas atmosphere

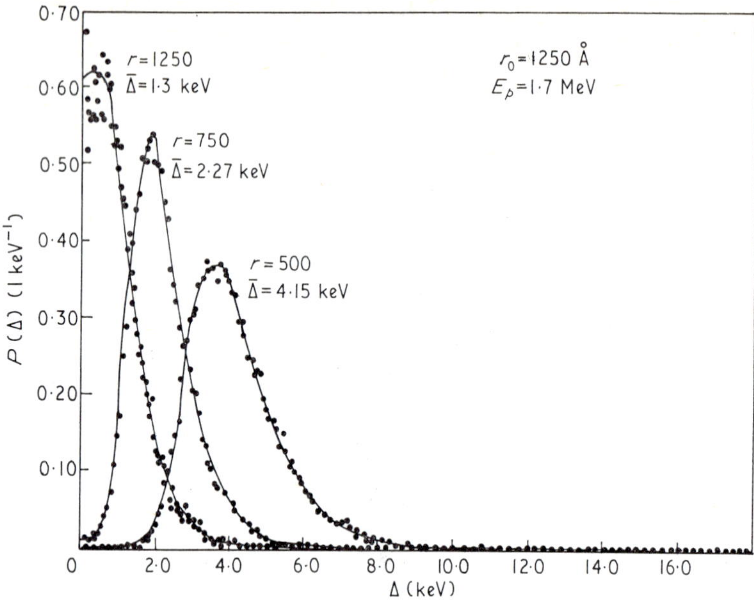

Figure 2. Representative energy absorption spectra for protons passing at various distances from the absorber axis.

necessary for reliable counter operation. The gas mixture consisted of methane, carbon dioxide, and nitrogen with proportions adjusted so the gas had the same stopping power as soft tissue. The absorption site simulated by these conditions had a radius equal to 1250 Å of unit density tissue.

Typical energy deposition spectra are shown in figure 2. The proton energy was 1·7 MeV for all data. For proton paths 500 Å from the axis the chord length is about 2250 Å and the mean observed energy absorbed was 4·15 keV. At 750 Å, the chord length is about 1980 Å and the mean energy deposited was 3·81 keV. At 1250 Å the chord length is zero, yet we observed a mean energy loss per proton of 1·32 keV. That is, when the protons passed tangent to the absorber the mean energy deposited in the site was about $\frac{1}{4}$ as great as found for a traversal across the diameter. Observed mean absorbed energies depart in a systematic way from the product of stopping power and chord length as shown in figure 3. This behaviour illustrates the relatively large quantities of energy carried away from the proton path by energetic δ-rays. The mean values of energy deposition observed in an experiment such as this are a combination of the radial energy distribution function for the proton track and the cylindrical geometry of the absorber.

The next figure shows the observed mean energy deposited per proton for protons passing at various distances outside the absorber. When the path is far outside the absorber, many protons deposit no energy in the site. The mean values plotted here include those zero energy events. The frequency of zero energy events increases as the distance between proton path and absorber increases. However, at distances as great as 150 Å outside the absorber, our data indicate that 50% of the protons will, on the average, deposit more than 100 eV of energy.

Experimental data such as just presented on the stochastic properties of particle tracks need a stochastic theory for their analysis. Such stochastic theories have emerged in the past few years, (Roesch 1968, Kellerer 1968). Models derived from

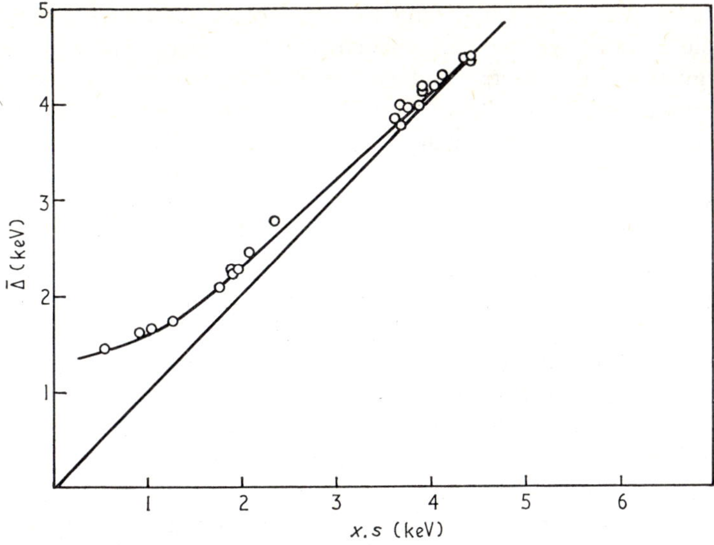

Figure 3. Observed mean energy absorbed versus the product of chord length and stopping power.

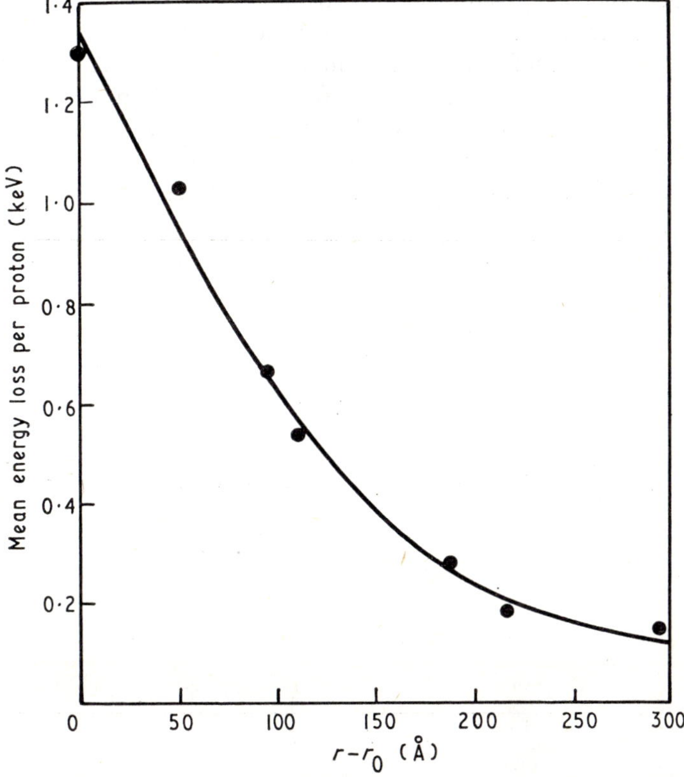

Figure 4. Mean energy deposited per proton for protons passing along various paths outside the absorber.

these theories are particularly applicable to the problem of energy-deposition spectra. Certain simple relations between the parameters of the primary interaction probabilities and the energy deposition spectra can be derived when these new methods are applied.

The distribution of energy losses by protons as they pass through a site can be calculated by combining the primary energy-loss distribution (for single primary interactions of the proton with the atoms of the medium) to give the distributions of energy losses for 2, 3, etc. interactions. These are then combined in proportion to the Poisson probabilities of 0, 1, 2, etc. atomic interactions occurring. It can be shown that both the variance and the mean of the resulting energy-loss distributions are proportional to the average number of primary events in the site; hence, their quotient is independent of the number of atomic interactions contributing to the energy loss. Indeed, this quotient equals the quotient for the energy-loss distribution for primary interactions; the latter can be shown to equal the quotient of the second and the first moments of the primary (single event) energy-loss distribution.

The energy lost by the proton differs from that absorbed in the site, because energy lost outside the site can be absorbed within it and energy lost inside can be absorbed outside because of transport by secondary particles. For each position along the path of the proton there will be a different primary energy-absorption distribution, i.e. a distribution of energy absorbed within the site due to single primary interactions of the proton with atoms of the medium at that position. These distributions could be combined, as above, to give the distribution for all primary events occurring at that position and then the resulting distributions for different positions along the path combined to give the energy-absorption distribution for the site. It can be shown, however, that one can first average the primary energy-absorption distributions along that part of the proton path from which secondary particles can reach the site and then construct the energy-absorption distribution from the average in the same way that the energy-loss distribution was constructed above. Therefore, the quotient of the variance by the mean of the energy absorption distribution gives the ratio of the second and first moments of the track-average primary energy-absorption distribution.

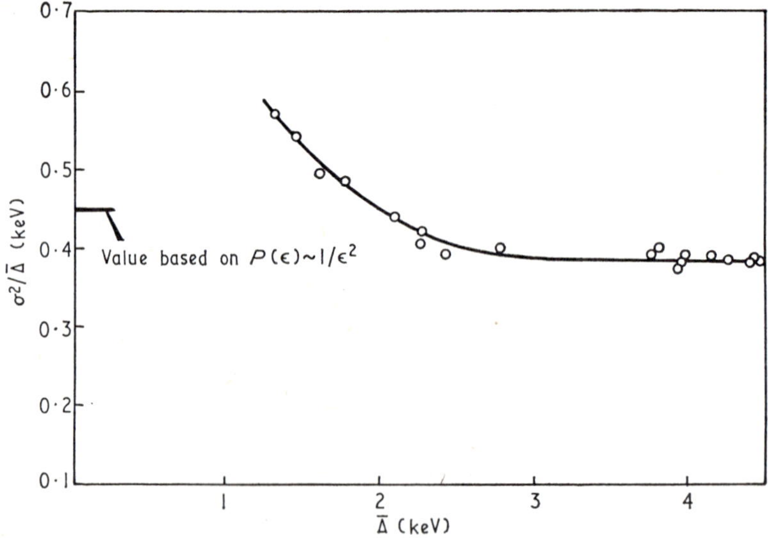

Figure 5. Experimental values of $\sigma^2/\bar{\Delta}$ for energy absorption spectra produced by protons passing along various chords within the absorber.

Figure 5 shows the quotient of the variance by the mean, $\sigma^2/\bar{\Delta}$, for the measured distributions described above. For large mean energy-absorption the quotient is independent of the mean energy absorbed. These large depositions are for tracks near the centre of the site. For these central tracks, energy transfer by secondary particles is relatively unimportant for a site of the size considered in this experiment and $\sigma^2/\bar{\Delta}$ is approximately equal to the quotient of the second and first moments of the primary energy-loss distribution. For smaller average energy absorptions which occur nearer the edge of the absorber, energy transfer by δ-rays becomes more important; it is the sole means of energy deposition for paths outside the site, of course. The quotient (which is equal to the quotient of the second moment of the primary energy absorption distribution by its first moment), $\sigma^2/\bar{\Delta}$, increases because the primary energy-absorption distribution must be broader than the primary energy-loss distribution due to the increased number of stochastic processes introduced by the boundary.

These data must be considered as preliminary and are presented here only to illustrate an approach to the study of the stochastics of energy desposition in small bounded volumes subjected to ionizing radiation. It is our hope that the new stochastic models of energy transport coupled with experimental data similar to that presented here will result in a tractable description of energy deposition in sites of chemical and biological interest.

References

Kellerer, A. M., 1968, *Biophysical Aspects of Radiation Quality*. Second Panel Report, IAEA, Vienna, 1968.

Roesch, W. C., 1968, *Energy Absorption Straggling*, Annual Report to USAEC.

Effect of chemical composition and of initial energy on the physical track structure for fast electrons†

I. SANTAR and J. BEDNÁŘ

Department of Radiation Chemistry, Institute of Nuclear Research, Czechoslovak Academy of Sciences, Řež, Czechoslovakia

Abstract. General formation developed earlier to interrelate the physical structure of the track of fast electrons and the concept of degradation spectra is summarized and applied to four representative light media. Extended computations of the degradation spectra are used for the discussion of extraspur structure and energy partition among track entities, and influence of the molecular nature and the initial electron energy is examined. The only major effect of the chemical nature of the irradiated medium concerns the yield and energy distributions of spurs, and is illustrated by combining the present method with recent optical-approximation data.

1. Introduction

We have recently suggested (Santar and Bednář 1969b—referred to in the text as Part V), that there exists a general relation between the average structure of the track of a fast electron and the degradation spectrum of this electron. We have shown that the skeleton of the track of an electron having a particular initial energy will be more or less the same provided the nuclei are not too different. The skeleton of the track will be composed of the main track of the primary electron together with the tracks of δ-electrons (e.g. blobs, short and branch tracks in actual models of the track) which are formed in hard collisions and depend only slightly on the particular chemical nature of the medium. This is significant only with respect to the number, distribution and average energy of spurs formed along all higher energy entities constituting the track skeleton. On the other hand, both the extraspur structure of the track, and the partition of absorbed energy among various types of entities in a particular medium depend on the initial energy of the primary electron, this being the result of the differences in the degradation spectra of variously energetic primary electrons.

In illustrating these conclusions in Part V, we limited ourselves to the track of a 1 MeV primary in water. Here we extend the calculations to a number of light media (helium, polyethylene, water, and air) and a number of initial energies of the primary electrons between 3 MeV and 1 keV. This extension should encompass the majority of systems studied by radiation chemistry.

For this purpose we use degradation spectra recalculated by the Spencer–Fano–McGinnies method and extended, in accordance with this method, to electron energies as low as 100 eV.

† This forms part VI of a series of papers on 'Theory of the Radiation Chemical Yield', the preceding part being published by Santar and Bednář (1969b) and referred to in the text as Part V.

The principles of the use of degradation spectra for the construction of the initial track structure were discussed in detail in Part V and, therefore we shall only summarize the important features of the method. As a slight modification of this approach, a compact procedure for the evaluation of extraspur structure will be introduced, and the spur distribution will be discussed independently.

2. Theoretical part

It is known (Spencer and Fano 1954) that during β or γ-radiolysis, a stationary degradation spectrum is formed in the irradiated medium, containing electrons of all energies from the maximum energy $T_{\max}$ of the incident radiation, down to thermal energies. This spectrum is formed by the gradual slowing-down of both the primary electrons and the whole cascade of secondaries, and it is through its action that the energy of radiation is transferred to the medium.

In general, the stationary degradation spectrum and the effects brought about by it, can be regarded as a superposition of independent processes each initiated by individual primary electrons. Thus, without restricting the generality of the following analysis, we need only consider the case of monoenergetic primary electrons.

In order to describe the degradation spectrum it is useful to choose the differential track length $y(T, T_0)$ as defined by Spencer and Fano (1954) and McGinnies (1959), which has the dimension of reciprocal stopping power (length per energy) and is normalized to one primary electron of initial energy T_0. This normalization leads (Santar and Bednář 1967, 1968a, 1969b) to simple expressions for the usual radiation-chemical quantities, such as yields and absorbed energy (dose), and displays most clearly the above interrelation between the degradation spectrum and the average track of the primary electron.

The yield of a particular collision process i is generally given (Platzman 1961, Durup and Platzman 1961, Platzman 1967) by simple averaging of the corresponding molecular cross section $\sigma_i(T)$ for an electron of energy T over the actual degradation spectrum

$$g_i = \frac{100}{Q_{\text{tot}}} \int_{E_i}^{T_{\max}} N\sigma_i(T) y(T, T_0)\, \mathrm{d}T \tag{1}$$

where N is the molecular density of the medium and E_i is the threshold energy of process i.

With the normalization adopted here, $T_{\max} = T_0$, and the total energy Q_{tot}, imparted to the medium during the complete absorption of the primary electron, equals its initial energy T_0, and the following identity is fulfilled

$$Q_{\text{tot}} = T_0 = \int_0^{T_0} \zeta(T) y(T, T_0)\, \mathrm{d}T. \tag{2}$$

Here $\zeta(T)$ is the net stopping power (cf. equation (6) in Santar and Bednář 1968a) involving only the energy transferred to the medium in irreversible collisions; i.e. not including any energy of the secondary electron.

The integrands in equations (1) and (2) have the important property that they display, over the whole integration range, the relative contributions of electrons of each particular energy T to g_i or Q_{tot}, respectively. Clearly, the stopping power is proportional to the density N, while the degradation spectrum is inversely proportional to it.

Therefore, density drops out from equations (1) and (2) as well as from subsequent considerations in so far as other effects of condensation are disregarded.

For the time being it is impossible to utilize fully the most instructive general equations (1) and (2). This is unfortunately due to the lack of quantitative information concerning the formation and action of slow ($T \lesssim 100$ eV) secondary electrons.†

Nevertheless, their use in the theoretical description of the initial track structure is justifiable owing to the following favourable circumstances.

(i) In all models of the track used hitherto, it is assumed that secondary electrons, with energies below 100 eV, lose their energy locally, in single spurs formed by faster ($T \gtrsim 100$ eV) electrons. The fast electrons interact with the medium either in infrequent hard collisions which lead to further extraspur entities, or in much more frequent optical collisions in which isolated or overlapping spurs are formed.

This scheme makes it possible to limit the problem of energy transfer between the incident radiation and the irradiated matter to the fast region of the degradation spectrum. Present knowledge of the spectrum of fast electrons, and of their action, then permits the use of equations (1) and (2) in the calculation of the yield of spurs and of the relative contributions of various portions of the fast electron spectrum to this yield.

In order to calculate the yield of spurs from equation (1), the optical approximation (Platzman 1962), i.e. the Bethe cross section for optical collisions, can be introduced and the lower integration limit put equal to 100 eV (cf. Part V). A similar limitation in equation (2) requires the simultaneous use of a restricted stopping power including all energy absorbed locally, i.e. in collisions involving the transfer of energy of less than 100 eV and leading to spur formation. Thus, a modified identity may be written

$$Q_{\text{tot}} = T_0 = \int_{100\text{ eV}}^{T_0} S(T, 100\text{ eV}) y(T, T_0)\, \mathrm{d}T + 100\mathcal{N}(100\text{ eV}). \tag{3}$$

This equation is a particular case of a general identity

$$T_0 = \int_{\Delta}^{T_0} S(T, \Delta) y(T, T_0)\, \mathrm{d}T$$

used by Spencer and Attix (1955), in the theory of cavity chambers. The only difference is that their $S(T, \Delta)$ also included in the region between $T = \Delta$ and $T = 2\Delta$, the fraction of absorbed energy deposited at the ends below Δ of the paths of fast electrons (Spencer 1965). On the other hand we express this fraction explicitly by the second term on the right involving $\mathcal{N}(100\text{ eV})$, the total number (per one primary) of fast electrons of all generations. Examination of the integrand in equation (3) permits an analysis of the gradual deposition of energy into spurs during the degradation of the primary electron to be made.

(ii) The other track entities (e.g. blobs, etc.) are defined by energy limits lying inside the region of fast electrons. These limits should be such that they schematize the varying degree of isolation of neighbouring interactions (spurs) inside the individual entities. [In the model of Mozumder and Magee (1966), proposed for a condensed medium of unit density, and used here, these limits are $T = 100$–500 eV for blobs,

† It is mainly this deficiency which, irrespective of the method of approach, greatly restricts all present efforts to present a complete picture of the effects of ionizing radiation on matter (cf. Platzman 1967, Santar and Bednář 1968a, b, 1969a). One cannot expect that either the theoretical or the experimental difficulties will be overcome for some time yet.

$T = 500$–5000 eV for short tracks, and $T > 5000$ eV for branch tracks and main track composed of isolated spurs. This particular division of track entities is, however, quite arbitrary from our point of view.]

The number and energy distributions of fast electrons (the extraspur structure of the track), are fully encompassed in the degradation spectrum as a substantial part of its calculation and can be deduced therefrom. The fractions of the total energy absorbed in particular types of track entities can then be obtained by an appropriate subdivision of the integral in equation (3).

3. Method of calculation

In this paper we systematically follow the model approximation in which the formation of fast secondary electrons in the whole region of energy transfers $E \geqslant 100$ eV is described by the Møller cross section $\sigma^*(E, T)$ for hard collisions with free electrons. Thus, the binding energy of molecular electrons is neglected throughout and the transferred energy E is identified with the kinetic energy of the ejected electron. The changes in the low energy part of the spectra of fast electrons, which occur even in light media owing to inner electron ionizations and the associated Auger effect, are also neglected.

A formal integration of the Møller cross section corresponding to the subtraction of the contribution of hard collisions with $E > \Delta$ from the total stopping power $S(T)$, is then used to derive the formula for the restricted stopping power $S(T, \Delta)$ required in equation (3) with $\Delta = 100$ eV.†

By itself, this model approximation may already be rather poor for T and E in the vicinity of 100 eV and represents the weak point of many present radiation-chemical and radiobiological applications and inferences so long as they require a detailed description of the energy deposition even in this low energy region. Nevertheless, we use it here to preserve the consistency with the degradation spectra calculated, on exactly the same basis, by the method of Spencer and Fano (1954).‡

McGinnies (1959) tabulated selected results of calculations by this method, of degradation spectra $y(T, T_0)$ for various energies $T > 450$ eV and $T_0 > 6440$ eV in different media. Bruce *et al.* (1963) published graphical results for water, obtained using the original program of McGinnies, but extended downwards to $T = 100$ eV and $T_0 = 1$ keV. This information was used in Part V to exemplify the calculation of track structure for a 1 MeV electron in water.

† We do not give the explicit expressions for $\sigma^*(E, T)$ and $S(T, \Delta)$ which can be readily found in the literature (see e.g. Spencer and Attix 1955 or Berger 1963). The total stopping power $S(T)$ is the limiting case of $S(T, \Delta)$ with $\Delta = T/2$. In contrast with Spencer and Attix however, we put $S(T, \Delta) = S(T)$ for all $T \leqslant 2\Delta$, (cf. below equation (3)). The use of $S(T, \Delta)$ in equation (3), instead of $S^0(T)$ defined in Part V, maintains the overall consistency of the calculations. As discussed in Part V, our understanding of S^0 is limited by the unknown parameter $\bar{c}$ in the optical cross section σ^0. Calculations showed that both stopping powers are approximately identical for reasonable values of $\bar{c}$.

‡ The influence of this approximation need not be too important, especially as far as total effects like yields are concerned. The total error introduced by neglecting the energy consumed irreversibly in hard collisions, should not exceed 1–2% Q_{tot} (Santar and Bednář 1968a). The effect of Auger electrons appears as a discontinuity in the degradation spectrum near the K-edge of light atoms of the medium. However, the overall shape of the degradation spectrum is not changed overwhelmingly (cf. Santar and Bednář 1969b, Klots 1968). These effects will presumably be taken into account in future attempts to solve the problem of obtaining reliable degradation spectra in the low energy region.

For the present work we have written a program which closely follows the numerical method of McGinnies (1959). Application of this program yielded in tabular form, the degradation spectra for all selected media in energy ranges $T_0=3\cdot3$ MeV to 1·0 keV and $T=T_0$ to 100 eV. In addition, we have obtained as output data, the quantities $\mathcal{N}(T, T_0)$ and $\mathcal{N}_1(T, T_0)$ which are intermediate products of spectra calculations. They represent the total numbers of electrons (including the primary one) produced with initial energy higher than T during the degradation, and take into account, respectively, all generations of δ-electrons, or just the first generation of short tracks and blobs. Together with the degradation spectra, these are the basic input data in the calculation of the initial track structure. (For details of our method see the Appendix.)

For the track structure calculations, another program has been written which, starting from the input data, produces the spectrum of gradual deposition of energy along the track and also, the quantitative description of the extraspur structure of the track. [This information corresponds, in principle, to figure 2 and table 2 (Santar and Bednář 1969b) in Part V.]

By analogy with the discussion of the problem in Part V, the calculation can proceed on the following levels:

(i) *The detailed structure* involving all generations of δ-electrons.

The differential yield of δ-electrons of energy E is given by

$$g^*(E)=\frac{100}{T_0}\,n^*(E)=\frac{100}{T_0}\left\{-\frac{\mathrm{d}\mathcal{N}(T, T_0)}{\mathrm{d}T}\right\}_{T=E} \tag{4}$$

and the total yield g^* within a particular range of energy E can be directly calculated as the difference of values of $\mathcal{N}(E)$ at both limits of the interval.

In the corresponding partition of absorbed energy, we must ascribe to the δ-electron of a particular initial energy E, only that part of E which is deposited in spurs during its slowing down. Otherwise the rest of the energy, which is consumed in the formation of higher generations of δ-electrons, and therefore taken into account at appropriate lower E's, would be counted twice thus violating the energy balance.

This aim is achieved by using in equation (5) for the resulting energy distribution, a dimensionless correction factor $\kappa(E)$, given by (6), and representing a number between 0 and 1:

$$Q(E)=g^*(E)\kappa(E)E \tag{5}$$

where

$$\kappa(E)=E^{-1}\int_0^E \frac{S(T, 100\text{ eV})}{F(T)}\,\mathrm{d}T. \tag{6}$$

The argument here is that the ratio of the restricted stopping power $S(T, 100\text{ eV})$ to the effective stopping power $F(T)$ (see equation (A8) in the Appendix) gives just the fraction of energy consumed on average in spur formation during the degradation at given T.†

† Below 200 eV all energy goes into spurs so that in the part between 0 and 200 eV of the integral in (6) this fraction equals unity by definition. Substituting from (4) and (6) into (5) and integrating by parts it may be shown readily that the integral of equation (5) from 100 eV to T_0 is equivalent to that displayed in equation (3) so that individual fractions of absorbed energy obtained by partial integrations of $Q(E)$ are nothing other than a specific subdivision of the integral on the right-hand side of equation (3). Other such subdivisions may also be considered, if more relevant to a particular application.

Note also that in the view of detailed structure, we include in each type of entity the end contribution from the nearest higher entity, conforming therefore to the strict definition of entities as parts of the track corresponding to instantaneous T within the prescribed limits.

(ii) *The gross structure* (including only the first generation of blobs and short tracks, formed along main track and branch tracks).

Here, this first generation is not developed any further and each of the entities is regarded as a single final entity in the scope of which the whole initial energy of the δ-electron is consumed.

The expressions for differential yields follow from equations (A3) or (A5), by restricting the lower integration limits for T' to $2E$ (or $2T$)$=5000$ eV. The corresponding quantity $\mathcal{N}_1(T)=1+n_1^*(E\geqslant T)$ is also obtained from the calculation of degradation spectra.

By analogy with equation (4), the differential yield of blobs and/or short tracks in the first generation is then given by

$$g_1^*(E)=\frac{100}{T_0}\,n_1^*(E\geqslant T)=\frac{100}{T_0}\left\{-\frac{\mathrm{d}\mathcal{N}_1(T)}{\mathrm{d}T}\right\}_{T=E} \tag{7}$$

and for the energy distribution we have

$$Q_1(E)=g_1^*(E)E. \tag{8}$$

Since $g_1^*(E)$ closely obeys the E^{-2} law, it is possible to obtain integrals of (8) within an energy interval (E_1, E_2) by applying a mean energy conforming to this law. $\bar{E}=E_1E_2\ln(E_2/E_1)/(E_2-E_1)$, to the corresponding yield, $\mathcal{N}_1(E_2)-\mathcal{N}_1(E_1)$.

4. Discussion of the results

The numerical calculations have been programmed in Algol and performed on a Gier computer. The computer has printed out tabulations of results for helium ($I=42{\cdot}0$ eV), polyethylene ($I=54{\cdot}9$ eV), water ($I=74{\cdot}1$ eV), and air ($I=94{\cdot}9$ eV), and for the following primary energies $T_0=3{\cdot}296$ MeV, 1·038 MeV, 327·0 keV, 103·0 keV, 32·45 keV, 10·22 keV, 3·219 keV, and 1·038 keV. Only selected results are presented here in graphical form.

The only way in which the chemical composition entered all calculations of the extraspur structure was through the mean excitation potential I which affected the various stopping powers involved. This influence is through logarithmic terms and therefore slight, although it becomes more important as the electron energy decreases. This common surmise is fully corroborated by the present results. Indeed, it was found that even with relatively strongly influenced quantities, such as the total yield g^* of δ-electrons, the dependence on I is slight and regular, so that linear interpolation with respect to I may be applied successfully (figure 2). The results for the four media considered here may therefore be regarded as typical for the radiolysis of any chemical system consisting of light elements, e.g. tissue and the like.

On the other hand, many of the calculated quantities exhibit characteristic dependence on the primary energy T_0. Both these aspects are now to be discussed on a few examples illustrating the conclusions and surmises made in this respect in Part V.

As far as the degradation spectra are concerned, the agreement with the results of McGinnies (1959) for cases and energy regions covered by this author is to about 1%. Some of the features of the spectra noted already by McGinnies manifest themselves distinctly in the present results.

4.1. *Extraspur structure*

An important byproduct of the present work are tables of quantities $\mathcal{N}(T)$ and $\mathcal{N}_1(T)$ defined above and in the Appendix. These describe the production of δ-electrons of all generations or of the first generation only along the track. The former of these quantities $\mathcal{N}(T)$ is shown in figure 1 for water and all the eight primary energies considered.

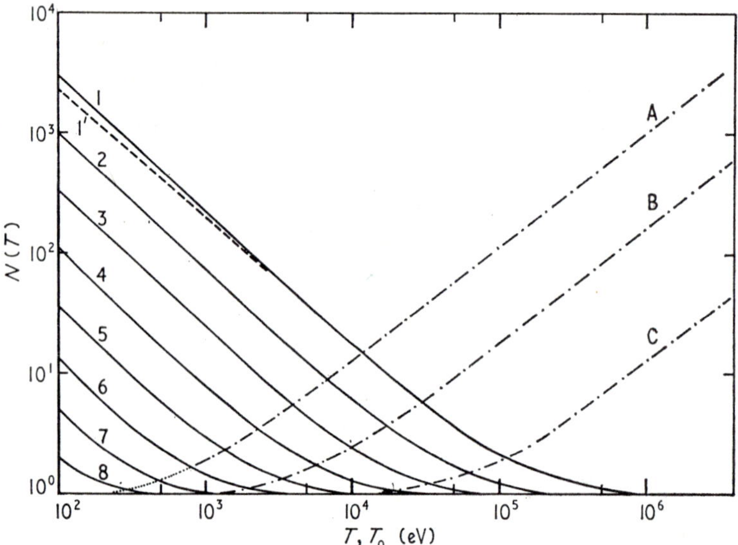

Figure 1. Production of extraspur entities—δ-electrons in water (I=74 eV). Primary energies T_0: 1, 3·3 MeV; 2, 1·0 MeV; 3, 330 keV; 4, 100 keV; 5, 32 keV; 6, 10 keV; 7, 3·2 keV; 8, 1 keV.

It is seen that $\mathcal{N}(T)$ behaves very regularly and obeys the $T^{-1\cdot 12}$ dependence, mentioned in Part V, as soon as $T \lesssim T_0/100$. This in turn means (equation (4)) an $E^{-2\cdot 12}$ limiting dependence for the all-generation yield of extraspur entities $g^*(E)$. Curve 1′ is drawn to indicate the departure which arises from taking into account only the first generation of blobs and short tracks, for which the energy dependence of $g^*(E)$ is close to the E^{-2} law.

Curves A to C in figure 1 display the T_0 dependence of $\mathcal{N}(T)$, i.e. of the total numbers of all extraspur entities, all short and branch tracks, and all branch tracks, respectively, for T=100, 500 and 5000 eV. We see that the increase with the primary energy T_0 is again very regular, roughly as $T_0^{-0\cdot 95}$ for $n^* = \mathcal{N}(100 \text{ eV})$ and high T_0. A consequence of the exponent being lower than 1 is a slight decrease of the total yield of δ-electrons, $g^* = 100n^*/T_0$, with increasing T_0.

This decrease is clearly seen at the top of figure 2, where the dependence of g^* on T_0 is displayed for all four media under consideration (curves 1 to 4). We note here the relatively strong influence of chemical composition already referred to, causing g^* to increase with increasing I. Yet the absolute values are in the neighbourhood of 0·1 in all cases at high primary energies, as surmised previously (Santar and Bednář 1968a, 1969b). Note that the curves for the detailed structure are identical with those for total g^* since the former also include the end contributions from higher entities.

The same trend with respect to T_0 and I is observed for the first generation of blobs

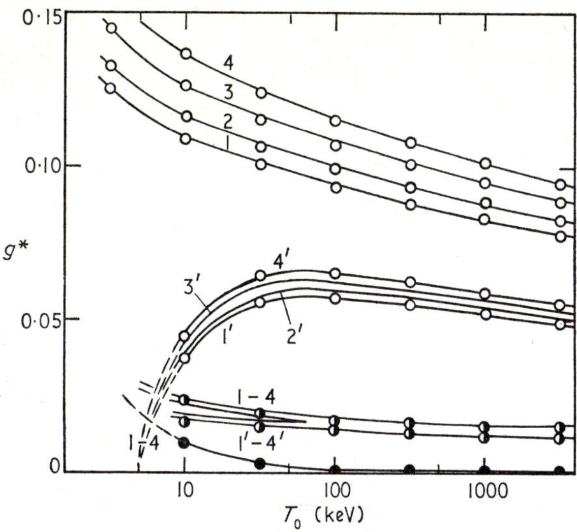

Figure 2. Yields of extraspur entities in various media for various primary energies T_0. ○ blobs, ◑ short tracks, ● main track+branch tracks; 1,1′ He ($I = 42$ eV), 2,2′ polyethylene ($I = 55$ eV), 3,3′ H_2O ($I = 74$ eV), 4,4′ air ($I = 95$ eV). Unprimed and primed numbers refer to all generation yields including the end contributions (detailed structure) and first-generation yields (gross structure), respectively.

in figure 2. The only difference here is the fall off of the yield below 10 keV due to the fact that these 'isolated' blobs are formed along branch tracks and the main track only, i.e. they disappear when T_0 drops below 5000 eV. It is seen that the yield of first-generation blobs typically amounts to 0·05–0·06 so that there is a factor of two between the yields of the two types of blobs. This might affect eventual discussions of the minor track effects associated with blobs.

The corresponding distinction between all-generation and first-generation yields is almost negligible for short tracks; the low values of about 0·01 to 0·02 may make their contribution to specific minor yields quite unimportant. To complete the comparison, figure 2 finally shows the yields of branch tracks and main track. Their increase at low T_0 is merely an artificial consequence of the factor $100/T_0$ in the yield formulae; note that it also appears in the other curves for all-generations yields and stems from the view of track entities we have adopted, including end contributions.

4.2. *Energy partition*

Trends similar to those found for the yields are also found in graphs of energy deposition in spurs along the track and its partition between the track entities, though different aspects are emphasized in this case.

Figure 3 is an example of the energy-deposition spectra, displaying the integrand of equation (3) (in polyethylene) for four different primary energies. For other media the picture is very similar. The curves are not normalized; in all cases studied, the total area agreed with $Q_{tot} = T_0$ to better than 2%. This represents a useful verification of the applicability of equation (3) even for Δ as low as 100 eV. Previously equation (3) was known to be valid only if Δ was of the order of 1 to 10 keV (Spencer and Attix 1955). The spectra of figure 3 may therefore be considered to describe correctly the relative contributions to spur formation of different portions of the degradation spectra of the average track.

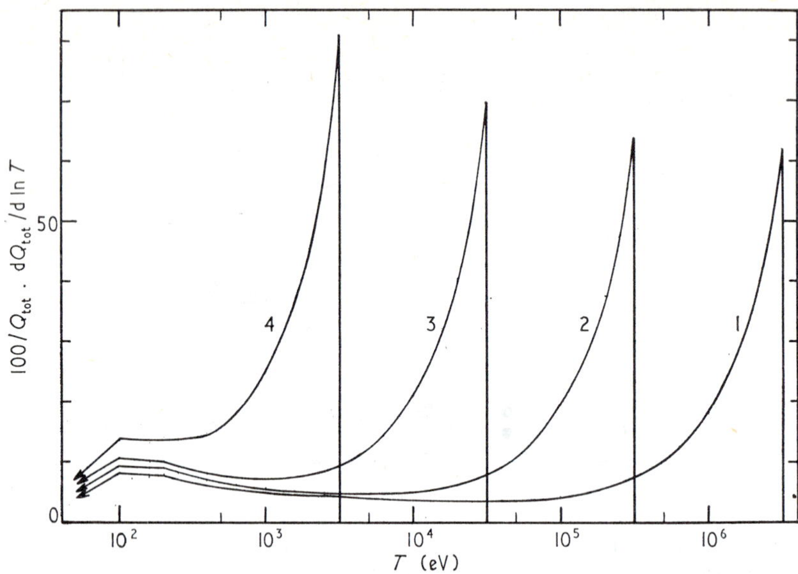

Figure 3. Energy deposition in spurs as function of instantaneous electron energy T. Medium: polyethylene ($I = 55$ eV); primary energies T_0: 1, 3·3 MeV; 2, 330 keV; 3, 3 2 keV; 4, 3·2 keV.

A characteristic feature noticeable in figure 3 is the great contribution from the high energy region. In fact, it is found that the contribution of the area above 5000 eV, i.e. that of the isolated spurs, is fairly constant and amounts to about 2/3 of total Q_{tot} for sufficiently high T_0, declining only as T_0 approaches 5000 eV.

In general, it may be pointed out that the overall relative apportionment of absorbed energy remains almost constant for T_0 greater than some 50 keV. This is most clearly seen from figures 4 and 5 which illustrate the T_0 dependence of energy partition in the two views of the track structure. Figure 4 corresponds to the detailed structure, i.e. the definition of track entities by instantaneous T, and is directly derived from the energy-deposition spectra. Characteristic values of energy fractions in various media in the high T_0 region are $Q_{is} \approx 60$–70% for the isolated spurs, $Q_{st} \approx 10$–15% for short

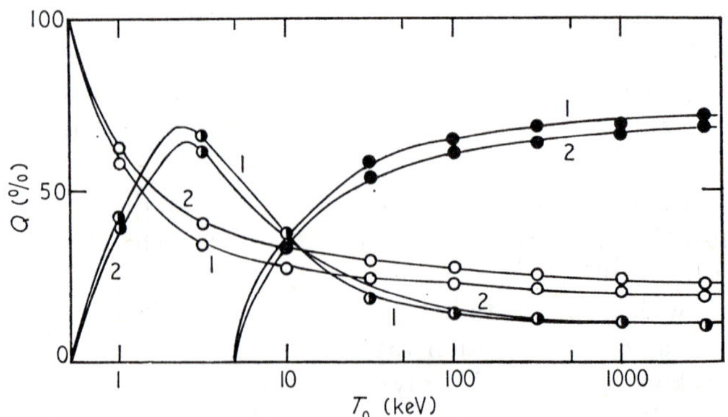

Figure 4. Dependence on T_0 of the energy partition among track entities in the view of detailed structure. ○ blobs, ◑ short tracks, ● isolated spurs; 1, He ($I = 42$ eV); 2, air ($I = 95$ eV).

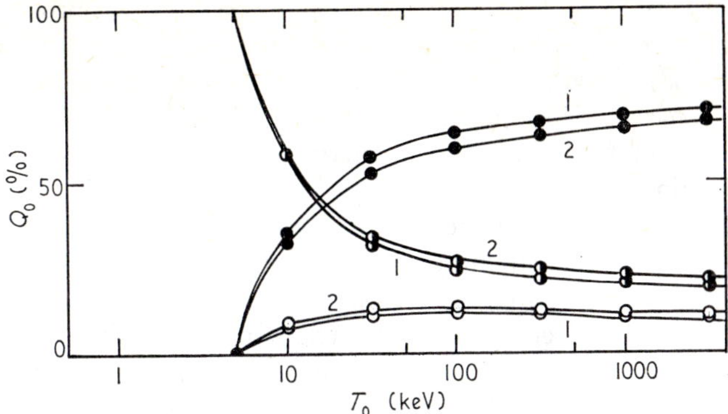

Figure 5. Dependence on T_0 of the energy partition among track entities of first generation—in the view of gross structure. ○ blobs, ◑ short tracks, ● isolated spurs; 1, He ($I=42$ eV); 2, air ($I=95$ eV).

tracks, and $Q_{bl} \approx 20$–30% for blobs. Below 50 keV, the influence of the energy limits defining the track entities starts to prevail, of course.

The curve for isolated spurs is the same also in figure 5 corresponding to the view of gross structure. In very good agreement with the previous finding of Mozumder and Magee (1966) we see an analogous situation with high T_0 limiting values of $Q_{is} \approx 60$–70%, $Q_{st} \approx 20$–30% and $Q_{bl} \approx 10$–15% for isolated spurs, short tracks, and blobs, respectively. Comparison of figures 4 and 5 clearly shows the pouring over of energy from blobs to short tracks when passing from detailed to gross structure (cf. Part V).

A qualitative appraisal of the effect of chemical composition may be inferred from the dependence on I of the individual factors in equation (3), where we resolve $y(T, T_0)$ according to (A8). Quite generally, an increase of I diminishes the two stopping powers $S(T, 100 \text{ eV})$ and $F(T)$ in such a way that their ratio S/F, or its integral κ from equation (6), determining the fraction of absorbed energy going to spurs, is also diminished. A concommitant trend is the increase in the number $R(T) \approx \bar{R}(T)$ of δ-electrons contributing to $y(T, T_0)$ at given T, already seen above. Both these factors are now seen to interact in determining the shape of the energy-deposition curves of figure 3. While the influence of the former prevails at high energies, where R is of the order of unity, the opposite is true at low T.

Thus, going to media with larger values of I we see the energy deposition change from the high T region of isolated spurs into the region of short tracks and blobs. An inspection of figures 4 and 5 clearly demonstrates this fact. The absolute effect is minor, however, as illustrated in both figures by the small vertical distances between the curves for helium and air, the two limiting representatives of light media considered here. The differences never exceed a few per cent on the absolute scale of Q. It is thus again seen that the results given in figures 3 to 5 are also of rather general validity. Moreover, other distributions may also be obtained from the ensemble of computed data corresponding with the requirements of particular applications.

4.3. *Formation and distribution of spurs*

So far, we have only encountered the relatively weak effect of the chemical composition, exerted through the value of I upon the number and energy content of the extra-spur entities—fast electrons forming the track skeleton. The original ideas of

Platzman (1962) as well as all later applications of the optical approximation in radiation cheimstry (Platzman 1967, Santar and Bednář 1968a, b, 1969a, Hatano *et al.* 1968, Hatano 1968) provide evidence that there must exist another, much more pronounced, primary effect of the chemical nature of the irradiated medium; namely, an effect upon the yield and energy distribution of the most important elementary activations—spurs.

It is now widely accepted that the low energy activations, induced by fast electrons and ranging from the threshold energy E_0 of electronic excitations in a given medium up to approximately 100 eV, are governed in the first instance by the optical spectrum of the medium (expressed conveniently in terms of the optical oscillator strength). The fortunate coincidence, at E and $T \sim 100$ eV, of characteristic energy limits of the optical spectra of light media in the theory of primary yields and of spur formation in the track theory introduces a wide parallelism between the theories.

Put briefly (cf. Part V), the situation is as follows. Along the path of any fast electron, whether primary or secondary, numerous spurs are formed in frequent optical collisions. Their energy and number distributions are determined by the optical spectrum $\mathrm{d}f/\mathrm{d}E$ and the excitation spectrum $(1/E)(\mathrm{d}f/\mathrm{d}E)$ of the medium, respectively, and are independent of the kinetic energy of the fast electrons. The same applies to the total yield of spurs determined essentially by the reciprocal of the mean energy of the excitation spectrum, $\bar{E}^0 = RZ_{\mathrm{eff}}{}^0/M^2$, where R is the Rydberg energy. The quantities $Z_{\mathrm{eff}}{}^0$, the effective number of valence-shell electrons per molecule of the medium, and M^2, the square of the total matrix element for optical transitions, represent integrals of the optical and excitation spectra, respectively, over the valence-electron transitions.

Attempts for an improved diffusion-kinetic solution of the radical-track problem led Mozumder and Magee (1966) to distinguish between isolated spurs formed along the main track and branch tracks and more or less overlapping spurs inside the blobs and short tracks. For this purpose it is now sufficient to apply the data on energy partition obtained in the present work to optical-approximation formulae for the optical primary yields, derived previously (Santar and Bednář 1968a, 1969a).

For instance, we may subdivide the differential yield of spurs at energy E according to

$$g_{\mathrm{spur}}(E)\,\mathrm{d}E = \frac{Q_{\mathrm{is}}+Q_{\mathrm{bl}}+Q_{\mathrm{st}}}{Z_{\mathrm{eff}}{}^0}\,\frac{1}{E}\,\frac{\mathrm{d}f}{\mathrm{d}E}\,\mathrm{d}E \tag{9}$$

where the fractions Q_{is}, Q_{bl}, Q_{st}, expressed in per cent, sum up to $Q_{\mathrm{tot}} = 100$. Similarly, the total yield of spurs may be written as

$$g_{\mathrm{spur}} = \frac{Q_{\mathrm{is}}+Q_{\mathrm{bl}}+Q_{\mathrm{st}}}{RZ_{\mathrm{eff}}{}^0}M^2. \tag{10}$$

An example of such a procedure is shown, for purposes of orientation, in table 1. This table incorporates some of the data on optical yields g^0 published earlier (Santar and Bednář 1969a), but modified by the application of equations such as (9) or (10). As the result we obtain illustrative data on the yields of spurs classified, on one hand, according to whether they are isolated or overlap, and, on the other hand, according to the type of elementary acts involved.

The yields of spurs involving only one act of primary excitation (S^{e}) or of superexcitation (S^{se}) have been obtained directly from the earlier results on $g_{\mathrm{exc}}{}^0$. A comparison of the optical ($g_{\mathrm{ion}}{}^0$) and total ($g_{\mathrm{ion}} = g_{\mathrm{ion}}{}^0 + g_{\mathrm{ion}}{}^{\mathrm{s}}$) yields of ionizations has yielded an approximate estimate of the relative contributions of spurs involving one or more ionizations. Both latter cases obviously refer to spurs with a higher energy content, where a part of the activation and subsequent chemical action is due to one or more slow secondary electrons. The latter may produce additional excitations in the first case, while in the second they also produce the extra ionization. If it is then assumed that at most one additional secondary ionization per spur occurs, one obtains $S^{i}=g_{ion}{}^{0}-g_{ion}{}^{s}$ and $S^{i'}=g_{ion}{}^{s}$ as estimates for the minimum value of the first and the maximum value of the second of these contributions. All of these data are included in table 1; may they inspire the radiation chemist undertaking the difficult task of identifying the precursors of the chemical changes observed after irradiation.

Table 1. Yields of spurs along the track of a 1 MeV electron in various media Based on data from Santar and Bednář (1969a).

Medium	Yield of spurs											
	Isolated				Overlapping				Total			
	S^{e}	S^{se}	S^{i}	$S^{i'}$	S^{e}	S^{se}	S^{i}	$S^{i'}$	S^{e}	S^{se}	S^{i}	$S^{i'}$
Helium	0·7	—	0·9	0·3	0·3	—	0·4	0·2	1·0	—	1·3	0·5
Polyethylene	0·5		2·4	0·3	0·2		1·1	0·2	0·7		3·5	0·5
Water	0·3	0·5	1·3	0·5	0·2	0·1	0·7	0·2	0·4	0·7	2·0	0·7
Air	0·5	0·1	1·1	0·4	0·2	0·1	0·6	0·2	0·7	0·2	1·7	0·6

Spurs containing: S^{e} excitation below I_0, S^{se} neutral decomposition from superexcitations, S^{i} one ionization, $S^{i'}$ one extra ionization from slow secondary electron.

5. Concluding remarks

Effects of condensation other than density of the medium were disregarded in the present paper. The density itself entered the model of track only indirectly through the choice of energy boundaries defining the track entities such as isolated spurs, etc. so as to correspond to common organic liquids, water, or tissue under normal conditions.

The condensation effects are presumably unimportant for the discussion of track skeleton except at very high electron energies where the 'density–effect' correction to stopping powers would have to be applied. However, in the treatment of spur formation by optical collisions of fast electrons we have used optical spectra for isolated molecules (including a tentative spectrum for the polyethylene unit —CH_2—). These spectra are valid for the gaseous state and neglect the actual changes in the distribution of optical oscillator strengths brought about by condensation; these affect especially the low energy excitations. We believe, however, that the error thus introduced cannot seriously invalidate our general conclusions concerning the effect of chemical composition on the track structure.

Two rather extreme views of the electron tracks have been considered, both intended in principle, to provide some initial information to insert into the diffusion-kinetics model of the radical track. One is the discontinuous model of Mozumder and Magee, introducing the sharp energy boundaries in the degradation spectrum of the primary electrons to distinguish the track entities which one subconsciously tends to associate with paths of the primary and its δ-electrons, visible in cloud or bubble chambers. These entities are then to be treated as having different properties during the development of the physicochemical stage of radiolysis. The other model, which uses the spectrum of energy deposition in spurs similar to that shown in figure 3 and which may be designated as continuous, has no energy boundaries and virtually disregards the geometry of the track. It is analogous to and might easily be converted into the 'Dose–LET spectra' used commonly in radiobiology. The argument in its favour would be its ability to give a quantitative description of instantaneous LET effects along the track of a fast electron.

The weak point in both these points of view as well as in any intermediate ones is our present inability to describe comprehensively the action of slow electrons with energies below about 100 eV. It is therefore an advantage of the model of Mozumder and Magee that it completely separates the effects of slow electrons by confining them within the particular entity called spur. It thus leaves open the possibility of almost independent solution of the slow electron problem by different theoretical or experimental means. Moreover, the notion of spur has been accepted by many radiation chemists as a fruitful basis for the interpretation of the subsequent physicochemical and chemical effects of radiolysis.

A final point which must be mentioned is that the present models represent essentially the *average* track and neglect the fluctuations due to the actual random sequence of elementary acts of energy deposition. If one were to take into account the latter, perhaps along the lines developed recently in 'microdosimetry', this might modify the present results quantitatively. The final goal—a comprehensive description of energy deposition in the track—must however wait for the resolution of the problem of slow electrons. It is therefore probable that the simple average models will retain their interpretative role in radiation chemistry for some time to come.

Appendix

Deduction of the extraspur structure from calculations of the degradation spectra

To derive the equations for the distribution of extraspur entities we start from the general equation (1) for yields. As adopted in the main text we use throughout the region of hard collisions of fast electrons, i.e. for all energy transfers $E \geqslant 100$ eV, the Møller cross section $\sigma^*(E, T')$ for collisions of free electrons, and identify the transferred energy E with the kinetic energy of the ejected δ-electron.

The expression

$$N\sigma^*(E, T')\,\mathrm{d}E \tag{A1}$$

gives the probability of δ-electron formation with energy between E and $E+\mathrm{d}E$ per unit path of the incident electron of energy T'. The whole integrand of (1) then reads

$$N\sigma^*(E, T')\,\mathrm{d}E \times y(T', T_0)\,\mathrm{d}T' \tag{A2}$$

and gives the number of δ-electrons of energies between E and $E+\mathrm{d}E$ formed by the action of the differential region T' to $T'+\mathrm{d}T'$ of the degradation spectrum.

The differential yield of δ-electrons of energy E is obtained by an integration of (A2) over all kinetic energies T' higher than $2E$†

$$g^*(E)\,\mathrm{d}E = \frac{100}{T_0}\,n^*(E)\,\mathrm{d}E = \frac{100}{T_0}\left\{\int_{2E}^{T_0} N\sigma^*(E, T')y(T', T_0)\,\mathrm{d}T'\right\}\mathrm{d}E. \tag{A3}$$

† We recall that the convention requires the slower from the two electrons after a hard collision to be the secondary when using the Møller cross section. An electron may therefore lose at most half of its energy in a single collision and thus also generate a secondary of, at most, that energy. Hence the lower integration limit in (A3) and similar consequences in a number of subsequent equations.

Overall yields of δ-electrons within particular energy regions then follow from integrations of the differential yield (A3) within corresponding limits of E. Equation (14) in Part V for the total yield g^* of all δ-electrons, where the integration over E ran from 100 eV up to $T_0/2$, may serve as an example of this procedure.

In fact, however, it is not necessary to perform calculations along these lines with the degradation spectrum since in the present approximation the whole extraspur structure is *a priori* entirely contained in it. The required numerical results for the distribution of the yields $g^*(E)$ can be directly extracted from the computation of the degradation spectra, where they play the role of basic data building up the spectrum.

The order of the double integration (over E and T') of (A2) can be reversed; in this way one can calculate the degradation spectrum. Integrating (A1) over E first, the total probability can be found for δ-electron formation with energy higher than a given T per unit path of the breeding electron of energy T'

$$K(T', T) = \int_{T}^{T'/2} N\sigma^*(E, T')\, \mathrm{d}E. \tag{A4}$$

Then, integrating over T' we get directly the number of δ-electrons of energies higher than or equal to T which have been formed by the whole degradation spectrum:

$$n^*(E \geqslant T) = \int_{2T}^{T_0} K(T', T) y(T', T_0)\, \mathrm{d}T'. \tag{A5}$$

Indeed, the integration indicated in (A5) is an important step (cf. McGinnies 1959) in the calculation of the degradation spectrum at given T. The quantity

$$\mathscr{N}(T) = 1 + n^*(E \geqslant T) \tag{A6}$$

appears there as the source of electrons and it consists of contributions from the single primary electron (within the accepted method of normalization) and of all secondaries formed with initial energies higher than or equal to T. It is closely connected with the yield distribution of extraspur entities by equation (4) of the text.

Since we have repeated, more extensively, the machine computations of McGinnies (1959), we have at our disposal the actual values of $\mathscr{N}(T)$. Otherwise, it is possible to approximate $\mathscr{N}(T)$ by the quantity $R(T, T_0)$ which, for a particular T, gives the ratio of the flux of all electrons to the flux of the primaries. Thus, $R(T, T_0)$ is defined by an important method of resolving the degradation spectrum, viz

$$y(T, T_0) = R(T, T_0) y_{\mathrm{prim}}(T, T_0). \tag{A7}$$

In the continuous slowing-down approximation, where $y_{\mathrm{prim}}(T)$ is given by the reciprocal stopping power $S(T)^{-1}$, the quantities $R(T, T_0)$ and $\mathscr{N}(T, T_0)$ would be identical (Spencer and Attix 1955). However, when one takes account of the fact that the energy of electrons changes from time to time by relatively large discrete jumps due to hard collisions, the two quantities turn out to be slightly different. A useful approximation to $R(T, T_0)$ is represented by the quantity $\bar{R}(T, T_0)$ defined by the resolution of $y(T, T_0)$ according to equation (A8) (McGinnies 1959) which represents a convenient substitute for (A7).

$$y(T, T_0) = \frac{\bar{R}(T, T_0)}{F(T, T_0)}. \tag{A8}$$

The function $F(T, T_0)$ which is available in analytic form represents an effective stopping power which allows for the discontinuous slowing-down as well as for the transient effect at the beginning of the degradation near T_0. The reciprocal of $F(T, T_0)$ provides a good approximation to the actual $y_{\rm prim}(T)$.

In our calculations of the degradation spectra we have obtained all the above mentioned quantities and found that, over a predominant part of all spectra, R and $\bar{R}$ differ by less than one per cent and they approach $\mathcal{N}$ to within a few per cent. At the low energy end of the spectra—for T approaching I—$y(T, T_0)$ behaved irregularly because of sharply declining accuracy of the computation method. We have therefore relied upon computed values of $y(T, T_0)$ above 200 eV only, while $\mathcal{N}(T)$, which receives contributions from energies larger than or equal to $2T$, was apparently correct down to 100 eV. In all applications we defined, for consistency, $F(T)$ by

$$y(T)=\mathcal{N}(T)/F(T)$$

except below 200 eV where we put $F(T)=S(T, 100\ \text{eV})=S(T)$, so that

$$y(T)=\mathcal{N}(T)/S(T).$$

Acknowledgment

The authors are grateful to Dr Cornelius E. Klots for sending them a preprint of his paper. They also thank the staff of the Computation Centre of the Institute of Nuclear Research for assistance in the computations.

References

BERGER, M. J., 1963, *Methods in Computational Physics*, Vol. 1 (New York: Academic Press), p. 135.
BRUCE, W. R., PEARSON, M. L., and FREEDHOFF, H. S., 1963, *Radiat. Res.*, **19**, 606.
DURUP, J., and PLATZMAN, R. L., 1961, *Discuss. Faraday Soc.*, **31**, 156.
HATANO, Y., 1968, *Bull. Chem. Soc. Japan*, **41**, 1126.
HATANO, Y., SHIDA, S., and INOKUTI, M., 1968, *J. chem. Phys.*, **48**, 940.
KLOTS, C. E., 1968, *Transition Yields in Irradiated Gases*, Preprint, abstracted in Adv. Chem. Ser., **82**, 539.
MOZUMDER, A., and MAGEE, J. L., 1966, *Radiat. Res.*, **28**, 203.
MCGINNIES, R. T., 1959, *U.S. Natn. Bur. Stands Circ. No.* 597.
PLATZMAN, R. L., 1961, *Int. J. appl. Radiat. Isotopes*, **10**, 116.
—— R. L., 1962, *The Vortex*, **23**, 372.
—— R. L., 1967, *Radiation Research*, Ed. G. Silini (Amsterdam: North-Holland), p. 20.
SANTAR, I., and BEDNÁŘ, J., 1967, *The Chemistry of Ionizations and Excitations*, Eds G. R. A. Johnson and G. Scholes (London: Taylor & Francis), p. 217.
—— 1968a, *Colln Czech. Chem. Commun.*, **33**, 1.
—— 1968b, *Adv. Chem. Ser.*, **81**, 523.
—— 1969a, *Colln Czech. Chem. Commun.*, **34**, 1.
—— 1969b, *Int. J. Radiat. Phys. Chem.*, **1**, 133.
SPENCER, L. V., 1965, *Radiat. Res.*, **25**, 352.
SPENCER, L. V., and ATTIX, F. H., 1955, *Radiat. Res.*, **3**, 239.
SPENCER, L. V., and FANO, U., 1954, *Phys. Rev.*, **93**, 1172.

Parameters of track structure

A. M. KELLERER

Institut für Biologie der Gesellschaft für Strahlenforschung, Neuherberg bei München and Radiological Research Laboratories, Columbia University, New York

Abstract. Linear energy transfer (or collision stopping power) and energy straggling along the tracks of charged particles are both relevant to the effectiveness of ionizing radiation. Energy straggling is the dominant aspect whenever one is concerned with small energy depositions (<1 keV) and a correction to LET is necessary in these cases. The correction term and its relation to the spectrum of energy transfers in primary collisions is derived. Other parameters of track structure are discussed, and an analysis is mentioned which can substitute the Landau- and Vavilov-theory in the analysis of the collision spectrum.

1. Introduction

Track structure is an aspect of radiation quality disregarded in conventional LET theory but indispensable for an understanding of radiation effects. D. Lea's and L. H. Gray's pioneering works have made this clear. These works have also shown that the random patterns of primary events are a fairly complex result of different stochastic factors. A limited number of parameters cannot possibly represent the full picture.

A complete description involves the spatial patterns of all different quantum-mechanical interactions of radiation and irradiated medium. At present such a description is out of the question, and it is therefore necessary to simplify the problem. A crude but practicable approximation is to assume that the distribution of energy deposition is the same as for the spatial distribution of primary events and their reaction products. This assumption is, of course, untenable on a molecular scale up to distances of about 100 Å. Over a larger scale, however, the spatial patterns of energy deposition are an important and characteristic index of radiation quality. Experimental determination and theoretical analysis of these microscopic patterns are the essential points of Rossi's concept of microdosimetry (Rossi 1959, 1967).

Microdosimetry is a synthesis of LET theory and straggling. But the cases of greatest biological interest are those where one deals with energy depositions of less than a few keV; and in those cases straggling, i.e. the fluctuations of energy loss along the track of an ionizing particle, is the dominating factor (Kellerer 1968b). This is why the theory of microdosimetry is by and large a theory of track structure.

While explicit analysis is rather involved (Kellerer 1967, 1968a) some aspects of track structure can be discussed in a simplified way. This is done in the first part of this paper. Specifically it will be shown that the LET concept can be modified in such a way that it takes into account the 'clustering' of energy deposition along the track. In a second part of this paper different parameters of track structure are compared and it is shown how straggling experiments can be used to obtain information on these parameters and on the delta-ray spectrum.

2. LET and linear energy concentration

The effectiveness of primary events and of their reaction products depends on their local concentrations. Due to the discontinuous nature of energy deposition the term 'local concentration' must refer to track intervals of finite extension. The relevant length of these intervals is determined by the range of interaction of ionizations or radiation-induced free radicals. It may also be related to the size of sensitive structures in the cell. The assumption of a critical distance is reflected in the choice of an energy cut-off in LET theory and in the use of the corresponding 'restricted' LET values.

The reaction probabilities of the irradiation products in the trail of an ionizing particle depend on the whole spectrum of their mutual distances. In other words, one must know the probability distribution of local energy concentration around the activated molecules. But whenever a single index of radiation quality is necessary, one has to substitute the probability distribution by its mean value. This mean value will be derived here, and it will be seen that LET in its present form is not the relevant parameter.

When an ionizing particle of linear energy transfer L traverses a certain distance d in tissue it loses an energy E. The local concentration of energy over the interval can then be measured by the 'linear energy concentration' E/d. This quantity is a random variable; while its expected value is equal to L, its actual values deviate considerably from L. In order to indicate the somewhat loose relation to L, the linear energy concentration will in the following be designated* by the Greek letter Λ.

Let the discussion be limited to intervals which are small enough that the LET of the incoming particle does not change appreciably over the interval. The track segment can then be assumed to be a straight line. Under these conditions the probability distributions of E or Λ can be calculated. In fact this is the object of the classical 'thin foil' straggling theory of Landau (1944), Simon (1948) and Vavilov (1957) and of a recent more general solution of the straggling problem (Kellerer 1968a). An analysis of this problem and its implications in a wider context is found in Fano's paper of 1953. We will, for the moment, not be concerned with the explicit distributions, but only with some of their basic properties.

The probability distribution of energy transfer E can be designated by $f(E; n)$ the parameter n is the expected number of collisions along the interval d. The fraction of absorbed energy associated with values of Λ between E/d and $(E+\mathrm{d}E)/d$ is then equal to:

$$E\ f(E; n)\ \mathrm{d}E/\int E\ f(E; n)\ \mathrm{d}E.$$

In order to derive the value of Λ averaged over all elements $\mathrm{d}E$ of absorbed energy, one has to evaluate the integral:

$$\overline{\Lambda}_{\mathrm{D}}=\frac{\int \Lambda\ E\ f(E; n)\ \mathrm{d}E}{\int E\ f(E; n)\ \mathrm{d}E}=\frac{1}{d}\frac{\int E^2 f(E; n)\ \mathrm{d}E}{\int E f(E; n)\ \mathrm{d}E}=\frac{1}{d}\frac{\overline{E^2}}{\overline{E}}. \tag{1}$$

This expression can be transformed to:

$$\overline{\Lambda}_{\mathrm{D}}=\frac{1}{d}\left(\overline{E}+\frac{\overline{E^2}-\overline{E}^2}{\overline{E}}\right)=L+\frac{\sigma^2(E)/\overline{E}}{d}. \tag{2}$$

* Note that Λ, while being related to Rossi's y, has a much more restricted meaning.

Because 'successive intervals of energy degradation contribute additive amounts to the mean square fluctuation of energy loss' (Fano 1953) the term $\sigma^2(E)/\bar{E}$ is independent of n. Its value is most easily derived in the limiting case of a very small collision number, $\epsilon \ll 1$. The distribution $f(E;\epsilon)$ is then a mere superposition of a delta function in $E=0$, and the collision spectrum (or delta-ray spectrum) $w(E)$

$$f(E;\epsilon)=(1-\epsilon)\ \delta(E)+\epsilon\ w(E). \tag{3}$$

If M_1 and M_2 are the mean and the second moment of the collision spectrum, $w(E)$, one obtains:

$$\bar{E}=\epsilon M_1 \quad \text{and} \quad \overline{E^2}=\epsilon M_2$$

and therefore if one disregards terms of the order of ϵ:

$$\sigma^2(E)/\bar{E}=M_2/M_1. \tag{4}$$

This quantity is the 'energy mean' of the collision spectrum and may be designated by δ_2:

$$\delta_2=\frac{\int E^2\, w(E)\ \mathrm{d}E}{\int E\, w(E)\ \mathrm{d}E}. \tag{5}$$

The index in δ_2 is chosen to distinguish this quantity from the 'number' mean:

$$\delta_1=\frac{\int E\, w(E)\ \mathrm{d}E}{\int w(E)\ \mathrm{d}E}. \tag{6}$$

δ_1 is not used in the present context.

In the usual approximation the collision spectrum $w(E)$ is assumed to be proportional to $1/E^2$, so that:

$$\delta_2=\frac{\int E^2k/E^2\ \mathrm{d}E}{\int Ek/E^2\ \mathrm{d}E}=\left[\frac{E}{\ln E}\right]_{E_{\min}}^{E_{\max}}\sim E_{\max}/\ln\left(\frac{E_{\max}}{I^2}\right) \tag{7}$$

where $E_{\max}$ is the maximum energy transfer in a collision, and $E_{\min}$ is equal to $I^2/E_{\max}$. The contribution of the lower limit of the integration has been neglected in equation (7). Actually the value of δ_2 is somewhat larger due to the contribution of the resonance collisions which are not properly represented by the $1/E^2$-spectrum (Fano 1953):

$$\delta_2=\tfrac{1}{2}E_{\max}/\ln\left(\frac{E_{\max}}{I}\right)+\frac{4}{3}T \tag{8}$$

where $E_{\max}$ is the maximum collision energy transfer and T the mean kinetic energy of electrons in the material.

Whenever $E_{\max}$ exceeds the proper cut-off energy, E_Δ, one has to break off the integration at this value. The formula for δ_2 which can be derived from equation (7) then contains both $E_{\max}$ and E_Δ. Because in this case $E_{\max}$ appears only in the argument of the logarithm, it is a reasonable first approximation to merely substitute E_Δ for $E_{\max}$ in order to account for the cut-off. As far as the relation between the interval length, d, in tissue and the cut-off energy, E_Δ, is concerned, it seems appropriate to assume that E_Δ is the energy of a delta-ray with a mean projected range equal to $d/2$.

Under these conditions the values of δ_2 for intervals from 100 Å to 1 μm range from about 150 eV to 500 eV.

If δ_2 is inserted into equation (2) one obtains:

$$\overline{\Lambda}_D = L + \delta_2/d \quad (9)$$

and with the assumptions mentioned above this is represented by the curves of figure 1.

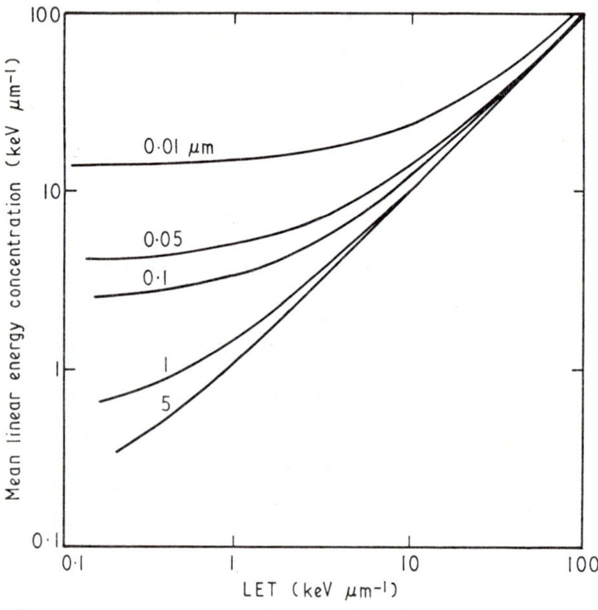

Figure 1. Mean linear energy concentration as a function of ∞. Parameter is the diameter of the region of interest.

The mean value of the linear energy concentration is always larger than LET or its mean value. The excess is due to the clustering of energy deposition along the track; it is most marked for sparsely ionizing radiation and for intervals of less than 1 μm in tissue. In fact in this region the effect of clustering (or straggling) is dominant.

LET is almost irrelevant if it is less than 10 keVμm^{-1} and if one deals with intervals of 100 Å or less. This is borne out in experiments on the inactivation of DNA, and it is generally true for radiation effects on dry materials.

With mammalian cells, on the other hand, there is an appreciable increase of RBE even below 10 keVμm^{-1}. Interaction of energy imparted along the track must therefore extend over fractions of a micrometer. This is in keeping with an analysis of sigmoidal inactivation curves, which shows that neighbouring tracks interact over distances of the order of a micrometer in the inactivation of mammalian cells (Hug and Kellerer 1966).

Figure 1 is also relevant for an understanding of the quality factor, QF, as an approximation for the effectiveness of radiation of different LET. The fact that QF has been chosen independent of LET below 3·5 keVμm^{-1} is a reflection of the fact that $\overline{\Lambda}_D$, not $\overline{L}_D$, is the parameter which determines the biological effect. A comparison of figures 1 and 2 shows that QF is roughly proportional to the mean linear energy concentration over distances of a fraction of a micrometer. This makes sense because the QF is related to overall effects on the cellular or intercellular level.

It may be mentioned parenthetically that recent attempts to determine the QF with microdosimetry instruments (Baum *et al.* 1969, Bengtsson 1969) should not be looked

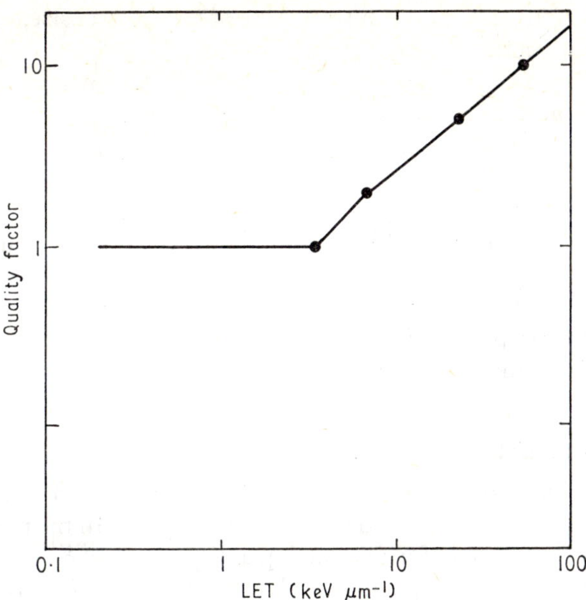

Figure 2. Quality factor as a function of ∞.

upon as approximate methods to derive the LET distribution in order to infer the quality factor. Λ or $\overline{\Lambda}_D$ is the quantity of interest; LET, though formally connected to QF, is merely an approximation to the actual energy concentration. Only Λ is experimentally observable and biologically relevant. It is appropriate to determine the local energy concentration over tissue equivalent distances of less than 1 μm in these experiments and simply derive $\overline{\Lambda}_D$ according to equation (1). This mean value can be obtained experimentally if one sums the squares of the pulse sizes in the proportional counter and divides this sum by the sum of the pulse sizes. The effective segment length d is equal to the second moment of the chord length distribution divided by the mean chord length. (This can be shown on the basis of formulae derived on the relative variance of the spectra of energy density (Kellerer 1967, 1969).

While the clustering of energy along the tracks is essential for the biological effect, it does not invalidate the LET concept as such. If one includes the appropriate term δ_2/d the LET values are, indeed, meaningful. LET values refer to a specific cut-off energy, E_Δ, thereby they also refer to corresponding values of δ_2 and d. If the interval size d is taken to be twice the projected range of the electron of energy E_Δ, one roughly obtains 0·01 μm, 0·1 μm, 1 μm, and 5 μm as intervals corresponding to the cut-off energies 100 eV, 1 keV, 4 keV, and 10 keV. Using equations (8) and (9) one finds that for the different cut-off energies one has to replace LET by the following values:

$$\overline{\Lambda}_{D,\,100\,\mathrm{eV}} = L_{100\,\mathrm{eV}} + 15\ \mathrm{keV}\ \mu\mathrm{m}^{-1}$$
$$\overline{\Lambda}_{D,\,1\,\mathrm{keV}} = L_{1\,\mathrm{keV}} + 2{\cdot}4\ \mathrm{keV}\ \mu\mathrm{m}^{-1}$$
$$\overline{\Lambda}_{D,\,4\,\mathrm{keV}} = L_{4\,\mathrm{keV}} + 0{\cdot}5\ \mathrm{keV}\ \mu\mathrm{m}^{-1}$$
$$\overline{\Lambda}_{D,\,10\,\mathrm{keV}} = L_{10\,\mathrm{keV}} + 0{\cdot}23\ \mathrm{keV}\ \mu\mathrm{m}^{-1}. \quad (10)$$

Note that on the right-hand side of these equations one may as well insert the energy mean, $\bar{L}_{D,\Delta}$, instead of L_Δ. The 'clustering term' depends, at least in our present approximation, only on the cut-off value of the interval size, not on the LET or the nature of the particle.

Instead of introducing the new quantity 'mean linear energy concentration' one might even be tempted to change the definition of L_Δ according to equation (9):

$$L_\Delta = \left.\frac{\mathrm{d}E}{\mathrm{d}x}\right|_\Delta + \delta_2/d \qquad (11)$$

where $(\mathrm{d}E/\mathrm{d}x)|_\Delta$ is the reduced collision stopping power. That would make L_Δ more than just another symbol for reduced collision stopping power. It would also give some degree of practical usefulness to the concept of energy cut-off in LET theory. The problem of the proper relation between E_Δ and d must, of course, be looked into more closely. The numerical values in equation (10) are only preliminary.

3. Parameters of track structure and the collision spectrum

Considering parameters of track structure, one may tend to think of the average energy, W, expended per ion pair and of δ_1 rather than δ_2. In the preceding section δ_2 has been shown to be the more fundamental quantity. δ_2, not δ_1, determines the mean square fluctuations of energy deposition. A few explanatory remarks on the meaning of the different parameters may help to clarify the relation between W, δ_1, and δ_2.

W together with LET determines the mean number of ionizations produced along a certain interval of the track. In dealing with a condensed medium this is, of course, an ill-defined statement. A sharp distinction between ionizations and excitations is highly questionable in this case. So is the distinction between events of more or less biological effectiveness. W values for a condensed medium are therefore to be used with caution. Moreover W is rather independent of the nature and energy of the ionizing particle, and is therefore not a very significant parameter of track structure.

For a given LET the mean energy transfer in a collision, δ_1 (see equation (6)), determines the mean number of collisions of the primary particle in a certain interval. It also determines the mean spacing, $\bar{l} = \delta_1/L$, between successive collisions. It is, however, important to note that δ_1 depends critically on the initial part of the collision spectrum. This initial part is not well-known at present. The experiments of Rauth and Simpson (1964) indicate that δ_1 has a value of approximately 60 eV for 20 keV electrons. Whether the same value can be assumed for electrons of higher energy or for heavy charged particles is an open question and has to be decided by further experiments on the collision spectrum. In radiobiology appplications δ_1 is not the physical parameter of chief interest. The plasma oscillations induced by the ionizing particle have supposedly little biological effect, while the value of δ_1 is, however, strongly influenced by these events of small energy transfer. Let us specifically assume that only 'ionizations' are considered as events. A conservative estimate is that half of all primary collisions do not lead to ionizations. The value of δ_1 measured in gas on the basis of ionizations alone must therefore be at least twice as large as the true value δ_1 in the condensed medium. This is in agreement with values of about 120 eV given by Hutchinson and Pollard (1961), while measurements by Sommermeyer and Dresel (1955, 1956) in the gas phase indicate lower values of δ_1.

Apart from being somewhat poorly defined δ_1 may also depend appreciably on the velocity of the ionizing particle. Calculated collision-spectra (Bichsel 1969, Choi and Merzbacher 1969) clearly indicate this dependence. The applicability of δ_1 is further limited by the fact that one can distinguish primary collisions only when their mean spacing, δ_1/L, is large as compared to the mean range of delta-rays or the mean cluster size. This requires that L is well below 10 keV μm^{-1}.

The bearing of δ_2 on track structure has been discussed in detail in the preceding section. The underlying fact may be illustrated by a simple statement. For larger intervals the distribution of energy deposition by a heavy charged particle is Gaussian. A Gaussian distribution of the same width results if the particle transfers energy in statistically independent events all of size δ_2. The value of δ_2 is rather insensitive to the initial part of the collision spectrum. It also makes little difference whether one deals with all collisions or with ionizations alone. From Fano's considerations on the fluctuation in the number of ions (Fano 1947) one derives that δ_2 increases only by $W/2$ (≈ 17 eV) if one determines its value on the basis of ionizations alone. Thus δ_2 in contrast to δ_1 can be determined very well from experiments in the gas phase. Since δ_2 depends strongly on the cut-off at high energies (see equation (8)) one must use devices which measure the energy imparted to the volume of interest, not the energy lost by the ionizing particle. Straggling experiments with thin foils usually yield values of $\delta_2 = \sigma^2(E)/\bar{E}$ which do not reflect a cut-off since one measures the energy loss of the primary particle, not the energy locally imparted. Straggling results from semiconductors (Maccabee 1968) are at present of limited significance to radiation biology since the equivalent layers are still too thick. Moreover the geometry of a foil cannot readily be compared to that of a cylindrical or spherical volume. For this reason the experiments with wall-less proportional counters by Rossi and Gross and by Glass, Roesch and Braby, which are being reported at this conference, are most useful for a determination of the effective δ_2 in microscopic volumes of different sizes. W. Glass *et al.* have computed numerical values of $\delta_2 = \sigma^2(E)/\bar{E}$ from experiments with protons of a few MeV; these values are consistent with the values derived in the preceding section.

The track of an ionizing particle is, of course, much too complicated to be described by a few parameters. The next step is therefore the question of the complete collision spectrum, $w(E)$. One answer to this question is given by observations on the electrons ejected from very thin foils. The answer, however, is incomplete, since one observes the kinetic energy of the electrons, not the total energy transfer including the binding energy. Thus it is not an indication of the applicability of the $1/E^2$ model, if one finds that, with the exception of the Auger peaks, the energy of the ejected electrons is distributed as $1/E^2$. On the contrary, this experimental finding supports the assumption that the collision spectrum is steeper than $1/E^2$ at low energies. Another approach is the observation of relative frequencies of ion clusters of different size in cloud chambers. The possibilities and the limitations of these observations are discussed in this volume in detail.

A third and very significant method is the use of straggling experiments in thin foils or in small gas gaps or volumes. The application of this method to the analysis of collision spectra has been limited in the past, because the corresponding mathematical problem, the compound Poisson process (Kellerer 1968a), had only been solved in the limit of many collisions and under the assumption of the $1/E^2$-spectrum or its relativistic modification. In the remaining cases rough numerical estimates or Monte Carlo methods had to be applied.

These restrictions have contributed to the fact that straggling distributions (i.e. distributions of the energy loss) have normally been determined for high collision numbers only, where the classical theory with its corrections according to Blunck and Leisegang (1950) and to Shulek *et al.* (1967) is adequate. Very little information on the collision spectrum can be derived from these many-collision experiments. On the other extreme one has tried to work with foils so thin that multiple collisions could be disregarded. This method is severely limited by the difficulty of controlling the purity and thickness of very thin foils and by the infrequency of near collisions and the resultant poor statistics in the high energy part of the collision spectrum. In the experiments of Rauth and Simpson (1964), for example, no significant information is available in the important region between 100 eV and 1 keV.

Straggling experiments with intermediate collision numbers, however, can now be analysed. Recently a computer programme has been developed which derives straggling distributions for arbitrary collision spectra and arbitrary collision numbers. The only condition is that the collision spectrum $w(E)$ is constant throughout the interval of interest. The programme (a new realization of an attempt which has been made here in Cambridge 40 years ago by E. J. Williams 1929) is based on the fact that energy loss in successive intervals is statistically independent. Starting from the approximation for extremely thin layers (see equation (3)) straggling distributions are derived in a series of successive convolutions; each of the successive distributions belongs to doubled thickness and doubled mean energy loss. The convolutions are executed on a logarithmic scale, since in certain cases the collisions spectrum covers so many orders of magnitude that use of a linear scale is not only inefficient but out of the question. This programme is available in Algol 60 and in Fortran IV. It has been published together with a theoretical analysis of the problem and its implications to radiobiology (Kellerer 1968a). An approach on the same line has independently been chosen by W. Roesch (1968).

The computer programme derives straggling distributions from collision spectra; it does not derive collision spectra from straggling distributions. One can obtain useful information on the collision spectrum by comparative curve fitting (Kellerer 1967). But ideally one would like to reverse the process. This is indeed possible. In mathematical terminology: a solution of a compound Poisson process uniquely determines the characteristic spectrum of the compound Poisson process. The reason can for example be understood from Fano's remarks on the Laplace-transform and the additivity of straggling (Fano 1953). These remarks hold for the characteristic functions (Fourier transforms) of a distribution as well as for the Laplace transform. It is therefore not necessary to repeat the argument here, in much detail.

If $\phi(\tau)$ is the characteristic function of the straggling distribution $f(E; n)$, then the characteristic function to the distribution $f(E; k.n)$ which belongs to a mean energy loss k-times as large is equal to $\phi(\tau)^k$; convolution reduces to mere multiplication for the characteristic functions. The relation holds for arbitrary non-negative values of k. Therefore it can also be applied to reduce the mean value. The only condition is that one must choose a proper continuous representation of $\phi(\tau)$ in the complex plane instead of deriving it modulo $\exp(2\pi i)$. Otherwise the function $\phi(\tau)^k$ would not be uniquely defined. This idea has been used to develop a computer programme for 'devolution' as well as convolution of given distributions. The programme (available in Fortran IV) uses the fast Fourier transform algorithm of Cooley and Tuckey (1965). Given an experimental straggling distribution to a certain mean energy loss (layer thickness) the straggling distribution to a larger or smaller mean energy loss (layer thickness) is computed. In choosing this mean value sufficiently small one directly obtains the collision spectrum.

With increasing mean values the solutions of the compound Poisson process converge to normal distributions regardless of the characteristic spectrum. In view of this fact it may seem unlikely that the underlying spectrum can be recovered from an arbitrary solution. Indeed the method is limited by the fact that small errors in the input data lead to considerable blurring of the resulting curve as the computer programme goes back to smaller mean values. Thus the method is applicable only if one either starts with moderate mean collision numbers or with extremely well-defined distributions, which are indeed 'infinitely divisible' (Gnedenko and Kolmogorov 1949, Linnik 1964). This is not a limitation of the mathematical procedure. It is, in fact, one of the main advantages of the programme that the results clearly indicate the range of different functions equally compatible with a particular result. Though it is beyond the scope of this paper to go into details of technique the procedure has been mentioned because it is felt that its existence may stimulate further experimental work on the analysis of the collision spectrum of charged particles.

References

Baum, G. W., Littlefield, P. S., and Scalsky, E. D., 1969, *Hlth. Phys. Res. Abstr.* IAEA, Vienna, **2**, 10.

Bengtsson, L. G., 1969, *Estimation of Dose Equivalent from Proportional Chamber Current Variation*, A. Rep., Radiation Research Laboratory, Columbia University, New York.

Bichsel, H., 1969, *Proc. Conf. Microdosimetry*, Stresa, Euratom, Ed. H. G. Ebert.

Blunck, O., and Leisegang, S., 1950, *Z. Phys.*, **128**, 500.

Choi, B.-H., and Merzbacher, E., 1969, *Phys. Rev.* A, **1**, 299.

Cooley, I. W. and Tuckey, J. W., 1965, *Math. Comp.*, **19**, 297.

Fano, U., 1953, *Phys. Rev.*, **92**, 328.

Fano, U., 1947, *Phys. Rev.*, **72**, 26.

Gnedenko, B. V., and Kolmogorov, A. N., 1949, *Limit Distributions for Sums of Independent Random Variables* (Moscow: G.T.T.I.); Engl. Transl. (1954) (Reading MA 01867: Addison-Wesley).

Hug, O., and Kellerer, A. M., 1966, *Stochastik der Strahlenwirkung* (Berlin-Heidelberg-New York, Springer-Verlag.)

Hutchinson, F., and Pollard, E., 1961, Physical Principles of Radiation Action, *Mechanisms in Radiobiology*, Vol. 1, Eds. M. Errera and A. Forssberg (New York: Academic Press).

Kellerer, A. M., 1967, *Proc. Conf. Microdosimetry*, Ispra, Euratom, Ed. H. G. Ebert.

—— 1968a, *Mikrodosimetrie, Grundlagen einer Theorie der Strahlenqualität* B1, GSF (Neuherberg bei München).

—— 1968b, 'Biophysical Aspects of Radiation Quality'. Second Panel Report, IAEA, Vienna.

—— 1969, *Proc. Conf. Microdosimetry*, Stresa, Euratom, Ed. H. G. Ebert,

Landau, L., 1944, *J. Phys. Moscow*, **8**, 201.

Linnik, Yu. V., 1964, *Decomposition of Probability Distributions* (Edinburgh: Oliver and Boyd).

Maccabee, H. D., 1968, *Phys. Rev.*, **165**, 2, 469.

Rauth, A. M. and Simpson, J. A., 1964, *Radiat. Res.*, **22**, 643.

Roesch, W. C., 1968, *Energy Absorption Straggling*, A. Rep. to USAEC.

Rossi, H. H., 1959, *Radiat. Res.*, **10**, 522.

—— 1967, *Advances in Biological and Medical Physics*, Vol. 11 (London: Academic Press).

Shulek, P. *et al.*, 1967, *Sov. J. nucl. Phys.*, **4**, 3.

Sommermeyer, K., and Dresel, H., 1955, *Z. Phys.*, **141**, 307.

—— 1956, *Z. Phys.*, **144**, 388.

Simon, K. R., 1948, *Fluctuations in Energy Loss by High Energy Charged Particles in Passing through Matter*, Thesis, Harvard University.

Vavilov, P. V., 1957, *Zh. eksp. teor. Fiz.*, **32**, 920, transl. *Sov. Phys.-JETP*, **5**, 749.

Williams, E. J., 1929, *Proc. R. Soc. Lond.* A, **125**, 420.

The measurement by electron spin resonance spectroscopy of local concentrations of radiation-produced radicals

S. J. WYARD

Physics Department, Guy's Hospital Medical School, London, SE1

Abstract. Electron spin resonance measurements of local concentrations of radicals, following low LET radiation, are reviewed; and are discussed in terms of current theories of radiation chemistry.

1. Introduction

It is well known that different varieties of ionizing radiation produce varying chemical and biological effects, and that these variations are usually explained by differences in the spatial deposition of energy. In the case of chemical damage an important intermediary step between the initial deposition of energy and the final chemical change is the appearance of unstable radicals. The different yields of stable products from different varieties of radiation can be explained by different local concentrations of the radical precursors. In the case of biological damage the greater complexity of the material makes it more difficult to trace the mode of action of the radiation, but it is reasonable to suppose that here, too, radicals play an important part, and that some at least of the differences between the effects of different varieties of radiation may be due to different local concentrations of radicals.

Fortunately, electron spin resonance spectroscopy (ESR) provides methods whereby direct measurements can be made of local concentrations of radicals, and such measurements have now been made in a number of laboratories. This article describes the methods, reviews the results obtained and discusses their interpretation and their relevance to the general problem of radiation damage in chemical and biological systems. The article is mostly confined to radiations of low LET, since this was the theme of the Second L.H. Gray Conference, but reference is also made to radiations of high LET and to ultraviolet radiation.

2. Methods

Electron spin resonance spectroscopy is well established, and has been described in a number of books; see, for example, Ingram (1958) or Wyard (1969). It detects paramagnetic species, i.e. those containing an unpaired electron; hence positive and negative ions, trapped electrons and free radicals will all give spectra. The sample can be in any state, but because of technical difficulties connected with the use of liquid samples, nearly all studies of radiation damage have been made with solid samples. From the spectrum it is a straightforward matter to obtain the overall concentration of paramagnetic species in the sample; in many cases the paramagnetic species can be identified from the spectrum, although this is not always possible. This article is concerned with a third kind of information which can sometimes be obtained from the spectrum, namely the local concentration of the paramagnetic species. If the distribution of paramagnetic species is homogeneous throughout the sample, then the local concentration will be the same as the overall concentration; but if the distribution is inhomogeneous due to the species being grouped in pairs, clusters, columns, etc., then the local concentration will be greater than the overall concentration.

There are three methods for measuring the local concentration of paramagnetic species, each based on a different effect in the spectrum. All three effects derive from the interaction between neighbouring magnetic centres, so all three will be present together, although in particular cases one or the other may be more prominent.

The first and most straightforward method is based on an increase in line-width of the spectrum. The line-width contribution from interaction with neighbouring magnetic centres can be written

$$\Delta H_L = \alpha C \tag{1}$$

where C is the concentration of magnetic centres and α is a constant. In a typical case the value of α is $9{\cdot}3 \times 10^{-20}$ gauss centre^{-1} cm^{-3} (or 55M gauss, where M is the molar concentration of magnetic centres) (Wyard 1965). It is important to realize that, because the magnetic field from a dipole falls off very rapidly, the line-width contribution comes almost entirely from near neighbours (Wyard 1965). It follows that when the distribution is inhomogeneous the method gives the local concentration, C_{loc}. The average concentration for the whole sample, C_{av}, can be obtained separately from the area under the spectrum.

In the special case where the sample is a single crystal and the magnetic centres occur in pairs, the interaction results not in line broadening, but in a splitting of the spectrum into a symmetrical doublet. The distance separating the centres in a pair can be calculated from the splitting, in gauss, in the spectrum.

The second method is based on the observation of a line in the spectrum at half the normal magnetic field. By comparing the intensity of the half-field line with that of the normal absorption line, one can obtain an average distance between nearest neighbours and hence derive the concentration. As with the first method, if the distribution is inhomogeneous, it is the local concentration which is measured.

The third method depends on measurements of saturation, i.e. the rate at which energy, absorbed by the magnetic centres from the microwave radiation, is handed over to the bulk of the sample. The theory of saturation involves two parameters; T_1, the spin-lattice relaxation time, and T_2, the spin-spin relaxation time. From T_2 one can obtain the line-width contribution from magnetic dipolar interaction by the formula

$$\Delta H_L = \frac{1}{\gamma T_2} \tag{2}$$

where $\gamma = 1{\cdot}76 \times 10^7$ when ΔH_L is in gauss and T_2 is in seconds. Then using equation (1) the local concentration can be found as in the first method.

There are two ways of performing the saturation measurements. One is *progressive saturation* in which the spectrum is recorded at a series of increasing power-levels. The other way employs very short and intense pulses of microwave power. The *spin-echo* technique is the most commonly used pulse method. Pulse methods are probably more accurate since they give T_2 directly; however, they require special apparatus, whereas progressive saturation can be measured on an ordinary spectrometer.

Table 1 summarizes the three methods.

Table 1. ESR methods of measuring local concentrations of magnetic centres

1 (a) Line-width $\Delta H_L = \alpha C_{loc}$
Typically $\alpha = 9{\cdot}3 \times 10^{-20}$ gauss centre^{-1} cm^{-3}
or $\alpha = 55$M gauss
(b) Line-splitting—generally requires a single crystal
2 Half-field line Intensity ratio gives average separation in a pair
3 Saturation $\Delta H_L = \dfrac{1}{\gamma T_2}$ $\gamma = 1{\cdot}76 \times 10^7$
(a) Progressive saturation
(b) Pulse methods, e.g. spin-echo

3. Results

The first ESR measurements of local concentrations of radicals in irradiated materials were made by Smith and Wyard (1961) and by Wyard (1962, 1964, 1965), although the possibility of such measurements had been earlier suggested by Livingston (1958). The measurements were made on frozen solutions of hydrogen peroxide in water, and mostly used the line-width method, although it was verified that the half-field line and the saturation methods gave the same results. The difficulty of separating out from the total line-width the contribution from the local concentration was overcome by the use of annealing. In this material annealing produces a homogeneous distribution of radicals, so that annealed samples can be used to calibrate the spectra for radical concentration. With low LET radiation (X-rays, γ-rays, β-rays and 6 MeV electrons) there was an initial local concentration of $3 \pm 1 \times 10^{19}$ radicals cm^{-3}. With high LET radiation (α-rays, deuterons and heavy ions from HILAC) the initial local concentration was somewhat greater. The greatest initial line-width was found in UV irradiated samples in which the radicals occur as closely spaced pairs.

Additional information about the distribution of radicals is given by the line-shape. A uniform distribution, either from a completely filled lattice, or from pairs with a constant separation, gives a spectrum with Gaussian tails. A random distribution, either from a partly filled lattice, or from pairs with variable spacing, gives a spectrum with Lorentzian tails. With every type of radiation used, including UV, the tails were Lorentzian for frozen solutions of hydrogen peroxide in water.

A number of measurements have been made on plastics. Using the method of progressive saturation, a local concentration of $1{\cdot}8 \pm 0{\cdot}2 \times 10^{19}$ radicals cm^{-3} was measured in syndiotactic polymethacrylate (PMMA), γ-irradiated at room temperature (Bullock and Sutcliffe 1964, Bullock Griffiths and Sutcliffe 1967). The local concentration was independent of dose, over a range for which the overall concentration varied from 10^{17} to 10^{18} radicals cm^{-3}. With the same method a local concentration of 0·037M was measured in ethylene glycol dimethacrylate (EDMA), γ-irradiated at 35 °C (Zimbrick Hoecker and Kevan 1968). This was also independent of the dose, which varied from 0·5 to 2 Mrads, corresponding to overall radical concentrations of 0·0021 to 0·0084M. A half-field line has been observed in a number of polymers, γ-irradiated at 77 K (Iwasaki and Ichikawa 1967). This was ascribed to pairs of radicals, but the separation distance was not given.

Further measurements have been made on organic glasses, γ-irradiated at 77 K. Smith and Pieroni (1965) investigated 2-methyltetrahydrofuran (MTHF). After irradiation this material contains both radicals and trapped electrons. Progressive saturation measurements were made on the trapped electrons, from which was calculated a local concentration of radicals and trapped electrons of $0{\cdot}02 \pm 0{\cdot}01$M. The overall concentration was 0·0018M, or eleven times smaller. Kevan and Chen (1968) confirmed the existence of high local concentrations of radicals and trapped electrons in MTHF, and also in methanol; but they did not give values for the concentrations.

Several measurements have been made on ices, γ-irradiated at 77 K. In alkaline ices Zimbrick and Kevan (1966, 1967a) made progressive saturation measurements on the trapped electron spectrum, and concluded that there was a local concentration which was independent of dose up to a dose of about 3 Mrads. Although the local concentration was not stated, the value of the parameter T_2 was given, from which the local concentration can be calculated as 0·04M. On the other hand, in acidic and oxyanion ices, the same workers found no evidence for a high local concentration of the hydrogen atoms which are trapped in these materials (Zimbrick and Kevan 1967b, 1967c). They concluded that hydrogen atoms are trapped uniformly throughout the sample. Other workers, however, also working with acidic ices and using the line-width method, reached a different conclusion (Bugaenko *et al.* 1967, Zador 1967). They found a local concentration of hydrogen atoms which varied with the strength of the acid, being lower in concentrated acid solutions. Both sets of workers agree that the trapped hydrogen atoms come from a reaction of mobile electrons, which are set free by the radiation; and Bugaenko *et al.* (1967) suggest that the higher the acid concentration the greater the distance travelled by the electron before reacting to give a trapped hydrogen atom. Since Zimbrick and Kevan used acid strengths of about 10M, the local concentration of hydrogen atoms may have been too small to detect. The largest local concentration measured by Bugaenko *et al.* was 0·024M in a 2·5M perchloric acid sample.

Pulse methods have been developed by Voevodsky (1967) and Salikhov *et al.* (1968). A number of substances were studied, including amino acids, nucleic acids and bases, and alcohols, but no evidence was found of local concentrations of radicals arising directly from the radiation. It was concluded that, in the materials studied, the distribution of radicals was either uniform throughout the sample, or that any inhomogeneity was due to inhomogeneity in the distribution of trapping sites, so that the distribution of radicals did not correspond with the distribution of the primary products of the radiolysis.

The single crystal results form a separate group, because the spectra are interpreted as due to pairs with a certain separation distance, rather than to clusters with a certain local concentration. Radical pairs were first observed in a number of oximes X-irradiated at 77 K (Kurita 1964, Kurita and Kashiwagi 1966, Kashiwagi and Kurita 1966). In each case the pairs were of similar radicals, and the separation distance was between 5 and 8 Å. Some samples contained two sets of pairs with different separation distances, as well as single radicals. Radical pairs have also been observed in oxalic acid, γ-irradiated at 77 K (Moulton Cernansky and Straw 1967). The separation distance was 6·33 Å in the dihydrate, and 9·93 Å in the anhydrous oxalic acid.

Dissimilar radical pairs have been observed in monofluoroacetamide, γ-irradiated at 77 K (Iwasaki and Toryama 1967). The suggested mechanism was

$$CH_2\,FCONH_2 \xrightarrow{\gamma} .CH_2\,CONH_2 + .F$$

$$.F + CH_2\,FCONH_2 \longrightarrow .CH\,FCONH_2 + HF.$$

The separation distance was 7·03 Å.

Pairs consisting of a positive and a negative ion radical have been observed in maleic anhydride, γ-irradiated at 77 K (Iwasaki and Eda 1968). The suggested mechanism was

$$\underbrace{\text{CO—CH=CH—CO—O}} \xrightarrow{\gamma} \underbrace{\text{CO—CH=CH—CO—O}^{+}} + e^{-}$$

$$\underbrace{\text{CO—CH=CH—CO—O}} + e^{-} \longrightarrow \underbrace{\text{CO—CH=CH—CO—O}^{-}}.$$

The separation distance was about 6 Å.

Pairs consisting of H and CH_3 have been observed in methane, γ-irradiated at 4·2 K (Gordy and Morehouse 1966). In this case the spectrum of a pair was observed, although the sample was not a single crystal. The separation distance was 6·76 Å.

The results are summarized in table 2.

Table 2. Values of local concentrations measured by ESR

Material	Authors	Method	Local concentration cm^{-3}; or separation distance
H_2O_2 in H_2O at 77 K	Smith and Wyard, 1961; Wyard, 1962, 1964, 1965.	Mainly line-width	$3 \pm 1 \times 10^{19}$(0·05M)
PMMA	Bullock and Sutcliffe, 1964; Bullock *et al.*, 1967	Progressive saturation	$1·8 \pm 0·2 \times 10^{19}$(0·03M)
EDMA	Zimbrick *et al.*, 1968	Progressive saturation	$2·8 \times 10^{19}$(0·037M)
Polyethylene etc. at 77 K	Iwasaki and Ichikawa, 1967	Half-field line	Value not given
MTHF at 77 K	Smith and Pieroni, 1965	Progressive saturation	$1·2 \pm 0·6 \times 10^{19}$(0·02M)
Methanol and MTHF at 77 K	Kevan and Chen, 1968	Progressive saturation	Value not given
Alkaline ices at 77 K (electrons)	Zimbrick and Kevan, 1966, 1967a	Progressive saturation	$2·4 \times 10^{19}$(0·04M)
Acidic ices at 77 K (H atoms)	Zimbrick and Kevan, 1967b, 1967c	Progressive saturation	Not detected
Acidic ices at 77 K (H atoms)	Bugaenko *et al.*, 1967 Zador, 1967	Line-width	$1·4 \times 10^{19}$(0·024M)
Variety of materials	Voevodsky, 1967; Salikhov *et al.*, 1968	Pulse methods	Not detected
Oximes at 77 K	Kurita, 1964; Kurita and Kashiwagi, 1966; Kashiwagi and Kurita, 1966	Pair spectrum	5 to 8 Å
Oxalic acid at 77 K	Moulton *et al.*, 1967	Pair spectrum	6·33 and 9·93 Å
Monofluoroacetamide at 77 K	Iwasaki and Toriyama, 1967	Pair spectrum	7·03 Å
Maleic anhydride at 77 K	Iwasaki and Eda, 1968	Pair spectrum	6 Å
Methane at 4·2 K	Gordy and Morehouse, 1966	Pair spectrum	6·76 Å

4. Discussion

It is clear from table 2 that the distribution of magnetic centres following low LET radiation is inhomogeneous in a number of solid materials. Similar or dissimilar radicals, ions or electrons are trapped in pairs, or in clusters with a local concentration considerably higher than the overall concentration. (There is also a larger number of materials studied by ESR for which no evidence was obtained of pair or cluster formation.) The interpretation of these results and a consideration of their relevance to current theories of radiation chemistry is not a simple matter because the only species observed by ESR are those which can be trapped, so that what is observed depends as much on the properties of the material as on the radiation. In particular the local concentration clearly cannot be greater than the concentration of trapping sites. Thus each material needs to be discussed separately. For this reason the discussion here will be confined to ice and to frozen solutions of hydrogen peroxide, two materials which are closely related and have been extensively studied, but which are quite different in respect of local concentrations of radicals.

In ice, γ-irradiated at 77 K, the main component of the ESR spectrum is due to OH radicals (Dibdin 1967, Gunter 1967), with an initial yield of 0·6 radicals per 100 eV (Siegel Flournoy and Baum 1961). Since most energy transfers are of less than 100 eV (Hutchinson and Pollard 1961) the majority of OH radicals should occur singly, which would account for the lack of any local concentration. Moreover the maximum concentration of OH radicals in heavily irradiated ice is 5×10^{18} cm^{-3} (Gunter 1965), and a local concentration of this magnitude, if it existed, would be very difficult to detect. At 4·2 K, both OH radicals and H atoms are trapped, with yields of 0·8 and 0·9 per 100 eV, respectively (Siegel Flournoy and Baum 1961). No measurements have been reported for the distance separating the H and OH radicals; presumably they are not very close, or else it is likely that some effect would have been observed in the spectrum.

In peroxide solutions, irradiated at 77 K, the main component of the ESR spectrum is due to HO_2 radicals (Wyard Smith and Adrian 1968). In contrast to ice, the initial yield is high, at 10 radicals per 100 eV; the maximum concentration is also high, at 5×10^{20} radicals cm^{-3}, and some radicals are trapped only 3·8 Å apart. Thus the conditions are favourable for detecting a local concentration and a value of 3×10^{19} radicals cm^{-3} was measured, as given in table 2. A possibility which must be considered is that the local concentration is higher when the radicals are first formed, and that they diffuse outwards before being finally trapped, owing to a local heating effect within the cluster. In the case of frozen solutions of hydrogen peroxide it is known from annealing experiments that HO_2 radicals diffuse freely at temperatures above 145 K, so that diffusion due to local heating is a possibility that has to be considered. However, after irradiation at 4·2 K the same local concentration was measured as at 77 K, which rules out local heating as a significant factor affecting the local concentration.

A commonly used model in radiation chemistry consists of a cluster of radicals formed from a 100 eV energy transfer. In frozen solutions of hydrogen peroxide, such a cluster would contain 10 HO_2 radicals. Assuming a homogeneous distribution within the cluster it would have a diameter of 90 Å. In the other materials listed in table 2, for which there is enough information to make a calculation, a 100 eV cluster would have a similar size, the diameters varying from 70 to 100 Å.

However, the ESR measurements do not in general discriminate between clusters of different sizes, so that the 'cluster' might, in fact, consist of a single pair of radicals. Moreover, a critical examination of the experimental evidence shows that in air rather more than 60% of all clusters consist of single pairs of ions (Ore and Larsen 1964). If the pattern of energy deposition in condensed media is similar to that in gases, the majority of clusters will consist of pairs of radicals, and it is of interest to calculate the separation distance. For frozen solutions of hydrogen peroxide an estimate can be obtained from the intensity of the half-field line. This gives an average separation distance of 6·6 Å, which is similar to the values listed for other materials in table 2.

In the cluster model there are two concentrations or separation distances to consider. Firstly, there is the separation between corresponding positive and negative ions, or from the pair of radicals formed from them (e.g. OH and H in water). Secondly, there is the separation between positive ions, or radicals formed from them (e.g. OH in water), within the same cluster. Of course, if the cluster consists of a single pair there is only the first separation to consider. One would like to measure both separations independently. This is not possible with hydrogen peroxide since only one radical is trapped, but it might be possible in materials in which radiation produces two radicals with distinguishable ESR spectra. It would also be of interest if the ESR methods could be further developed so as to distinguish between single pairs and larger clusters.

Acknowledgment

This work was supported by a grant from the British Empire Cancer Campaign.

References

BUGAENKO, L. T., BELEVSKY, V. N., ZADOR, E., and GOLUBEV, V. B., 1967, *The VIIIth Int. Symp. on Free Radicals*, p. 145.
BULLOCK, A. T., GRIFFITHS, W. E., and SUTCLIFFE, L. H., 1967, *Trans. Faraday Soc.*, **63**, 1846.
BULLOCK, A. T., and SUTCLIFFE, L. H., 1964, *Trans. Faraday Soc.*, **60**, 2112.
DIBDIN, G. H., 1967, *Trans. Faraday Soc.*, **63**, 2098.
GORDY, W., and MOREHOUSE, R., 1966, *Phys. Rev.*, **151**, 207.
GUNTER, T. E., 1965, *University of California, Berkeley Rep. UCRL*-16613, p. 82.
GUNTER, T. E., 1967, *J. chem. Phys.*, **49**, 2780.
HUTCHINSON, F., and POLLARD, E., 1961, *Mechanisms in Radiobiology*, Eds M. Errera and A. Forssberg (London: Academic Press) p. 1.
INGRAM, D. J. E., 1958, *Free Radicals as Studied by Electron Spin Resonance* (London: Butterworth).
IWASAKI, M., and EDA, B., 1968, *Chem. Phys. Lett.*, **2**, 210.
IWASAKI, M., and ICHIKAWA, T., 1967, *J. chem. Phys.*, **46**, 2851.
IWASAKI, M., and TORIYAMA, K., 1967, *J. chem. Phys.*, **46**, 4693.
KASHIWAGI, M., and KURITA, Y., 1966, *J. phys. Soc. Japan*, **21**, 558.
KEVAN, L., and CHEN, D. H., 1968, *J. chem. Phys.*, **49**, 1970.
KURITA, Y., 1964, *J. chem. Phys.*, **41**, 3926.
KURITA, Y., and KASHIWAGI, M., 1966, *J. chem. Phys.*, **44**, 1727.
LIVINGSTON, R., 1959, *Radiat. Res. Suppl.*, **1**, 463.
MOULTON, G. C., CERNANSKY, M. P., and STRAW, D. C., 1967, *J. chem. Phys.*, **46**, 4292.
ORE, A., and LARSEN, A., 1964, *Radiat. Res.*, **21**, 331.
SALIKHOV, K. M., YUDANOV, V. F., RAITSIMRING, A. M., ZHIDOMIROV, G. M., and TSVETKOV, YU. D., 1968, *Proc. XVth Colloque Ampère*, 278.
SIEGEL, S., FLOURNOY, J. M., and BAUM, L. H., 1961, *J. chem. Phys.*, **34**, 1782.
SMITH, D. R., and PIERONI, J. J., 1965, *Can. J. Chem.*, **43**, 876.
SMITH, R. C., and WYARD, S. J., 1961, *Nature*, Lond., **191**, 897.
VOEVODSKY, V. V., 1967, *The VIIIth Internat. Symp. on Free Radicals.* Plenary lecture.
WYARD, S. J., 1962, *Proc. XIth Colloque Ampère*, 388.
WYARD, S. J., 1964, *University of California, Berkeley Rep. UCRL*-11387, Suppl. p. 1.
—— 1965, *Proc. phys. Soc.*, **86**, 587.
—— 1969, *Solid State Biophysics*, Ed. S. J. Wyard, (New York: McGraw-Hill Book Company), p. 1.
WYARD, S. J., SMITH, R. C., and ADRIAN, F. J., 1968, *J. chem. Phys.*, **49**, 2780.
ZADOR, E., 1967, *KFKI Kozlemenyek*, **15**, 223.
ZIMBRICK, J., HOECKER, F., and KEVAN, L., 1968, *J. phys. Chem.*, Ithaca, **72**, 3277.
ZIMBRICK, J., and KEVAN, L., 1966, *J. Am. chem. Soc.*, **88**, 3678.
—— 1967a, *J. chem. Phys.*, **47**, 2364.
—— 1967b, *Nature*, Lond., **214**, 693.
—— 1967c, *J. chem. Phys.*, **47**, 5000.

Spatial distribution of free radicals in aqueous salt solutions gamma-irradiated at 77 K

B. G. ERSHOV and G. P. CHERNOVA

Institute of Physical Chemistry of the USSR Academy of Sciences, Moscow

Abstract. Information on the local concentrations of radicals within 'spurs' can be obtained from an analysis of line-broadening in EPR spectra as a function of the dose given. The line width of the H-atom spectrum reflects the local concentration of radicals in a spur and is found to remain constant until a dose is reached at which it is supposed that neighbouring spurs begin to overlap. The relevant theoretical relations are discussed and experimental data are presented for γ-irradiated frozen salt solutions.

At low temperatures the radicals, having lost their kinetic energy, are stabilized in the bulk of the matrix. Their spatial distribution may reflect the specific features of their formation mechanism in 'spurs'. The information obtained at present on the distributions of paramagnetic centres is based on the analysis of dipole–dipole interactions (B. G. Ershov, to be published). The basis of the most reliable method of estimating the local concentration of radicals ($C_{loc.}$) is the analysis of the concentration broadening of the electron paramagnetic resonance (EPR) line. The broadening of the EPR signal may be associated with the local radical concentrations by the relation (Lebedev and Dobryakov 1967).

$$\Delta H_0{}^2 = \Delta H_G{}^2 + \Delta H_0 \Delta H_s \simeq \Delta H_G{}^2 + A \,.\, C_{loc.} \Delta H_0 \qquad (1)$$

where ΔH_0 is the width of the EPR line measured at the points of maximum slope, ΔH_G is the effective width of the non-uniform distribution approximated by the Gaussian function, ΔH_s is the width of the spin packet and A is a coefficient equal to $5{\cdot}4 \times 10^{-20}$ Oe cm^{-3} for the random radical distribution and the Lorenz form of the spin packet.

This investigation has studied the radical distribution in frozen aqueous salt solutions γ-irradiated at 77 K. The investigation method is based on taking into account the contribution of the electron dipole–dipole interaction to the total width of the EPR line of a hydrogen atom.

The concentration of hydrogen atoms in glassy solutions of 8M $NaClO_4$ and 5M NaH_2PO_4 has been found to reach its maximum value (2–3×10^{18} cm^{-3}) at the absorbed dose $\sim 3 \times 10^{21}$ eV cm^{-3} and then to decrease. At the same time, the total concentration of radicals attains steady-state values ($\sim 10^{20}$ cm^{-3}) at the dose $\sim 3{\cdot}10^{22}$ eV cm^{-3}.

The width of the EPR line of a hydrogen atom in γ-irradiated frozen salt solutions, remains approximately constant up to the average radical concentrations $(1{\cdot}3$–$2{\cdot}0) \times 10^{19}$ cm^{-3} (the dose absorbed being 3–6×10^{20} eV cm^{-3}). Then it begins to increase, reaching maximum values 8 and 9 Oe for perchlorate and phosphate solutions, respectively. The analysis of the area of the concentration broadening shows good agreement with equation (1). Having determined ΔH_G graphically, on the basis of relation (1) one can calculate $C_{loc.}$ in the immediate environment of a hydrogen atom. In the region of small doses the local concentration has been found to be constant and to exceed considerably the average concentration (table 1).

This picture corresponds to the formation of radicals in a 'spur', while $C_{loc.}$ represents the average concentration of radicals in it. The broadening of the EPR line of a hydrogen atom takes place when the spurs begin to overlap, and the contribu-

Table 1.

System	Dose*, 10^{-20} eV g^{-1}	$C_{av.} \times 10^{-19}$ cm^{-3}	$C_{loc.} \times 10^{-19}$ cm^{-3}	$\rho = \frac{C_{loc.}}{C_{av.}}$
8M $NaClO_4$, glass	$0{\cdot}33 \div 5$†	$0{\cdot}15 \div 2{\cdot}7$	$2{\cdot}9 \pm 0{\cdot}2$	$18 \div 1{\cdot}1$
8M $NaClO_4$, polycryst.	$1{\cdot}3 \div 25$	$0{\cdot}18 \div 2{\cdot}3$	$5{\cdot}9 \pm 0{\cdot}4$	$25 \div 2{\cdot}7$
5M NaH_2PO_4, glass	$0{\cdot}33 \div 5$	$0{\cdot}13 \div 1{\cdot}8$	$1{\cdot}8 \pm 0{\cdot}4$	$11 \div 1{\cdot}3$

* The dose at which $C_{loc.}$ is approximately constant.
† The sign $\div$ separates the extreme values used.

tion of dipole–dipole interactions from neighbouring 'spurs' begins to be observed. In this case in the glassy samples their uniform distribution ($\rho = C_{loc.}/C_{av.} = 1$) is finally attained. However, in the sample of 8M $NaClO_4$ having non-uniform polycrystalline structure, $C_{loc.}$ and $C_{av.}$ increase in parallel but $C_{loc.}$ remains 2·7 times as great as $C_{av.}$.

This fact is probably caused by different compositions of microphases and by non-uniform distribution of the traps of radicals in the sample bulk. In this case, in each of these phases the distribution of radicals may be uniform, but may differ in $C_{av.}$. This is shown by the constancy of ρ.

The point of bending on the curves of the concentration broadening corresponds to the concentration of radicals ($\bar{C}_{av.}$) at which the overlapping of 'spurs' begins. On this basis it is not difficult to show that assuming spherical shape of a 'spur' its radius (r) is equal to

$$r = \left(\frac{3n}{4\pi \bar{C}_{av.}}\right)^{1/3} \times 10^8 \ (\text{Å}) \tag{2}$$

where the unknown value n is the average number of radicals in a 'spur'.

Table 2 gives the results of the calculation for a number of values of n. The radicals in a 'spur' are supposed to be distributed uniformly. For $n = 4$–8 the radius is seen to be 30–50 Å.

Table 2.

n	r, Å in 8M $NaClO_4$ glass	r, Å in 5M NaH_2PO_4 glass
2	26	31
4	33	39
6	36	45
8	41	49
10	45	53

In spite of the obvious simplification of the model used for the accumulation of radicals and track structure as well as for 'the spur' itself, the value obtained may evidently be considered reasonable enough. By comparing data derived from a study of $C_{loc.}$ in aqueous solutions of alkalis, acids and salts, one may observe the effect of the mode of formation of radicals on the form of their spatial distribution. In alkaline aqueous glasses the thermalized electrons are captured non-uniformly (Ershov and Pikaev 1967, Zimbrick and Kevan 1967a, Ershov *et al.* 1968) whereas, in acid glasses their reaction with hydrogen ions results in the formation of hydrogen ions that are already uniformly distributed (Ershov *et al.* 1968, Zimbrick and Kevan 1967b). This may be accounted for by the fact that electrons in acid 'glasses' have a long range due to the different nature and concentration of their traps. Moreover, it may be associated with the fact that when the thermalized electrons react with H_3O^+ ions there arise hot hydrogen atoms that can move some additional distance to the place where they are stabilized; one can then understand the non-uniformity of their distribution in salt solutions. In these systems, at least, a considerable proportion of the hydrogen atoms are formed when excited molecules of water decompose. It is quite possible, however, that the difference in the distribution of hydrogen atoms in acid and neutral media is due to the different distribution of H_3O^+ ions. In this case in neutral 'glasses' hydrogen atoms are formed by the reaction of electrons with H_3O^+ ions in 'spurs'.

References

Ershov, B. G., and Pikaev, A. K., 1967, *Zh. fiz. Chimii*; **41**, 2537.
Ershov, B. G., Chernova, G. P., Grinberg, O. Ya., and Lebedev, Ya. S., 1968, *Izv. Akad. Nauk. SSSR,–ser. chin.*, 2439.
Lebedev, Ya. S., and Dobryakov, S. I., 1967, *Zh. Strukt. Chimii*: **8**, 838.
Zimbrick, J., and Kevan, L., 1967a, *J. chem. Phys.*, **47**, 2364.
—— 1967b, *J. chem. Phys.*, **48**, 5000.

ESR spin echo method for determining the spatial distribution of free radicals

YU. D. TSVETKOV

Institute of Chemical Kinetics and Combustion, Academy of Sciences, Novosibirsk, USSR.

Abstract. The advantages of the electron spin echo (ESE) method of studying the spatial distribution of radicals are discussed and preliminary results on γ-irradiated organic glasses at 77 K are reported. The ESE method of studying radiation-produced radicals is at present limited by the sensitivity of the available spectrometers, but once this limitation has been overcome the method should be a powerful tool for studying track effects.

It seems worthwhile to draw the attention of scientists interested in track effects or other inhomogeneous chemical and physical effects to the possibilities of using the pulsed ESR method (or 'Electron Spin Echo'—ESE—method) for the investigation of the local free radical concentrations in irradiated materials.

The spin echo method is at present widely used in nuclear magnetic resonance. This phenomenon was first discovered by Hahn (1950) while studying the behaviour of nuclear magnetization under the action of two powerful radio frequency pulses separated by the time interval τ (the magnetic system being in the resonance condition). It turns out that after a time interval 2τ has elapsed, measured from the first pulse, a signal appears, which can be detected by means of a receiver. The amplitude of this signal depends on τ and it decays exponentially with characteristic relaxation time (T_2) which depends upon the dipole–dipole interaction in spin-system

$$T_2^{-1} \propto \frac{1}{\langle r^3 \rangle}$$

where $\langle r^3 \rangle$ is the average cubed distance between spins. The theory of ESE has some special peculiarities but in general the same principles hold as in ESR and these can be used to derive information about the distances between paramagnetic particles or about the local concentrations of free radicals in the specimen under investigation.

This type of measurement, which has been done mainly on γ-irradiated organic solids at 77 K, started in the Institute of Chemical Kinetics and Combustion quite recently in Professor V. V. Voevodsky's Laboratory and the aim of this report is to summarize briefly the main results.

The first group of results refers to the investigation of T_2 values during the accumulation of radicals. It was found for the γ-irradiated crystalline solids in the dose range 0·5–10·0 M rads that the ratio of the local to average radical concentration (the last may be determined by usual ESR methods) lies between 1 and 2. In cases where it is more than 1, for example in cyclohexane, methyl alcohol, hexyl alcohol, stearic acid, the concentration dependence of T_2 during the accumulation of free radicals under irradiation can be explained in terms of a non-uniform spatial distribution of free radicals, if radicals are assumed to be stabilized only in particular local regions or somewhere near crystal defects.

The size of these stabilization volumes may be estimated from the measured T_2 relaxation times and it can be found that they must be not smaller than 100 Å. The local concentration in these volumes grows in accordance with the growth of the average concentration of the radicals and the ratio of $[\dot{R}]_{loc}/[\dot{R}]_{av}$ in this range of dose does not change.

The other group of results which were obtained by means of the ESE method refer to the concentration dependence of relaxation times during thermal radical decay in irradiated organic solids. The experimental curves are usually more complicated in these experiments and in many of the cases examined they exhibit two types of region. The first type is linear, with $(T_2)^{-1}=K[\dot{R}]_{av}$ and in the second type the relaxation time for radical decay is independent of $[\dot{R}]_{av}$ in spite of changes of some 50% or more in $[\dot{R}_{av}]$. This kind of result was obtained, for example, for free radical decay in the temperature range 125–165 K in γ-irradiated ($\sim$20 Mrad) cyclohexane and in a number of other experiments of the same kind. These results can be understood only if radicals initially trapped in solids do not decay uniformly throughout the whole volume but disappear preferentially in some parts of it. This conclusion is of interest because of its bearing on theoretical models for radical recombination kinetics in irradiated solids.

The ESE method has a number of advantages over the usual measurements of ESR line-broadening and of progressive saturation of ESR signals. The most important of these, in relation to studies on free radical distribution, are:

(1) It is possible to measure accurately the dipole–dipole interactions in the spin system not only in the simple cases of ESR lines but with any type of paramagnetic species, since the other kinds of interaction—inhomogeneous broadening (anisotropic and g-value) become unimportant for ESE measurements.

(2) It is possible to measure spin–spin (T_2) and spin–lattice (T_1) relaxation times independently.

When doing the experiments on spatial distribution of free radicals by means of ESE, it is necessary to have in mind that the other types of broadening may occur in the spin system. For radiation chemistry experiments one important consideration may be the broadening of the relevant ESR lines by the other types of paramagnetic species which cannot be detected due to their short magnetic relaxation times, or indeed other types of broadening in the ESR spectrum. There may be, for example, paramagnetic ions present. These can affect the T_2 relaxation times and the local concentration of free radicals deduced may be higher than the real one. This remark refers, however, not only to ESE measurements but to all magnetic resonance measurements of the spatial distribution of paramagnetic species.

Systematic ESE investigation of free radicals distributions have commenced quite recently. It is believed that the real track effects in solids will be worth investigating by means of ESE at much lower doses of γ-radiation than can be done today owing to the comparatively low sensitivity of ESE spectrometers (about 10^{15} radicals in a normal ESR sample). But it seems that the results obtained up to the present indicate that very soon ESE may be one of the most direct methods for the investigation of free radical spatial distribution after or during irradiation, or during the subsequent chemical reactions in solids.

Reference

HAHN, E., 1950, *Phys. Rev.*, **77**, 746.
——, 1950, *Phys. Rev.*, **80**, 580.

Information about the structure of the track of a high energy electron in a liquid as obtained from electric conductance measurements

A. HUMMEL

Reactor Institute, Delft, Netherlands

Abstract. The spatial distribution of the charged species in the track of a fast electron in a dielectric liquid is considered. The ions will mostly be formed in groups of a few ion pairs. Most of them will recombine with ions of the same group, but some of them will escape and become homogeneously distributed throughout the liquid. The number of ion pairs that escape, which can be determined from conductance measurements, contains information about the spatial distribution of the ions in the track. The theoretical background is reviewed. Experimental methods are described and some typical results are discussed.

We will consider what information can be obtained about the spatial distribution of the charged species in the track of a high energy electron in a liquid from measurements of the electric conductance.

During thermalization of a high energy electron in a liquid, positive ions and secondary, tertiary, etc., electrons are found. The electrons as well as the positive ions eventually give rise to charge carriers in thermal equilibrium with the surrounding molecules, which we will call ions. These ions will be found in groups of different sizes, where they will be moving in one another's Coulomb field, and which often results in the (initial) recombination of positive and negative ions of the same group. Some of the ions however will diffuse out of the Coulomb field of the charge carriers of the group, they will escape initial recombination, and become homogeneously distributed throughout the liquid. The number of ion pairs that escapes initial recombination with and without an externally applied field, contains information about the spatial distribution of the ions in the track. Measurements of the electric conductance of liquids under irradiation may enable us to determine this yield of escaped ions as a function of the applied field.

Consider a liquid in a homogeneous dc field between two parallel plates of a conductivity cell under continuous irradiation, with an absorbed dose rate D. The yield of escaped ion pairs per 100 eV absorbed as a function of the external field will be expressed as $G(E)$. With a vanishingly small field this yield is not affected (zero field yield, $G(0)$), a steady-state concentration of ions n, recombining in a second order fashion with a rate constant k, is built up and given by

$$DG(0)=kn^2 \tag{1}$$

assuming one kind of ion of either sign to be present. If certain experimental conditions are fulfilled, Ohm's law is found to hold and it can be shown that for small fields the conductivity κ is given by

$$\kappa = eun = eu \left\{ \frac{DG(0)}{k} \right\}^{1/2} \tag{2}$$

where u is the sum of the mobilities of the positive and negative ion with a charge e and n is given by equation (1).

When we increase the field strength the number of ions attracted to the electrodes increases, and the steady-state concentration n is not given by equation (1) as Ohm's law is not obeyed, and the current as a function of applied field tends to 'saturate' (as in the gas phase); however, the current never reaches the 'saturation' value due to the fact that the yield of escape from the track increases with the field strength. It is possible to obtain a rough estimate of $G(0)$ from measurements at intermediate field strength, and by extrapolating to the 'saturation' value using the semi-empirical description of the current–voltage relationship for gases.

At even higher field strengths, no homogeneous recombination occurs any more, and the measured current is proportional to the rate of production of escaped ion pairs in the liquid, $DG(E)$. This makes it relatively easy to determine $G(E)$. Often $G(E)$ takes approximately the form of $G(E) = G_0 + aE$ for extended ranges of E, which makes it tempting to identify G_0 with $G(0)$. In some cases the two values correspond reasonably well, but there is no theoretical basis for the identification.*

Another way to determine $G(0)$ is to sweep out all the ions with a large field which is applied immediately after the irradiation is interrupted and before an appreciable amount of homogeneous recombination and diffusion towards the electrodes has taken place (Schmidt and Allen 1968).

The principles of the measuring techniques mentioned above have all been described in the beginning of this century by Langevin, Thomson and Jaffé. A great deal of information exists on measurement of conductance at higher field strengths in different liquids, but it is only in recent years that reliable absolute values for the yields of escaped ions as a function of the field have become available.

In table 1 some typical results are shown of measurements of $G(0)$ with (*a*) the steady-state method, using equation (2) and (*b*) by sweeping out the ions according to the method of Schmidt and Allen. As a comparison we have also given some values of G_0, the zero field yield as determined by extrapolation from $G(E)$ at high field strengths. The ratio $G(E = 10^4\ \text{V cm}^{-1})/G_0$ gives an impression of the effect of the field on the yield of escaped ions.

The first attempts to obtain information about the spatial distributions of the ions in the tracks of high energy particles in liquids from conductance measurements have been made by Jaffé (1913). Following a suggestion of Langevin he took cylindrical tracks, and assumed the density distribution to be gaussian at all times (prescribed diffusion approximation) and equal for both positive and negative ions. He ignored the interaction between the ions. Consideration of diffusion and recombination in the track leads to a correlation of the number of ions escaping recombination in the track as a function of E with the width of the track and the charge carrier density. Lea (1934) introduced a modification, especially meant for treatment of the tracks of electrons of intermediate energy. He assumed the track to consist of a string of spherical groups, merging into a column. The mathematical treatment here is quite similar to Jaffé's.

* For a discussion of the measuring techniques see Hummel (1967).

Table 1. Yields of escaped ions for some liquids under different experimental conditions

Radiation or radiation source	Liquid	Temperature K	$G(0)$ (b)	$G(0)$ (a)	G_0	$\frac{G}{G_0}(E=10^4\,\mathrm{V\,cm^{-1}})$	Ref.
Bremstrahhlung	n-hexane	296	0·131				S
1·5 MeV electrons	n-hexane	293		0·095	0·125	1·62	H
^{39}Ar	n-hexane	293		0·10			H
^{60}Co, ^{137}Cs, ^{65}Zn, ^{170}Tm	n-hexane	293			0·089		J
^{37}Ar	n-hexane	293		0·050			H
T	n-hexane	294		0·04			H
T	n-hexane	294			0·064	2·3	G
		324			0·087	1·96	G
Bremstrahhlung		371	0·256				S
1·5 MeV electrons		296	0·131				S
		293		0·095	0·125	1·62	H
		219		0·055	0·055	2·12	H
	3-methylpentane	296	0·146				S
	2,3-dimethylbutane	296	0·192				S
	2,2-dimethylbutane	296	0·304				S
	n-pentane	296	0·145				S
	neopentane	296	0·857				S
	diethylether	296	0·35				S
	diethylether	296			0·36	1·3	H
	1,4-dioxane	296	0·046		0·038	2·1	H
	CCl_4	296	0·096				S
	CCl_4	296		0·068			H

(a) Steady-state method; (b) sweep method; see text: G—Gzowski, O., Chybicki, M., 1963; H—Hummel, A., Allen, A. O., 1966, Hummel, A., Allen, A. O., Watson, Jr., F. H., 1966, Hummel, A., Allen, A. O., 1967, Hummel, A., 1967; J—Jauszajtis, A., 1963; S—Schmidt, W. F., Allen, A. O., 1968.

Kramers (1952) criticized the prescribed diffusion approximation and gave an improved treatment to describe the results of Gerritsen on liquid He, Ar, H_2 and N_2.

Measurements of absolute yields of escape as a function of the field strength and the temperature for high energy electrons in liquid n-hexane have shown that the Jaffé–Lea model is not applicable (Hummel and Allen 1967, Hummel 1967).

In the meantime Onsager (1933) had treated the diffusion of one positive and one negative ion in each other's Coulomb field. He calculated the probability of escape as a function of the external field. For $E=0$ this leads to an escape probability $\phi=\exp\{-r_c/r\}$ with $r_c=e^2/\epsilon kT$ with ϵ the dielectric constant, e the (electronic) charge of the ions, and k Boltzmann's constant. We will make use of this treatment.

Consider a track of a 1 MeV electron in n-hexane. When the primary fast electron is slowing down it will suffer losses of a wide range of energy, secondary electrons will be formed, which in turn may form tertiaries and so on. The majority of the losses will be small (<100 eV) and will result only in a few ionizations spaced closely together. Occasionally a larger energy loss occurs which gives rise to a more extended track of ions. After thermalization or trapping of the electrons the ions are found in groups, and the size of each group corresponds to a given energy loss. Each ion in the group moves in the field of all the others and the ions will recombine pairwise.

We will now consider the fate of the ion pair in the group that remains after recombination of all the others. In the case of very large groups, there may be two or more ion pairs left which are so far apart that cross-interaction between members of the separate pairs is negligible. In this case we consider both these separate pairs.

The number of such pairs of ions with separation between r and $r+\mathrm{d}r$ resulting from ionizations produced by an electron of energy W we call $g_W(r)\,\mathrm{d}r$. The number of escaping ion pairs per energy loss of W is then

$$\int_0^\infty g_W(r)\phi(r)\,\mathrm{d}r \tag{3}$$

where $\phi(r)$ is given by the theory of Onsager.

If $n(W)\,\mathrm{d}W$ is the number of energy losses between W and $W+\mathrm{d}W$ suffered by the primary electron during its thermalization, the total number of escaped pairs originating from the primary is

$$N=\int_0^{W_{\max}} n(W)\,\mathrm{d}W \int_0^\infty g_W(r)\phi(r)\,\mathrm{d}r. \tag{4}$$

If the external field is zero, this can be written as

$$N(0)=\int_0^{W_{\max}} n(W)\,\mathrm{d}W \int_0^\infty g_W(r)\exp\left(-\frac{r_c}{r}\right)\mathrm{d}r. \tag{5}$$

Electrons which are formed with such an energy that they separate from the parent ion by a distance which is large compared to r_c will have a large probability of escape. We therefore assign an escape probability of 1 to all losses >1000 eV. In this way we get

$$N(0)=\int_{1000\text{ eV}}^{W_{\max}} n(W)\,\mathrm{d}W+\int_0^{1000\text{ eV}} n(W)\,\mathrm{d}W \int_0^\infty g_W(r)\exp\left(-\frac{r_c}{r}\right)\mathrm{d}r. \tag{6}$$

The number of losses >1000 eV turns out to be 67 and contributes only little to $N(0)$ which experimentally is found to be of the order of 1000. If W is so small that only one ion pair is created, then $g_W(r)$ is the range distribution of the electron with energy W. For a somewhat larger W, corresponding to a few pairs, the order of magnitude of the width of the distribution $g_W(r)$ will still approximately correspond to the range of the electron. For $g_W(r)$ we will take

$$g_W(r)=\frac{4\pi r^2}{\pi^{3/2}b^3}\exp\left(-\frac{r^2}{b^2}\right) \tag{7}$$

which has the form of the probability distribution found for a free diffusing particle in 3-dimensions, starting at $r=0$. b is the most probable distance and is a function of W. We will use this expression for g_W for values of W below 1000 eV. Since $g_W(r)$ is normalized to one, the integral over r in equation (6) is always <1. This means that the yield of separated pairs from an event $W<1000$ eV cannot be larger than one. In view of the fact that we know from the ^{37}Ar experiment that the total yield for $W=2{\cdot}6$ keV is only 1·3, this assumption cannot introduce a large error.

After substitution of (7) in (6) the integral over r turns out to be a function of r_c/b only, so that (6) can be written as

$$N(0)=67+\int_0^{10^3} n(W)F\left(\frac{r_c}{b}\right)\mathrm{d}W \tag{8}$$

with $b=b(W)$; $n(W)$ can be estimated fairly well and $b(W)$ is now chosen so that $N(0)=1000$, the value for n-hexane at room temperature. The curves for $b(W)$ in figure 1 are arbitrarily made to coincide with the range at high energies, but this is of little importance, since the events around 1000 eV are few in number and the probability of escape is close to unity. Curve (A) is the range energy curve calculated by Magee *et al.* (1964) for energy losses down to subexcitation level; the yield computed with this curve is about 10 times lower than the experimental one. Curve (B) is obtained by adding a constant to the ranges of Magee and assuming b to stay constant below 80 eV, where this value is chosen such as to get $N(0) = 1000$. Curve (C) is simply an empirical curve drawn so as to give a value for $N(0)$ that fits the experiment. With this curve the temperature dependence of $N(0)$ can also be predicted quite adequately.

It should be realized that from this model one can at best get the separation of the last pair in each group. Only for a group of one ion pair will this correspond to the range of the electron; for small groups the separation may not be too different from the range. For large groups the physical significance of b is lost. Fortunately there are so many small losses that our lack of knowledge about the larger groups may not introduce too large an error in the case of n-hexane. Experimental values of $G(0)$ for electron energies below the 2·6 keV of the ^{37}Ar decay are needed to resolve this problem. Another difficulty with this model is the uncertainty about the probability that a given small energy loss in the liquid leads to a pair of charge carriers in thermal equilibrium with the surrounding. This probability may be dependent on the properties of the liquid, and also on solutes. Fast-pulse radiolysis techniques, where the ions can be observed spectroscopically with a time resolution of nanoseconds, may provide valuable information.

Mozumder and Magee (1967) have considered the movement of low energy electrons in the field of a positive ion during their thermalization in n-hexane. They take the distance at which the energy of the electron has decreased down to the lowest electronic excitation level (taken as 6 eV) to be given by the curve (A) in figure 1. The distance at which the electron energy has dropped under the lowest vibration level (taken as 0·4 eV) is now calculated, taking a mean free path of 5 Å, a cross section of 40 Å^2 and an average energy loss of 0·2 eV. The remaining kinetic energy (0·4 eV) is considered to be lost in N interactions, where an average energy of ΔW_s is dissipated, while the potential energy change of the electron in the field of one positive ion is now taken into account. Assuming the electron to have a random path, it is now possible to calculate the probability distribution $g_W'(r)$ for the electron with initial energy W to get thermalized at a distance r for given values of ΔW_s and the mean free path. By using $g_W'(r)$ like $g_W(r)$ in (5) values for the parameters can now be found by adjustment to experimentally determined yields. The distribution $g_W'(r)$ found by Mozumder and Magee for n-hexane is quite different from our $g_W(r)$. In order to compare the yields of escape calculated with both distributions, we put

$$\int_0^\infty g_W'(r) \exp\left(-\frac{r_c}{r}\right) \mathrm{d}r = \int_0^\infty \frac{4\pi r^2}{\pi^{3/2} b^3} \exp\left(-\frac{r^2}{r} - \frac{r_c}{r}\right) \mathrm{d}r. \tag{9}$$

The values found for b in this way have been plotted in figure 1, curve D, which shows that the effective yield of escape as a function of W for the model of Mozumder and Magee is quite close to the yield which is calculated with curve B in our treatment.

Mozumder and Magee have also made an attempt to calculate the yield of escape from tracks of electrons with an energy of a few keV. They find a temperature dependence of $G(0)$ which is much smaller than the fast electron tracks. As can be

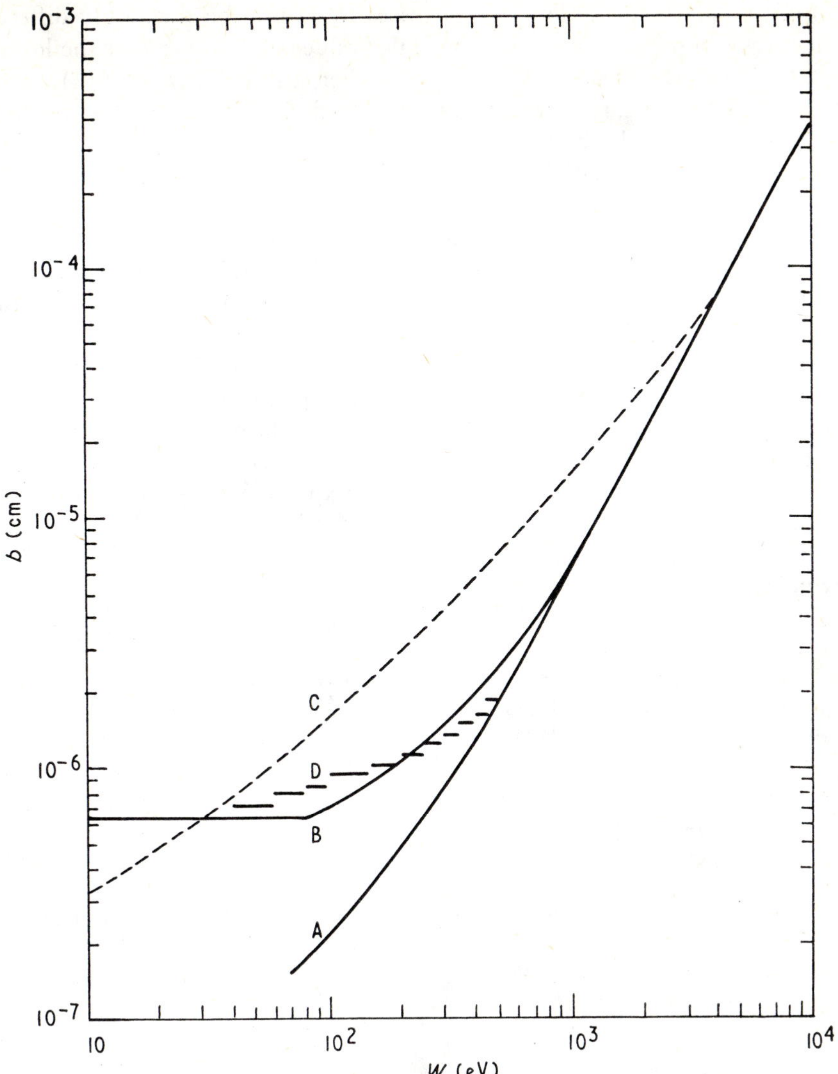

Figure 1. The separation parameter b against the energy loss W.

seen from table 1 Gzowski and Chybicki have found a large temperature dependence of G_0, which probably indicates that $G(0)$ will also change considerably with the temperature.

Schmidt and Allen (1968) have used the expression

$$G = G_g \int_0^\infty \frac{4\pi r^2}{\pi^{3/2} b^3} e^{-r^2/b^2} e^{-r_c/r} \, dr = G_g \, F\left(\frac{r_c}{b}\right) \tag{9}$$

with one value for b, and where G_g is the gas phase value for ion production (for hydrocarbons taken as 4·3). It is interesting to see that their value of $b = 67{\cdot}4$ Å for n-hexane (taking $G(0) = 0{\cdot}131$) corresponds closely to the value of b which we find in figure 1 for small energy losses ($b = 63$ Å, with $G(0) = 0{\cdot}1$). For neopentane ($G(0) = 0{\cdot}856$) they find in this way $b = 178{\cdot}4$ Å. If we assume the relative contribution to $G(0)$ of the high and the low energy losses to be not too drastically different for neopentane as compared to n-hexane, a similar value is found for b for the low energy region of W with the treatment presented earlier, using (5) ($b \sim 190$ Å). For 2,2-dimethylbutane and 2,2,4-trimethylpentane they find 92·0 and 95·0 Å, respectively. The large b-values for the branched hydrocarbons and especially for the neopentane are quite striking. Although they are not the ranges of the low energy electrons, they do indicate large thermalization distances.

The large differences are probably not caused by differences in cross section for electronic or vibrational excitation in these liquids. Neutron scattering experiments on n-pentane and neopentane do not show clear evidence of an appreciable difference in the intermolecular interactions (L. A. De Graaf 1969 private communication; Larsson and Dahlborg 1964). A possible explanation may be that a spherically symmetrical molecule like neopentane, showing no change in polarizability during rotation does not provide any drag to the passing electron due to energy loss to rotational motion of the molecule. The fact that CCl_4 does not show a large G-value and a correspondingly large b-value could be explained by capture of epithermal electrons.

References

GZOWSKI, O., and CHYBICKI, M., 1963, *Conf. on Electronic Processes in Dielectric Liquids,* Durham, England.

HUMMEL, A., 1967, Thesis Free University, Amsterdam.

HUMMEL, A., and ALLEN, A. O., 1966, *J. chem. Phys.*, **44**, 3426.

HUMMEL, A., ALLEN, A. O., and WATSON, JR., F. H., 1966, *J. chem. Phys.*, **44**, 3431.

HUMMEL, A., and ALLEN, A. O., 1967, *J. chem. Phys.*, **46**, 1602.

JAFFÉ, G., 1913, *Ann. der Phys.*, **42**, 303.

JANUSZAJTIS, A., 1963, *Conf. on Electron Process in Dielectric Liquids*, Durham, England.

KRAMERS, H. A., 1952, *Physica*, **18**, 665.

LARSSON, K.-E., and DAHLBORG, U., 1964, *Physica*, **30**, 1561.

LEA, D. E., 1934, *Proc. Camb. Phil. Soc.*, **30**, 80.

MAGEE, J. L., FUNABASHI, K., and MOZUMDER, A., 1964, *A.E.C. Rep.* No. COO-38-378.

MOZUMDER, A., and MAGEE, J. L., 1967, *J. chem. Phys.*, **47**, 939.

ONSAGER, L., 1938, *Phys. Rev.*, **54**, 554.

SCHMIDT, W. F., and ALLEN, A. O., 1968, *Science*, **40**, 301.

SCHMIDT, W. F., and ALLEN, A. O., 1968, *J. phys. Chem.*, **72**, 3730.

Information from induced electrical conductance and from the nonhomogeneous kinetics of charge scavenging reactions

G. R. FREEMAN

Chemistry Department, University of Alberta, Edmonton, Alberta, Canada

Abstract. The pairs of positive and negative charges generated during the radiolysis of a liquid either undergo geminate neutralization or they escape from each other by random diffusion. The competition between geminate neutralization and diffusive escape is strongly dependent on the initial distance between the energetically thermalized pair of charges. The ions that escape geminate neutralization can be measured by electrical conductance methods. Thus the magnitude of the radiation-induced conductance provides a measure of the competition between geminate neutralization and diffusive escape, and hence gives an indication of the magnitude of the separation distance between the charged species in spurs. More detailed information about the spacial distribution of charged species in spurs is obtainable by measuring the scavenger concentration dependence of charge scavenging reactions.

A heavy particle track consists of a single, long spur. A fast electron track consists of a series of small spurs of various sizes.* The studies of induced conductance and scavenging kinetics have provided information about the structure of spurs, and therefore of particle tracks, in liquids.

Consider the general reactions

$$\mathrm{M} \rightsquigarrow [\mathrm{M}^+ + e^-] \tag{1}$$

$$[\mathrm{M}^+ + e^-] \longrightarrow [\text{geminate neutralization}] \tag{2}$$

$$\longrightarrow \mathrm{M}^+ + e^- \text{ (free ions)} \tag{3}$$

$$\text{free ions} \longrightarrow \text{random neutralization.} \tag{4}$$

The square brackets about reactants or products indicate that the species are inside a spur. The competition between reactions (2) and (3) depends on the spacial distribution of the species in the spur, i.e. on the spur 'structure'.

The conductance method measures only the free ions and is mainly restricted to liquids of low polarity. The intrinsic conductance of highly polar liquids tends to obliterate the induced conductance in these substances. The scavenging kinetics method measures all of the ions and can be applied equally well to liquids of high and low polarity. The scavenging kinetics method is potentially more powerful than the conductance method, but the two complement each other.

Induced conductance measurements give a rough indication of the distance that the secondary electrons travel away from their parent ions during the slowing-down process. In any system there is a wide distribution of ion–electron separation

* According to the simplest definition, a spur is a group of reactive intermediates that are close enough together that there is a significant probability of their reacting with each other before diffusing into the bulk medium.

distances, but the exact distribution function is not known. In most liquids of low polarity, only a few per cent of the total number of ions formed become free ions. In these cases the free ions result from the long-range tail of the distribution, so little information is provided about the main part of the function. To illustrate the present level of development, the free ion yield in n-hexane, $G_{fi}=0{\cdot}1$, has been interpreted by different authors in terms of ion–electron separation distributions in which the average separation distances were roughly 25 (Freeman 1967), 60 (Hummel Allen and Watson 1966), 70 (Schmidt and Allen 1968), and 80 Å (Mozumder and Magee 1967).

Comparison of the induced conductance of various compounds (Schmidt and Allen 1968, Tewari and Freeman 1968) has shown that the magnitude of the free ion yield in a liquid hydrocarbon is dependent on the molecular structure (table 1).

Table 1. Free ion yields in X-irradiated hydrocarbon liquids at 22 ± 2°

Hydrocarbon	G_{fi} (Schmidt and Allen 1968)	G_{fi} (Tewari and Freeman 1968)	$G_{fi}/G_{total\ ions}$
Cyclohexane	0·15	0·11	0·03
n-Hexane	0·13	0·11	0·03
n-Pentane	0·15	0·12	0·03
Neohexane	0·30	0·40	0·08
Neopentane	0·86	0·81	0·20

A larger free ion yield indicates a larger average ion–electron separation distance. Thus an average secondary electron travels farther from its parent ion in neohexane than it does in cyclohexane, n-hexane or n-pentane, and it travels still farther in neopentane than in neohexane. The range of a secondary electron in a liquid hydrocarbon is greater when the molecules are more nearly spherical in shape. The energy distributions of the secondary electrons are probably very similar from one hydrocarbon to another, so the differences in range can be attributed to differences in energy loss processes and electron localization efficiencies (Tewari and Freeman 1968). This suggestion is supported by the fact that essentially all of the ions generated during the X-radiolysis of liquid argon become free ions (Tewari and Freeman 1969). Electrons do not form a localized state in liquid argon (Schnyders Rice and Meyer 1966). The properties of the hydrocarbon molecules that are most relevant to the present considerations and that correlate with molecular geometry are anisotropy of polarizability and dipole moment. Small dipole moments of n-alkanes have been proposed previously to explain the microwave spectra (Lide 1960) and the dependence of the Clausius–Mossotti function on density and temperature (Mopsik 1967). The anisotropic polarizability and the dipole moment probably facilitate both energy transfer from the secondary electron to the medium and the formation of a localized state of the low energy electron in the medium.

The average ion–electron separation distance in a newly formed spur in a fast electron track in neopentane appears to be about 170 Å (Schmidt and Allen 1968, G. R. Freeman unpublished), which is much greater than the 25–80 Å in n-hexane.

In the X-radiolysis of n-hexane 3% of the ion pairs become free ions (table 1). However, when n-hexane is irradiated with 5 MeV α-particles, only 0·2% of the ions escape geminate neutralization (Chybicki 1966). The concentration of charged species in the α-particle spur is much larger than that in the fast electron spurs, and the effectiveness of reaction (3) in competition with (2) is 15-fold smaller in the former than in the latter.

More information about the spacial distribution of the charged species in spurs is obtainable from scavenging studies. The addition of a charge scavenger changes the radiolysis product yields. The measurement of the product yields allows one to estimate the extent of occurrence of the scavenging reaction as a function of scavenger concentration. Kinetic analysis of the results gives information about the concentration of the charged species in the spurs.

Nitrous oxide reacts with electrons to form nitrogen.

$$e^- + N_2O \rightarrow N_2O^- \tag{5}$$

$$N_2O^- \rightarrow N_2 + O^-. \tag{6}$$

The nitrogen yields obtained from nitrous oxide solutions in cyclohexane (Robinson and Freeman 1968), methanol (Jha and Freeman 1968) and water (Head and Walker 1965, Dainton and Logan 1965, Russell and Freeman 1968) are shown in figure 1.

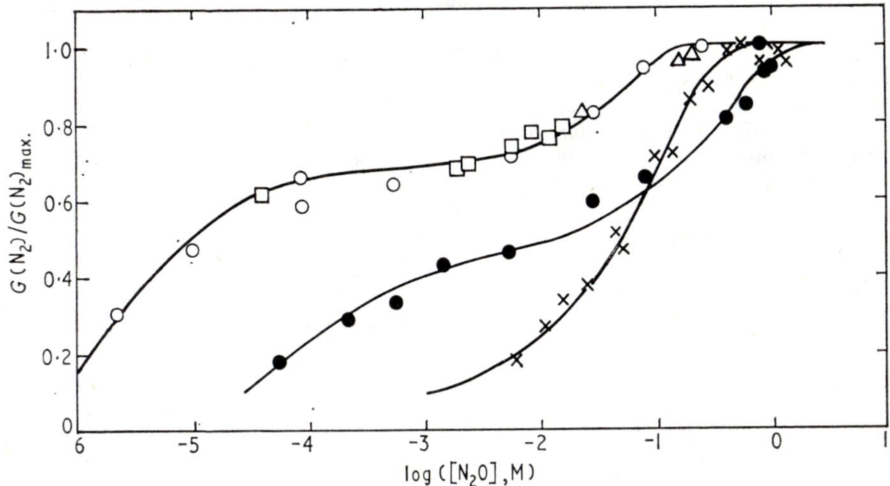

Figure 1. Nitrogen yields from the γ-radiolysis of solutions of nitrous oxide in the following liquids at 25°: cyclohexane (X, Robinson and Freeman 1968); methanol (●, Jha and Freeman 1968); water (□, Head and Walker 1965; △, Dainton and Logan 1965; ○, Russell and Freeman 1968). The curves were calculated from the model described by Freeman (1967), with the parameter β_- equal to 6×10^{14} V cm^{-2} in cyclohexane, 4×10^{13} in methanol and 8×10^{14} in water. The values of $G(N_2)_{max}$ were 5·1 in cyclohexane, 4·6 in methanol and 4·0 in water.

The solid curves in figure 1 were calculated from the model described by Freeman (1967). The distribution functions used for the ion–electron separation distances in the three liquids were related to each other by a slightly modified Bethe equation (we used the valence electron density and a value of I near to the ionization potential).

Curves such as these are relatively sensitive to the particular form of the ion–electron distribution function. The free ion yield by itself is much less sensitive. The functions used for the curves must lead both to the correct value of the free ion yield and to the correct dependence of the scavenging reaction on the concentration of scavenger. The average ion–electron separation distances in the functions used for the γ-radiolyses of cyclohexane, methanol and water were all in the vicinity of 20–25 Å. The general function leads to the correct dependence of the free ion yield on liquid properties such as dielectric constant and density, for liquids with distinctly nonspherical molecule.

The study of LET effects will also be fruitful. Increasing the value of LET from 0·03 to 30 eV $Å^{-1}$ decreased the free ion yield in methanol from 2·0 to 0·2 (Imamura and Seki 1967, Seki and Imamura 1968).

Ion scavenging and ion lifetime measurements will lead to information about ion–electron separation distances in spurs. However, independent information will be required to help transform the separation distance distribution into a secondary electron energy distribution and a relationship between electron energy and range at low energies.

References

CHYBICKI, M., 1966, *Acta phys. pol.*, **30**, 927.
DAINTON, F. S., and LOGAN, S. R., 1965, *Trans. Faraday Soc.*, **61**, 715.
FREEMAN, G. R., 1967, *J. chem. Phys.*, **46**, 2822.
HEAD, D., and WALKER, D. C., 1965, *Nature*, **207**, 517.
HUMMEL, A., ALLEN, A. O., and WATSON, F. H., JR., 1966, *J. chem. Phys.*, **44**, 3431.
IMAMURA, M., and SEKI, H., 1967, *Bull. Chem. Soc. Japan*, **40**, 1116.
JHA, K. N., and FREEMAN, G. R., 1968, *J. chem. Phys.*, **48**, 5480.
LIDE, D. R., 1960, *J. chem. Phys.*, **33**, 1514.
MOPSIK, F. I., 1967, *J. Res. Nat. Bur. Stand.*, A**71**, 287.
MOZUMDER, A., and MAGEE, J. L., 1967, *J. chem. Phys.*, **47**, 939.
ROBINSON, M. G., and FREEMAN, G. R., 1968, *J. chem. Phys.*, **48**, 983.
RUSSELL, J. C., and FREEMAN, G. R., 1968, *J. chem. Phys.*, **48**, 90.
SEKI, H., and IMAMURA, M., 1968, *J. chem. Phys.*, **48**, 1866.
SCHMIDT, W. F., and ALLEN, A. O., 1968, *J. phys. Chem.*, **72**, 3730.
SCHNYDERS, H., RICE, S. A., and MEYER, L., 1966, *Phys. Rev.*, **150**, 127.
TEWARI, P. H., and FREEMAN, G. R., 1968, *J. chem. Phys.*, **49**, 4394.
TEWARI, P. H., and FREEMAN, G. R., 1969, *J. chem. Phys.*, **51**, 1276.

Particle tracks in condensed matter*

R. KATZ and E. J. KOBETICH†

Behlen Laboratory of Physics, University of Nebraska, Lincoln, Nebraska 68508

Abstract. A proposed theory of track formation attributes observed effects to the spatial deposition of energy by secondary electrons. The theory is compared to experimental observations in a variety of fields, with good results. In contrast, linear measures of the interaction of charged particles with matter (LET, primary excitation, primary ionization, restricted energy loss) are unsuitable parameters through which to describe particle tracks, except in limiting cases, because many detection media are saturable. Energy deposited close to a particle's path often produces a disproportionately small response, because of 'overkill'.

1. Introduction

To a large extent, the damage to condensed matter from γ-ray irradiation and charged particle bombardment is due to the interaction of secondary electrons with the medium, so that differences in the observed effects arise from the time scale with which the secondary electrons are generated and from their spatial distribution. The energy deposition from secondary electrons gives rise to excitations, to bond rupture, to molecular and crystallographic rearrangements which are detected in different ways, of which the alteration of biological function may be the most specific. In small subvolumes near an ion's path we assume that the response of the medium to a passing ion is as if the subvolume were part of a larger system uniformly irradiated with γ-rays to the same average energy deposition. Thus, knowledge of the response function of a medium to γ-rays may be coupled with knowledge of the spatial distribution of the energy deposited by secondary electrons about an ion's path, to yield the spatial distribution of response, and therefore the total response of the medium to a single charged particle.

Several detection systems have been analysed on this basis.

An assumption of exponential response (one-or-more hit in the cumulative Poisson distribution) to dose has been used to describe particle tracks in emulsion, the scintillation pulse heights in NaI(T1), and the RBE for dry enzymes and viruses.

An assumption of threshold response (many-hit in the cumulative Poisson distribution) has been used to correlate observations of the formation of etchable tracks in dielectrics.

Since experimental data represent average responses of the medium to the passing charged particle, as when one measures the blackness of the particle's track in emulsion, or its grain count, theories of track formation must predict average response. In the present work, the average is taken at the level of the energy deposition by secondary electrons. We make no attempt to follow the paths of individual electrons as they meander through the medium. So long as the theory predicts correctly the mean probability for the production of detectable events as a function of radial distance from

* Supported by the US Atomic Energy Commission and the National Science Foundation.
† Present Address: H. H. Wills Physics Laboratory, University of Bristol.

the ion's path, the fact that these events really lie along the path of a secondary electron may be inconsequential.

2. Spatial distribution of ionization energy

To find the spatial distribution of ionization energy about the path of a charged particle, electron data have been combined with the δ-ray distribution formula and an assumed angular distribution of the ejected secondary electrons (δ-rays).

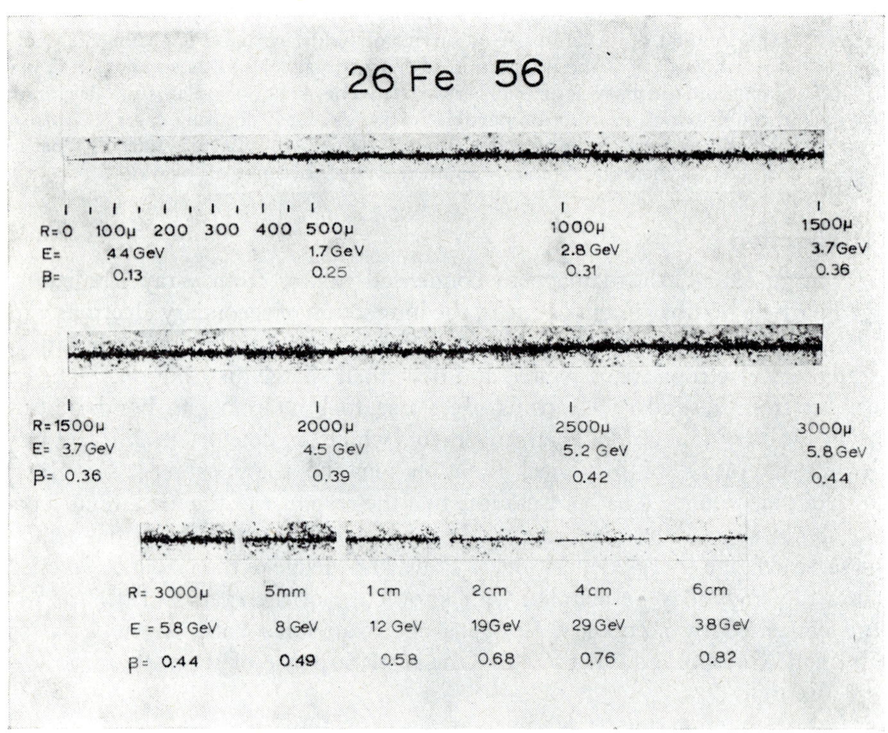

Figure 1. Track of an iron nucleus in G·5 emulsion, with residual range, energy, and $\beta\ (=v/c)$ shown. The maximum specific energy loss occurs at a residual range of about 25 μm. (Courtesy P. H. Fowler.)

Electron energy dissipation data for a range of materials and energies have been summarized into computer algorithms, in excellent agreement with experimental data, from 20 keV to 2 MeV (Kobetich and Katz 1968a, 1969). These algorithms have been used for extrapolation to low electron energies, as needed.

In all computations, the δ-ray distribution formula differential in energy, and calculated for the interaction of an incident charged particle with free electrons, is used. It is assumed that the energy which would be given to a free electron is the energy transferred to the bound electron.

Where interest centres on effects close to an ion's path, say within 1000 Å, all δ-rays are taken to be ejected normally, for most of the significant energy deposition is associated with low energy δ-rays, ejected in grazing collisions. Normal ejection has been used in the analysis of RBE (Butts and Katz 1967), NaI(T1) (Katz and Kobetich 1968a), and the formation of etchable tracks in dielectrics (Katz and Kobetich 1968b).

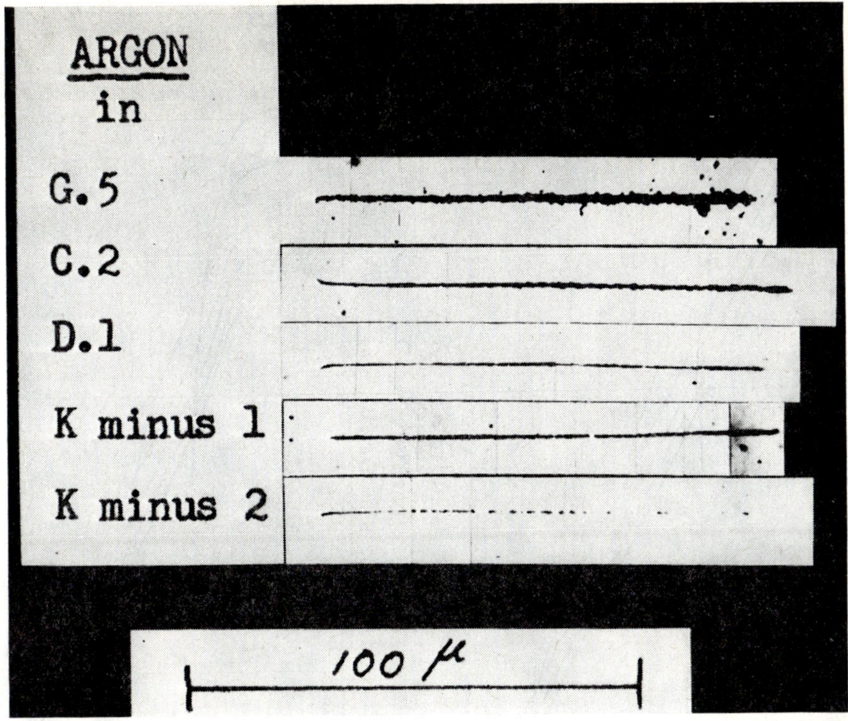

Figure 2. Tracks of 400 MeV argon ions in emulsions of different sensitivities. The silver deposited does not scale with sensitivity. There are qualitative differences in response between the most sensitive and the least sensitive emulsions. The Bragg peak occurs at a range of about 15 μm. Note that only in the least sensitive emulsion does the amount of deposited silver seem to follow the Bragg curve. (Barkas 1963.)

Where the events of interest are microns distant from the ion's path, the angular distribution has been adjusted to give best agreement with experimental data. Track formation in emulsion has been studied through the use of a distribution of the form $5 \cos^4 \theta$ (Kobetich and Katz 1968b), and the classical distribution for the encounter of a heavy particle with a free electron (Katz and Kobetich 1969). Fortunately the results of the theory are insensitive to the details of the angular distribution. Indeed the principal results seem to derive from the fact that under reasonable assumptions of angular distribution or energy dissipation by δ-rays, the spatial distribution of the energy is nearly inversely proportional to the square of the distance from the ion's path, and nearly inversely proportional to the square of the ion's speed, and directly proportional to the square of its effective charge (accommodating for electron capture and loss).

In all applications of the theory up to the present time, the direct excitation of the medium by the passing ion has been neglected. In essence the model was initially developed for the study of the tracks of energetic heavy ions in emulsion (Katz and Butts 1965) where this assumption is clearly valid, and has been applied to other situations where this neglect might be thought to generate difficulty. Nevertheless, in only two cases have possible difficulties appeared. The model does not predict the relativistic rise in grain counts or bubble counts from the tracks of singly charged

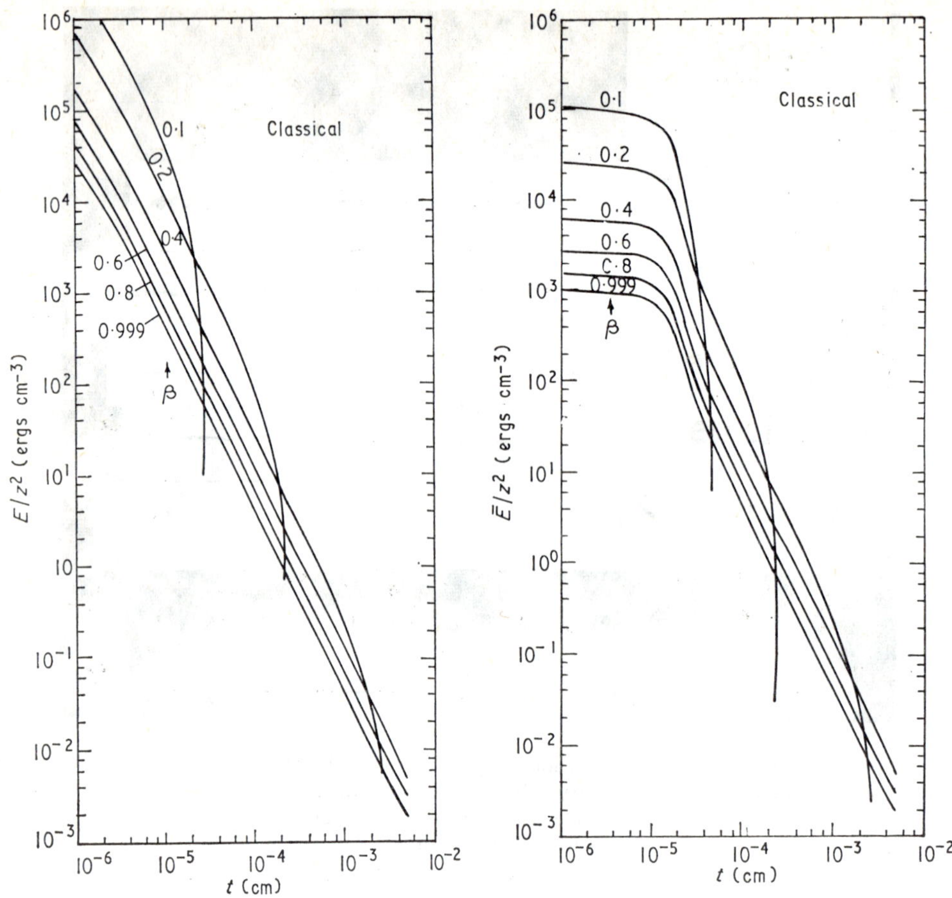

Figure 3. Point distribution of the energy deposition in emulsion by δ-rays, divided by the square of the effective charge, as a function of *t*, the distance from the path of an ion moving at speed βc. For this calculation the angular distribution implied by classical kinematics was used. (Katz and Kobetich 1969.)

Figure 4. Spatial distribution of the average energy deposited in spheres of radius 0·2 μm, divided by the square of the effective charge, as a function of the distance *t* of the centre of the sphere from the ion's path, in emulsion. (Katz and Kobetich 1969.)

particles, nor does it properly yield pulse height for proton or α-particle bombardments of NaI(T1). The neglect of direct excitations may be responsible for these failures. The validity of the neglect may be due to 'overkill', for the response of a medium is saturable, and a small fraction of the response may result from the large fraction of the energy loss to be associated with direct excitation.

3. Results

The present theory of track formation asserts that track formation must be understood from the spatial distribution of ionization energy deposited by secondary electrons about the path of a charged particle. Such an assertion implies that track formation

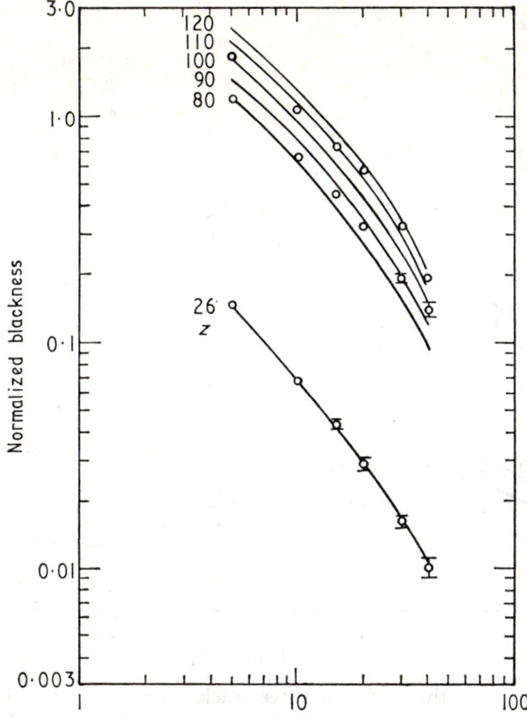

Figure 5. Blackness variation as a function of distance from the ion's path calculated for ions of atomic number 26, 80, 90, 100, 110, and 120, moving at $\beta=0{\cdot}95$. These calculations are normalized so as to agree with measurements (P. H. Fowler 1969, private communication) of relativistic iron tracks in emulsion. Measurements of the tracks of very heavy cosmic-ray particles are shown. Agreement in the slopes of the curves implies that the calculations of spatial distribution of the energy is in reasonable agreement with blackness measurement at very large distances from the path of an ion. Final Z identification of these particles depends on knowledge of E_0 and β. (Katz and Kobetich 1969.)

cannot be understood from linear measures of charged particle interaction, for these have no explicit knowledge of spatial effects. The LET, and similar parameters cannot properly describe track formation.

Data from several sources confirm this view. Thus the light pulse from a scintillation counter is not proportional to the specific energy loss, the specific energy loss is not a good parameter for describing the formation or non-formation of etchable tracks in dielectrics, but the most spectacular demonstration arises from particle tracks in emulsion, shown in figure 1, where for an energetic iron ion, the maximum silver development occurs at a range of around 1000 μm, but the maximum of the energy loss occurs at a range of about 25 μm. The appearance of the track simply does not follow from linear energy loss considerations. In figure 2 we note that the change in emulsion sensitivity changes the response of the emulsion to ions in a qualitative way. The response of the medium simply does not scale with its sensitivity; the very character of the response is altered.

To describe the formation of particle tracks in emulsion we first find the point distribution of energy, as in figure 3, and then average this distribution over the size of a sphere approximating an emulsion grain, as shown in figure 4. The two distributions

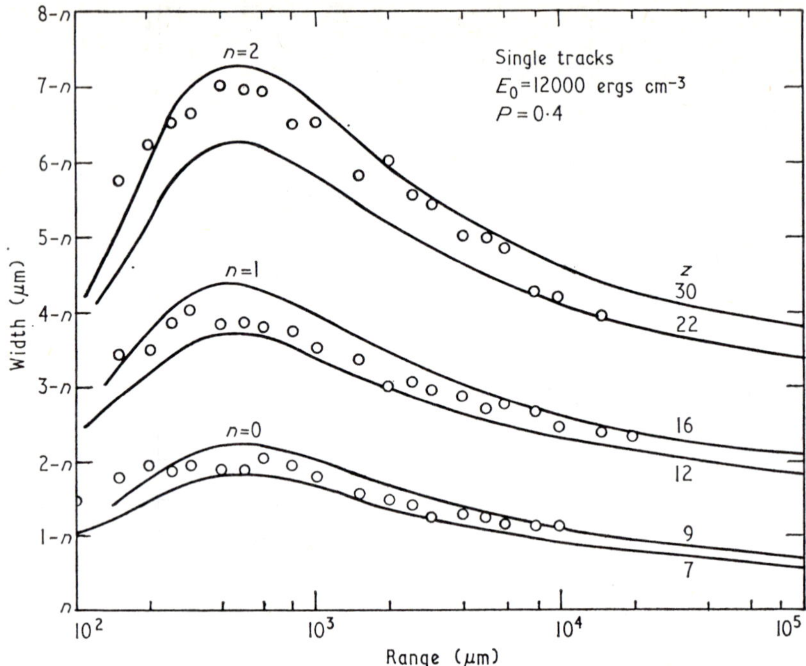

Figure 6. Measurements of the width of three tracks in a stack of G.5 emulsion exposed at balloon altitudes, bracketed between theoretical curves of the indicated Z, at indicated values of E_0 and P. The divergence between theory and measurement at low range is expected, for the tracks are very dense in this region, and it is not possible to locate the experimental value of t at which $P=0{\cdot}4$ (see figure 1). (Katz and Kobetich 1969.)

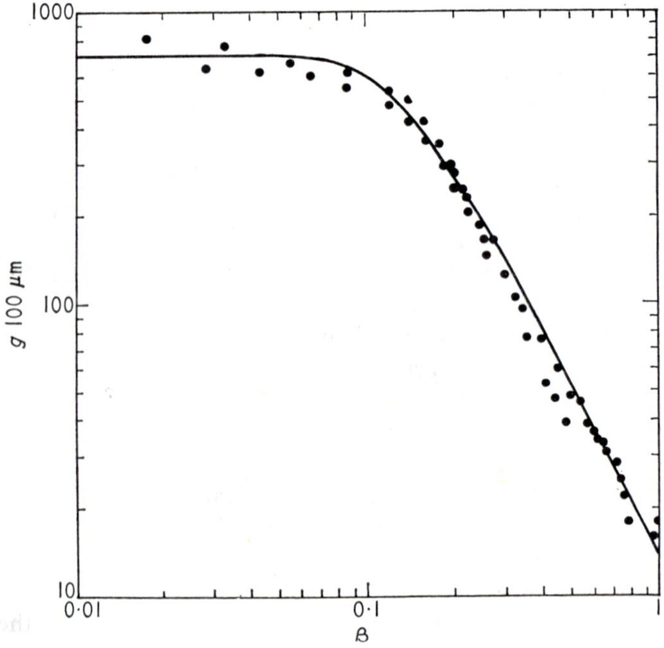

Figure 7. Measured grain count for singly charged particles in K.5 emulsion, shown against calculations from equation (2). (Barkas 1963, Katz and Kobetich 1969.)

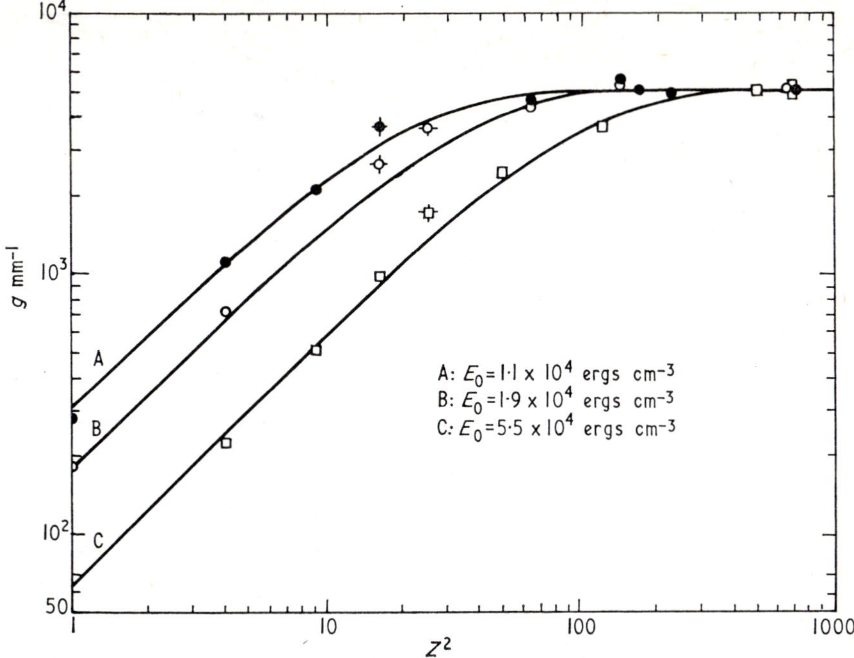

Figure 8. Grain counts for relativistic ions in G·5 emulsion as a function of Z^2 (Powell *et al.* 1959) shown against curves from equation (2) at $\beta=0{\cdot}95$. Solid circles *A* and hollow circles *B* are from emulsions representing the limits of normal development, while hollow squares *C* are from underdeveloped emulsion. Note that variations in processing are accommodated by variations in E_0 alone. (Katz and Kobetich 1969.) Points flagged with a cross, at $Z^2=16$ and 25 would fit the theoretical curves if experimentally assigned *Z* values were off by 1. Experimentally, the value of *Z* and β are unknown and interact in the *Z* assignment.

approach each other beyond two sphere radii from the ion's path, so that where grain sensitization at distances larger than a grain diameter is significant, figure 3 may be used instead of figure 4 to represent the mean dose received by the grain. In earlier work this has been called 'the point target approximation' (Butts and Katz 1967).

When the response of the medium is exponential (1-hit) we take the probability *P* for the sensitization of an emulsion grain to be represented by the expression

$$P(t)=1-\exp\{-\bar{E}(t)/E_0\} \tag{1}$$

where $\bar{E}(t)$ is the mean energy density deposited in a grain whose centre is at distance *t* from the ion's path (figure 4), and E_0 (also called the D-37 dose) is the characteristic dose for activation of 63% of the emulsion grains in an emulsion uniformly exposed to γ-rays. A single parameter, E_0, accommodates variations in emulsion and in processing. Equation (1) has also been used to describe NaI(T1) and RBE. Combined with figure 4, it yields the spatial distribution of developed grains, and can therefore be used as the basis of a calculation of the microdensitometry of particle tracks in emulsion, with results shown in figure 5.

The appearance of a heavy ion track in emulsion may also be characterized by its 'width', measured by manually tracing around track segments, and dividing the included area in a segment by its length. We compare the measured width to $2t$, found theoretically by taking $P=0{\cdot}4$, in figure 6.

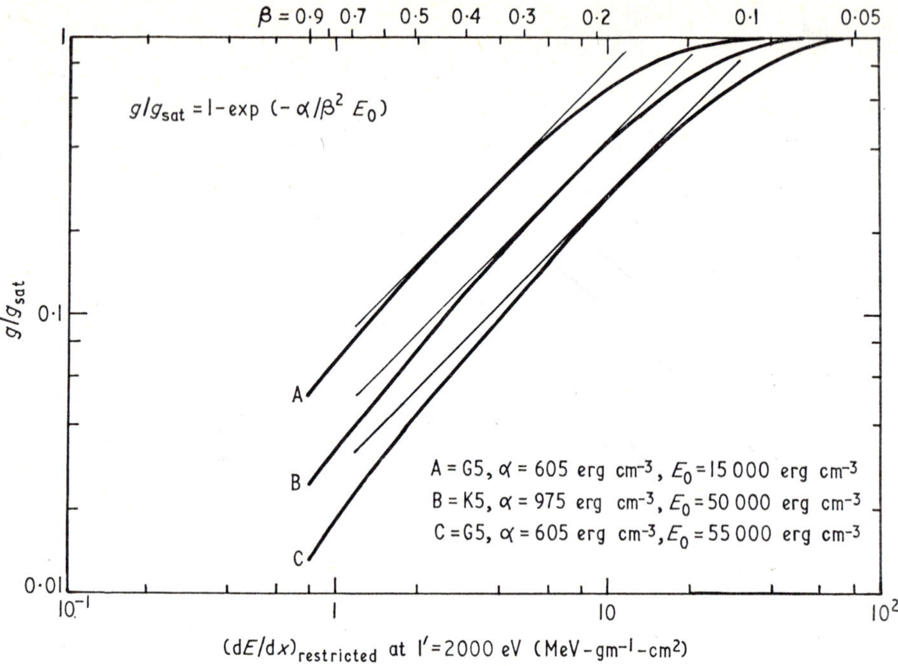

Figure 9. Grain count is plotted against restricted energy loss in AgBr (Barkas 1963), from equation (2). The grain count expected from average development of G.5 emulsion is shown in curve A, from average development of K.5 emulsion in curve B, and from underdeveloped G.5 emulsion in curve C. Lines at 45° tangent to these curves are shown, where grain count is proportional to restricted energy loss. Note that even where the track is linear, linear measures of the energy loss do not describe the observed effect well.

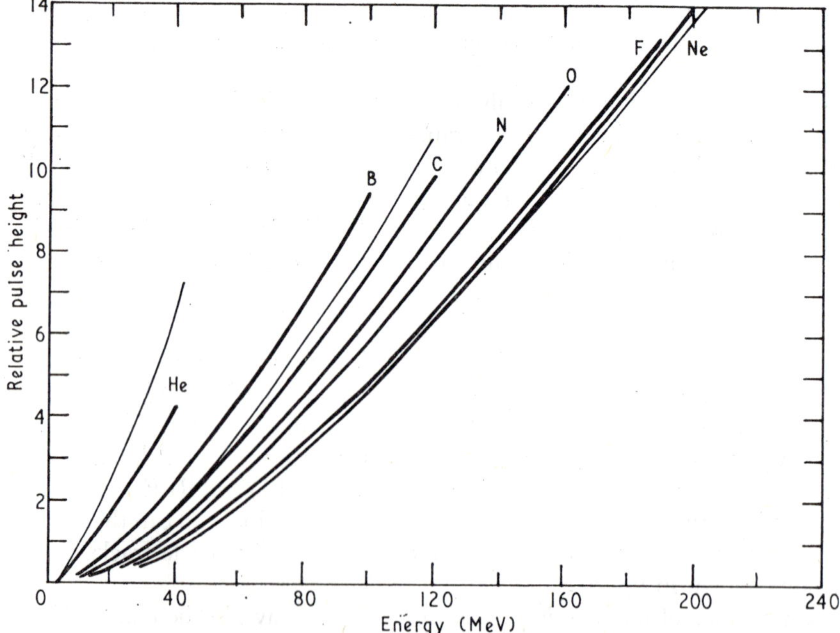

Figure 10. Experimental values of the relative pulse heights generated in NaI(T1) by ions of varying incident energies (light lines) are compared to theory (heavy lines). (Katz and Kobetich 1968a.)

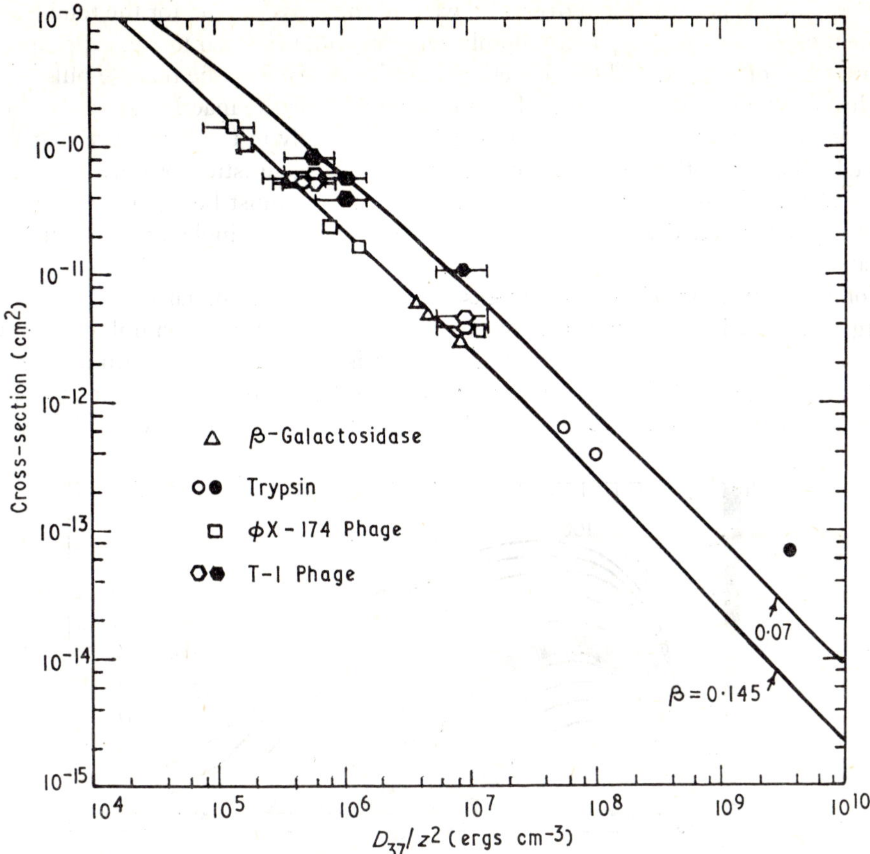

Figure 11. Theoretical relationship between the cross section for heavy ion inactivation and the D-37 dose for γ-rays, for dry enzymes and viruses. Unadjusted experimental data (references cited in Butts and Katz 1967) are plotted over theoretical curves derived from the energy dissipation algorithm of Kobetich and Katz 1968a. Horizontal error bars on T-1 phage points show the range of D-37 data for this material.

Rapidly moving singly charged particles do not form a closed track. Observers make grain or gap counts, which they convert into grain counts by statistical procedures. An observer must decide which grain belongs to a track and which is background, and typically decides that grains whose centre is beyond some characteristic distance, τ, do not belong to the track. We therefore integrate equation (1) from 0 to τ to find the theoretical grain count cross section, and compare these calculations to experimental data in figures 7 and 8. We find (Katz and Kobetich 1969) that the number of grains per unit length g is related to the maximum observable number, g_{sat}, by the expression

$$g=g_{sat}[1-\exp\{-\alpha z^2/\beta^2 E_0\}] \tag{2}$$

where $\alpha z^2/\beta^2$ is the mean value of $\bar{E}$ within the cylinder of radius τ, and βc is the speed of the ion. When we plot equation (2) against the restricted energy loss in silver bromide, as in figure 9, we see that the proportionality between grain count and restricted energy loss depends on the emulsion sensitivity and the particle speed. The curve bears an intimate relationship to a conceptual structure used in radiobiology, where it might be interpreted as a plot of 'efficiency factor' against LET.

If equation (1) is integrated over all t, we find the cross section for the total response of the medium to charged particle bombardment, and, using parameters appropriate to the medium of interest and the detection techniques, we find the relative pulse heights produced by heavy ions in NaI(T1), as in figure 10, or the inactivation cross sections for dry enzymes and viruses, as in figure 11. It is only in the last figure that experimental data for both the cross section and the characteristic dose are known from experiment. In all other cases the characteristic dose must be determined by trial.

The phenomena described above are treated with a single model, with minor variations.

For the formation of etchable tracks in dielectrics, the spatial distribution of the energy deposited by δ-rays again appears to be determining. A simple model which supposes that track formation or non-formation is to be associated with a criterion of minimal dosage at a minimal distance is found to be in good agreement with experiment, as shown in figure 12 (Katz and Kobetich 1968b). If the critical distance is

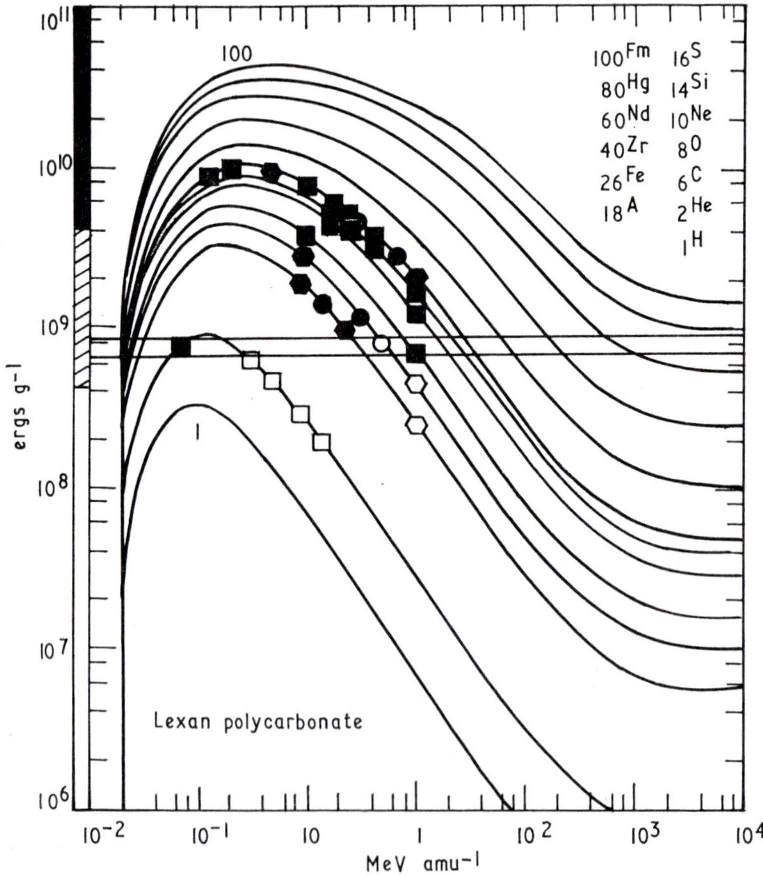

Figure 12. Dosage of ionization energy in Lexan polycarbonate at 2×10^{-7} g cm^{-2}, with superimposed data from two observers plotted as solid figures if etchable tracks are formed, and as hollow figures if not. The two adjacent horizontal lines are thresholds for track formation for the two sets of data. Shading along the dose axis is open if the indicated γ-ray dose gives negligible damage, is cross hatched if it produces moderate damage, and is solid if the indicated dose produces severe damage to a variety of physical properties of the bulk material. (Katz and Kobetich 1968b.)

taken as 2×10^{-7} g cm^{-2}, an energy deposition of about 8×10^{8} erg g^{-1} is the minimal dosage in Lexan polycarbonate. As consistent with views expressed earlier about the relationship between γ-ray and charged particle bombardment, the minimal dosage for track formation is a dosage producing macroscopic damage when this material is irradiated with γ-rays.

4. Discussion

From the variety of phenomena described by the present theory of track structure, and the quantitative agreement between theory and experiment, we must conclude that the theory encompasses many of the essential features of track formation, though details in the calculation of the energy deposition, or in the assignment of the critical energy dose will be altered with improved knowledge. The average energy deposition in small subvolumes (microdose) around the path of an ion is taken to be the significant parameter. The theory asserts that track structure can only be understood from the spatial distribution of the microdose, and from the response function of the medium.

Finally, the present theory of track formation enables us to compare the relative sensitivities of different detecting systems in terms of the critical energy dosage for track formation, and the estimated radius of a sensitive volume, as shown in table 1.

Table 1. Sensitivity of detecting systems

1-hit	E_0 erg cm^{-3}	a_0 cm
G.5 emulsion	$1{\cdot}5 \times 10^{4}$	$1{\cdot}5 \times 10^{-5}$
K.5 emulsion	$5{\cdot}0 \times 10^{4}$	$1{\cdot}0 \times 10^{-5}$
NaI(T1)	$4{\cdot}0 \times 10^{7}$	$3{\cdot}3 \times 10^{-7}$
T-1 Phage	$5{\cdot}7 \times 10^{7}$	6×10^{-7}
Trypsin	$3{\cdot}6 \times 10^{9}$	2×10^{-7}

many-hit	critical dose erg g^{-1}
cellulose nitrate	3×10^{8}
Lexan polycarbonate	7×10^{8}
mica	3×10^{9}
olivine	1×10^{10}

The critical dose varies over 5 or 6 orders of magnitude for the different detectors thus far studied, but when this dose is multiplied by the volume of the sensitive site, it appears that an energy deposition ranging from 3 to 300 eV is required to sensitize these detectors.

References

BARKAS, W. H., 1963, *Nucl. Res. Emulsions* (New York: Academic Press).
BUTTS, J. J., and KATZ, R., 1967, *Radiat. Res.*, **30**, 855.
KATZ, R., and BUTTS, J. J., 1965, *Phys. Rev.*, **137**, B198.
KATZ, R., and KOBETICH, E. J., 1968a, *Phys. Rev.*, **170**, 397.
—— 1968b, *Phys. Rev.*, **170**, 401.
—— 1969, *Phys. Rev.*, **186**, 344.
KOBETICH, E. J., and KATZ, R., 1968a, *Phys. Rev.*, **170**, 391.
—— 1968b, *Phys. Rev.*, **170**, 405.
—— 1969, *Nucl. Instrum. and Meths*, **71**, 226.
POWELL, C. F., FOWLER, P. H., and PERKINS, D. H., 1959, *The Study of Elementary Particles by the Photographic Method* (New York: Pergamon Press).

Track structure in relation to radiobiology

G. J. NEARY

Medical Research Council Radiobiology Unit, Harwell, Didcot, Berkshire, England

Abstract. Track structure is important in radiobiology because the highly localized pattern of energy deposition on both the cellular and molecular scales influences the nature of the molecular damage in critical cellular elements and the potential consequences for a cell of the production of correlated sites of damage. It is necessary to consider the spectrum of the various precise chemical changes that may result from any one individual energy absorption process (an activation event), the degree of localization of such events, and their possible interactions at the molecular and cellular levels of organization. A number of questions, physical, chemical, biochemical and biological, on which information is needed are listed.

1. Introduction

When we couple together the two topics of track structure and radiobiology, it is obvious that we are focusing attention on the basic mechanisms of the biological effects of radiation. Those of us concerned with these radiobiological problems will be only too well aware of the difficulty of finding any new approach by which to gain a deeper insight into them. Many promising lines have been tried but have generally come to a halt far short of the conclusive result hoped for, bogged down by the complexity of the problems, the lack of basic physical data concerning details of track structure, and uncertainty even about the identity and chemical nature of the cellular entities involved, and their structural and biochemical characteristics. It might be not unfair to say that at this basic level, the subject has been stagnating for some time now, though the achievements of molecular biology have helped to define in concrete terms some, at least, of the important cellular systems which must be considered. We are also beginning to learn something about the interactions between these systems and about the existence and nature of repair processes. At the same time there has been progress on the basic physics of track structure, and in radiation chemistry, for example, the direct demonstration of the occurrence of the hydrated electron in radiolysis (Boag and Hart 1963) and the investigation of its reactions.

It is thus appropriate that this volume on track structure should come at this time, and, in particular, that we should be trying to achieve some alignment of concepts deriving from the physical, the chemical and the radiobiological approaches. The difficulties of finding a new starting point for our radiobiological thinking are oppressively obvious, but if we are to break out of our present impasse, I think we radiobiologists must abandon some habits of thought, or more accurately lack of thought, about the nature of primary processes and products, and turn to the physicists and chemists for help in finding an adequate foundation on which to build our radiobiological theories.

Since this is an interdisciplinary gathering, let us start at the beginning and say explicitly why track structure should be important in radiobiology. The story goes back to the origins of the hit, or target, hypothesis of Dessauer, Crowther and others, notably developed by Lea (1946) and by Timoféeff-Ressovsky and Zimmer (1947). Quite simply, biological material has structure; structural entities may be damaged by an activation event, and such damage in a critical structure or structures may have observable biological effects; finally, activation events are not distributed in space purely at random but in relation to the paths of charged particles. Hence track structure influences the probable biological effects of radiation.

There have been two principal mathematical elaborations of these concepts. In the first, the multitarget formulation, it is assumed that several target structures must be damaged for the biological effect to develop; usually the targets are assumed to be damaged by independent tracks and in the commonest formulation, a single activation event in a target is considered to suffice. In the second, the multi-hit formulation, a certain minimum number of 'hits' in a particular target volume is thought to be required. In some treatments, a 'hit' is interpreted as the passage of a track through the target; the numbers of track intersections are distributed according to a Poisson distribution. Such a formulation seems logically incomplete since if it is thought that several tracks are required to provide some minimum number of activation events in the target, the possibility of fluctuations about the mean number of activation events by any one track should also be taken into account. In fact, the number of activation events in a target follows a double-Poisson, rather than a simple Poisson distribution. This mathematically correct formulation of the multi-hit hypothesis has seldom been used but it is presented clearly in an insufficiently known paper by Dittrich (1961).

The appropriateness of a multi-hit concept will be considered below but first it must be noted that in all these target theory treatments the notion of a track has been severely formalized. Even when the statistical distribution of collisions along the tracks and the separate identity of δ-ray tracks has been allowed for in principle, the physical input data have been approximate and of uncertain validity and, usually, scattering of tracks has not been taken into account. In view of the difficulties of making a more precise detailed picture of the relevant features of track structure, the attack on this problem from the side of the radiobiologists has come almost to a standstill, in the sense that no new ground has been broken in making a more precise specification of relevant primary events. We must take note, however, that an attempt is being made to bypass some of the difficulties of the detailed kinetic picture of tracks by means of the hypothesis that the relevant quantity for production of a biological effect is the total energy delivered within some particular volume. This is the basis of the concept of microdosimetry introduced by Rossi (1967), and its great advantage lies in the fact that in principle the energy can be measured rather than calculated. This is an important development which will be discussed later.

2. Activation events: Nature and consequences

Before we can profitably consider the relevance to radiobiology of track structure, that is the spatial and temporal relations between activation events in an irradiated system, it is necessary to scrutinize first the concept of an activation event itself. The term 'activation event' has been used by Platzman (1967) as a convenient inclusive term for any of the particular kinds of individual interaction events of the primary particle with an atom or molecule of the medium, for example ionizing and non-ionizing collisions. Thus, broadly speaking, it refers to the same thing as the term 'energy loss event' of Rauth and Simpson (1964), but from the standpoint of the medium rather than the primary particle.

A prerequisite for further progress in understanding radiobiological mechanisms at the molecular level is detailed information on the character of activation events and their chemical consequences. To go on thinking, in the rigid formal terms of the older target theory, of an average activation event and an average set of unspecified chemical changes with some undetermined probability of biological effect is an obstacle to further progress. We must recognize that the nature of the target molecules and environmental conditions such as temperature, concentration of water and other modifying substances are not universal factors common to all radiobiological situations; consideration of all the physical and chemical details seems to be unavoidable. For the activation event itself, is the amount of associated energy alone a sufficient specification, or is it necessary to know the spectrum of the different levels and kinds of activation (e.g. singlet or triplet states), each with its own associated spectrum of chemical changes? It is of particular importance to know whether there are qualitative differences in the spectrum of effects produced by radiations of different qualities. If so, then since the ability of cells to repair chemical or structural damage may be expected to vary according to the nature of the damage, the fact that the reparability of biological damage generally depends on the quality of the radiation used would be partly explicable at the level of the individual initial lesions. If, as seems likely, different kinds of activation events must be considered, which are the chemically important ones? Is ionization *per se* of a molecule particularly effective chemically, and is the pathway by which the ionized state is reached important, for example, by direct ionization or by autoionization (preionization) of a molecule in a superexcited state (Platzman 1962)?

Another general physical question is the degree of initial localization of an activation event and the possible spreading of its influence through energy or charge transfer or radical migration. There is also the chemical question of the reaction kinetics which may apply between activation centres in the physico-chemical and early chemical stages when the molecules still possess considerable excess energy. We are still left with the biochemical and biological questions as to the identity and physical state of the supposed target molecules or structures in cells, and the expectation of a particular biological effect consequent upon a set of specified chemical changes.

3. Interrelation of activation events: Track structure

In a given irradiation situation the distribution of activation events in space will be non-random and concentrated along tracks provided that the mean separation of tracks is greater than the mean separation of events along a track. The approximate numerical condition for distinguishable tracks is

$$D < (16L)^3$$

where D is the dose in rads, L the LET in keV μm^{-1} and a value of 60 eV is assumed for the mean size of events (Rauth and Simpson 1964). Even for a particle of minimum ionization, the dose limit is several rads and if the heterogeneity in track characteristics in any practical situation is taken into account, the limit becomes hundreds of megarads. Moreover, even for doses above the limit, the dose would have to be concentrated into a time interval shorter than the lifetime of the primary products in order to lose the identity of the separate tracks. In the vast majority of practical situations, therefore, track structure cannot be disregarded.

All the events along one track in a cell are produced within a time interval of the order of 10^{-13} s; thus, so far as the time factor itself is concerned, all these events coexist for any possible interactions in the chemical stage and even in the physico-chemical stage (Boag 1967). For any one track, therefore, the important feature of the interrelation of activation events is the spatial one. The mean separation of events along the track may be anything from the order of 1 μm to the order of 1 Å, according to the quality of the radiation. Initial events are unlikely to be produced by a particle at more than a few ångströms radial distance from the trajectory (Hutchinson and Pollard 1961). However, some of the collisions of the primary particle lead to energetic secondary electrons, the δ-rays, each of which can produce many activation events. Since about half of the collisions of a charged particle involve energy transfers greater than 100 eV, a considerable fraction of the total activation events is produced by δ-rays and their spatial distribution is of practical importance. This brings us to another area of considerable uncertainty at the physical level: the stopping power data for low energy electrons is not well-established and furthermore, it is difficult to allow for the effect of the scattering of the electrons on the radial distribution of activation events. Thus the calculations required even for *simple* target theory, the evaluation of the probability of occurrence of a given number of activation events in a target or targets of specified size and shape under particular irradiation conditions, cannot be carried out with any confidence. In any case, the quantitative specification of the statistical spatial distribution of activation events is severely limited by the already mentioned lack of definition in simple target theory of what constitutes an event and still more by the omission of a specification of different kinds of event and their chemical consequences.

In considering the fine-scale distribution of activation events in and around a track, information is also needed on the initial localization of any one event. For example is the mean free path for production of activation events by a very low energy electron greater or less than the linear dimensions of the possible region of localization of any one of these events in a condensed medium? In other words, is the concept of separate 'ionizations' in a cluster valid in a condensed medium, or is there simply an unresolvable local concentration of energy? What is the minimum amount of energy in a collision which can be regarded as giving rise to resolvable activation events? For collision events below this energy limit, how does the probability of chemical and biological change depend on the magnitude of the energy transferred? Are multi-hit calculations based on data for ion-cluster size distribution in gases valid (Pollard *et al.*, 1955, Howard-Flanders 1958)?

When one starts to analyse track structure in this degree of detail, it soon becomes obvious that one cannot proceed far without specifying the nature and geometry of the targets and interacting systems of targets which are thought to be concerned. Even for the simplest cells there are great gaps in our knowledge of these matters. Certainly, there are many general indications that the DNA is an important target for radiation damage in a cell (Hutchinson 1966, Kaplan 1968) but information on the details of its molecular organization, particularly in higher cells is very inadequate, and even if we accept that the DNA is the most important primary target, there is very little precise information about the nature and arrangement of the repair systems which are undoubtedly of great importance in determining the biological end-result of the initial radiation damage.

The provision of answers to these questions is obviously not primarily the responsibility of the physicists and radiation chemists in this interdisciplinary arena. They may be able, however, to give the radiobiologists some guidance on the possibilities of cooperation between the effects of activation events at different parts of a molecule, or in neighbouring molecules or structures. For example, Davy (1967, 1968) has used an *ad hoc* model to explain LET dependence, based on the perturbation of neighbouring C—H valency bonds in a molecule by separate activation events along the track. Likewise, there has been a general tendency to suppose that double-strand scission of a DNA molecule is likely to involve the occurrence of activation events in each strand. Are such models legitimate ways of thinking about intramolecular primary processes?

There is plenty of evidence from radiation chemistry of the importance of the interaction of events or their immediate products within regions measured on the atomic or molecular scale. Most of the data relate to the formation of specific products from relatively simple molecules. For the biochemical inactivation of a complex molecule, *any* change may be sufficient rather than damage of some specific variety. Since there is little evidence of energy transfer or radical migration in cells over distances much more than 50 Å (Guéron and Shulman 1968, Hutchinson 1961), it would seem that the distribution of activation events on the molecular scale is important. Of course, there could be an interaction of damaged molecules on a more extended space and time scale through later biological processes, for example the interaction of a damaged critical structure and a partly damaged repair system, the precise location of which is at present unknown. Similarly, chromosome aberrations are thought to result from the interaction of damaged chromosomes which may be separated by distances of the order of a micron (Lea 1946, Wolff *et al.* 1958, Neary *et al.* 1967). In such cases, although the overall biological mechanisms may involve cellular distances much greater than ordinary molecular dimensions, is there any reason to doubt that, for the component individual primary steps themselves, the distribution of activation events on the molecular scale is important?

This question is vital for microdosimetry which seeks to avoid the uncertainties about the nature and distribution of activation events, by determination of the total energy absorbed in a supposed critical volume through measurement of the ionization in an equivalent mass of gas. Techniques of measurement hitherto have imposed a lower limit of about one-tenth of a micron for the volume. Is the information attainable with this degree of resolution likely to be sufficient or must the details of track structure down to molecular dimensions be taken into account? There is also the general question of whether a description in terms only of total energy is adequate without a detailed specification of the kinds of activation events and the dependence of the likelihood of a required particular molecular change on the energy of the activation event directly responsible. These questions are particularly important for the practical problem of radiation of different qualities—are there qualitative differences in the cellular lesions produced by low and high LET tracks, either because of differences at the level of the individual activation events or because of differences due to interaction of events or their chemical products for different initial spatial distributions of events (Shalek Humphrey and Sedita 1963, Tobias and Manney 1964)?

It remains to note some relevant but almost incidental features or consequences of track structure. One of these is the transient local heating in a track (Norman 1967). It seems unlikely that all the effects of a charged particle could be explained solely in terms of thermal reactions from the local heating, but are the final chemical changes influenced by this heating? If so, there is another possibility of a qualitative difference in damage produced by radiations of different qualities.

There is also the transient electric field produced by the column of activation events. A possible radiobiological effect of this field has been suggested, namely an extensive disruption of hydrogen bonds (Platzman and Franck 1958). It is convenient to mention here the possibility that, for the important molecular targets in cells, the hydrated electron may play a less important role than in aqueous solutions, since there is evidence that in the frozen state recapture of the electron by the parent positive ion becomes more likely (Stein 1967).

Another possible consequence of track structure which, though almost incidental, would be important radiobiologically, may be the endogenous production of oxygen in the track, particularly at high LET (Shekhtman 1960, Swallow and Velandia 1962, Neary 1965, Alper and Moore 1967, Alper, Moore and Smith 1967, Hüber 1969). This process would provide an explanation of the decreasing influence of environmental oxygen on radiosensitivity as the LET is increased. The mechanism was originally proposed in relation to the radiation chemistry of aqueous solutions; it is not obvious that formation of molecular oxygen itself could occur where a track passed through certain of the critical structures in a cell. However, the general principle already noted could be expected to hold, namely, that differences in the spatial distribution of activation events by radiations of different qualities should influence the interaction of the chemical products of the activation events and the general chemical reaction sequence.

A general consequence of track structure for radiobiology is that two quite distinct effects may be separately produced in the same cell but with a high degree of correlation, and there is no way of avoiding this confounding of the two effects even by going to extremely low doses or dose rates. For example, a high LET track may have a high probability of producing a mutational change in a cell but a lethal effect is also likely to occur as well. This is another source of complication in the investigation and interpretation of the biological effects of radiation in cells.

4. Conclusions

Factual conclusions in the ordinary sense of the word could hardly be expected in this rather general survey of the relevance of track structure in radiobiology. On the other hand, there are a number of questions which are important for the further progress of radiobiology, and which lie within the scope of the discussions in this volume. First, the physical and chemical questions:

1. What should our mental picture of an activation event be, to take the place of the purely formal notion of the unit event of classical target theory?
2. Is there any simplifying broad classification of the possible chemical effects of an activation?
3. Is the energy of the activation event an important parameter for the chemical effects?
4. Is ionization resulting from an activation event of any special significance?
5. Is there any qualitative difference in the chemical effects of an activation event according to the quality of the radiation?
6. What is the degree of localization of an activation event?
7. In what way, and to what extent, do events cooperate (a) within a molecule (b) between molecules?
8. Is the microdosimetric specification in terms of gross energy within a volume of linear dimensions as large as one-tenth of a micron a complete one for determination of biological effects?

9. Are the chemical reaction kinetics greatly influenced by residual excitation energy or by local temperature rise?

10. What is the stopping power relation for electrons of low energy? At what energy do the events become unresolvable?

Some biochemical and biological questions are:

1. What is the chemical nature and physical state of the important targets in cells?
2. What is the size and disposition of the targets?
3. Is associated damage to separate targets important?
4. Are complex molecules such as DNA or protein inactivated biologically by any activation event or has intensity of damage to be taken into account?

References

ALPER, T., and MOORE, J. L., 1967, *Brit. J. Radiol.*, **40**, 843.

ALPER, T., MOORE, J. L., and SMITH, P., 1967, *Radiat. Res.*, **32**, 780.

BOAG, J. W., 1967, *Radiation Research*, Ed. G. Silini (Amsterdam: North-Holland), p. 43.

BOAG, J. W., and HART, E. J., 1963, *Nature, Lond.*, **197**, 45.

DAVY, D. R., 1967, *Hlth Phys.*, **13**, 1159.

DAVY, D. R., 1968, *Hlth Phys.*, **14**, 569.

DITTRICH, W., 1961, *Z. Naturf. B*, **16**, 398.

GUÉRON, M., and SHULMAN, R. G., 1968, *A. rev. Biochem.*, **37**, 571.

HOWARD-FLANDERS, P., 1958, *Advs biol. med. Phys.*, **6**, 553.

HÜBER, R. P. O., 1970, *Charged Particles Tracks in Solids and Liquids*, (London: The Institute of Physics and The Physical Society).

HUTCHINSON, F., 1961, *Science*, **134**, 533.

—— 1966, *Cancer Res.*, **26**, Part 1, 2045.

HUTCHINSON, F., and POLLARD, E., 1961, *Mechanisms in Radiobiology*, Vol. I, Eds M. Errera and A. Forssberg (New York: Academic Press), p. 1.

KAPLAN, H. S., 1968, in *Actions Chimiques et Biologiques des Radiations*, 12ième Série, Ed. M. Haissinsky (Paris: Masson).

LEA, D. E., 1946, *Actions of Radiations on Living Cells* (London: Cambridge University Press).

NEARY, G. J., 1965, *Int. J. Radiat. Biol.*, **9**, 477.

NEARY, G. J., PRESTON, R. J., and SAVAGE, J. R. K., 1967, *Int. J. Radiat. Biol.*, **12**, 317.

NORMAN, A., 1967, *Radiat. Res.*, **7**, 33.

PLATZMAN, R. L., 1962, *Radiat. Res.*, **17**, 419.

PLATZMAN, R. L., 1967, *Radiation Research*, Ed. G. Silini (Amsterdam: North-Holland), p. 20.

PLATZMAN, R. L., and FRANCK, J., 1958, *Symp. on Information Theory in Biology*, Eds H. P. Yockey, R. L. Platzman, and H. Quastler (New York: Pergamon) p. 262.

POLLARD, E. C., GUILD, W. R., HUTCHINSON, F., and SETLOW, R. B., 1955, *Prog. Biophys. biophys. Chem.*, **5**, 72.

RAUTH, A. M., and SIMPSON, J. A., 1964, *Radiat. Res.*, **22**, 643.

ROSSI, H. H., 1967, *Advs biol. med. Phys.*, **11**, 27.

SHALEK, R. J., HUMPHREY, R. M., and SEDITA, S. J., 1963, *Radiat. Res.*, **19**, 214 (Abstr. H10).

SHEKHTMAN, Y. L., 1960, *Role of Peroxides and Oxygen in the First Stages of the Radiobiological Effect* (Akad. Nauk. SSSR). (In Russian).

STEIN, G., 1967, *The Chemistry of Ionization and Escitation*, Eds G. R. A. Johnson and G. Scholes (London: Taylor and Francis), p. 25.

SWALLOW, A. J., and VELANDIA, J. A., 1962, *Nature, Lond.*, **195**, 798.

TIMOFÉEFF-RESSOVSKY, N. W., and ZIMMER, K. G., 1947, *Biophysik*, Band I, *Das Trefferprinzip in der Biologie* (Leipzig: S. Hirzel Verlag).

TOBIAS, C. A., and MANNEY, T., 1964, *Ann. N.Y. Acad. Sci.*, **114**, 16.

WOLFF, S., ATWOOD, K. C., RANDOLPH, M. L., and LUIPPOLD, H. E., 1958, *J. biophys. biochem. Cytol.*, **4**, 365.

Stochastic equations relating RBE to LET

P. R. J. BURCH

Medical Research Council, Environmental Radiation Unit, Department of Medical Physics, The University of Leeds, The General Infirmary, Leeds, 1.

Abstract. Stochastic equations are developed from the following basic hypothesis: A single ionizing particle may induce biological change through a 'hit'—direct or indirect—on a single target structure, or by independent 'hits' on each of multiple target structures. Experimental data from the literature for several systems (bacterial inactivation, mutations in yeast, chromatid aberrations, inhibition of clone formation by cells of human origin) are interpreted in terms of the basic hypothesis. The problems of distinguishing between 'complex'- and 'featureless'-target models are discussed. It is concluded that, in the analysis of the relative biological effectiveness of an irradiation, the detailed internal structure of biological target molecules is as important as the track structure of the ionizing particles.

1. Introduction

Any quantitative theory of radiation biology must be able to account not only for dose-response relations, but also for the dependence of relative biological effectiveness (RBE) on linear energy transfer (LET). When we take up this latter problem we are immediately confronted with some awkward decisions. How should we characterize the LET of a radiation? Will an average value suffice? If so, should we calculate a track average, an energy average, or something more complicated? And what cut-off energy, if any, should we adopt to allow for δ-rays?

To answer the foregoing questions we require a model of radiobiological action. But to arrive at such a model it is important to know the relation between RBE and LET. In other words, we need to know the answer to the problem before we can start to solve it. At first sight, the dilemma might appear to be unresolvable. However, provided our model does not depend *critically* on the choice of a particular form of LET, and provided we can obtain favourable experimental evidence for interpretation, we should be able to approach a quantitative solution by exploiting the method of trial-and-error.

The problem of characterizing the quality of a radiation can be minimized by relying on the results of the track-segment type of experiment. In such experiments the distribution of LET is narrow, and, for a very narrow distribution, track- and energy-average values of LET coincide. Even with track-segment experiments, however, we are often faced with the problem of δ-tracks that are effectively independent of the primary track. We may hope to progress by assuming initially that the effect of δ-tracks is small compared with that of the primary track. A preliminary analysis should show whether or not this assumption can be justified.

2. Stochastic models

Most dose-response relations in radiobiology are consistent with the thesis that biological change results from essentially *random* transfers of energy from the ionizing particle to one or more critical target structures. Transfers may occur directly through the ionization, for example, of a key macromolecule such as DNA, or they may occur indirectly by way of active chemical agents. Thus, radicals may be produced with a more-or-less random distribution along the track of the particle, and they may then diffuse short distances to damage key biological structures.

2.1. *Single-hit model*

Suppose a single random transfer of energy from the primary particle to a critical target region suffices to effect biological change. Let the effective area of the target normal to the path of the particle be A, and suppose the average length of track over which energy transfers can be effective is l. Assume the average total LET of the particle (sometimes symbolized by L_∞) is L, and that it stays effectively constant throughout the target region. If the average number of *biologically effective energy transfers* (or 'hits') per unit length of track is cL, where c is a constant, then the effective cross section of the target towards such primary particles is given by

$$A_L = A(1 - \exp - clL) \tag{1}$$

2.2. *Model postulating at least one independent hit, on each of* n *targets, from a single primary particle*

Suppose biological change is effected when a *single* primary particle produces at least one hit, in each of n targets. Assuming each hit conforms to the statistics of model (1) above, then the effective cross section for irradiation with particles of a total LET, L is given by

$$A_L = A(1 - \exp - clL)^n \tag{2}$$

(If the n targets are very close to one another, then for heavy primaries at ordinary dose levels, change is much more likely to be produced by one, than by multiple primary particles.)

2.3. *Model with multiple, complex* (r-*fold*) *targets and Weibull statistics*

For concreteness, suppose the overall target involved in the inactivation of the proliferative capacity of mammalian cells consists of a very large number of vulnerable sections of chromosomes. Suppose the cell will be inactivated if each and every one of r distinctive elements, in at least one vulnerable section of chromosome, receives at least one independent hit from the primary particle. (Target elements may consist, for example, of the two strands of DNA, and/or associated RNA and/or protein.) In traversing a vulnerable section, the probability that each and every one of the r elements will receive at least one independent hit from the primary particle will be

$$(1 - \exp - c_1 l_1 L)(1 - \exp - c_2 l_2 L) \ldots (1 - \exp - c_r l_r L) \tag{3}$$

Suppose $c_i l_i L \ll 1$ for all $c_i l_i L$ of interest. Equation (3) then approximates to $(bL)^r$, where b is a constant. Suppose the average number of vulnerable sections traversed by a primary particle which passes through the effective total area A, is N. Then the average number of vulnerable sections receiving r-fold hits, per particle transit through A, is $N(bL)^r$.

Hence, the probability that all r elements of at least one vulnerable section will receive a hit, per particle transit, is described by the Weibull expression

$$1 - \exp - N(bL)^r$$

The effective target cross section A_L at L is therefore given by

$$A_L = A\{1 - \exp - N(bL)^r\} \tag{4}$$

that is,

$$A_L = A(1 - \exp - kL^r) \tag{5}$$

When kL^r is unity, a cell will have received, on the average, one effective r-fold inactivating event in a vulnerable section, per primary particle transit through A. At the value of L making kL^r unity, the average spacing between successive hits will approximate to the average spacing, for random orientations, between successive elements of a vulnerable section.

In all the above expressions, A_L is proportional to the *relative effectiveness* per particle, and RBE is proportional to A_L/L.

The following single general stochastic equation contains each of the above models

$$A_L = A(1 - \exp - kL^r)^n \tag{6}$$

where n and r are positive integers, 1, 2, . . ., etc.

Interestingly, a stochastic equation of the same general form

$$P_t = S(1 - \exp - kt^r)^n \tag{7}$$

(or its differentiated version) describes the age-specific prevalence (P_t) of initiated disorders (or their age-specific initiation-rates, dP/dt) for many diseases in man (Burch 1968). The parameter S represents (or occasionally it is proportional to) the fraction of the community predisposed to the disorder; k is a constant describing the kinetics of initiation; t is the age at initiation; n and r are again positive integers. Equations (6) and (7) both reflect the high degree of specificity connected with damaging random events in biological systems.

3. Stochastic models and experimental findings

In considering the fit of equation (6) to experimental data, certain restrictions need to be borne in mind. If A_L (or A_L/L) is plotted on a log scale against L also on a log scale, the *shape* of each specific function (given n and r) remains invariant for all k and A. Thus, if each function, for all n and r of interest, is represented on tracing paper using a consistent log–log grid size, the curve may be moved up and down over the graph of experimental data—equivalent to a change of A—or left and right—equivalent to a change of k—but the size and shape of the theoretical curve itself remain invariant for all positions. Given suitably accurate experimental data, latitude of interpretation in terms of the integers n and r is therefore often absent. Of course, uncertainties in the numerical values of k and A must be present in all analyses.

3.1. *Bacterial inactivation*

Figure 1 shows results (Munson, Neary, Bridges, and Preston 1967, Alper, Moore and Bewley 1967) for the inactivation of two strains of bacteria. The relation between A_L/L and L cannot be represented by a single version of the general equation (6) and this indicates that two or more basic mechanisms of inactivation are implicated. Two versions of equation (6), the first with $n=1$, $r=1$; and the second with $n=2$, $r=1$ give a very satisfactory fit to the experimental data in figure 1. From the absolute values of D_{37} obtained by Munson *et al.* (1967) we find the value of A for the first mode is

0·32 μm^2. The value of L making kL unity is about 76 keV μm^{-1}. For the second mode ($n=2$, $r=1$), $A \simeq 0{\cdot}12$ μm^2, and the value of L making kL unity is about 20 keV μm^{-1} (see summary in table 1).

Table 1. Parameters connected with equation: $A_L = A(1 - \exp - kL^r)^n$

System and reference	Mode of action n	r	Value of L making $kL^r = 1$ (keV μm^{-1})	l; or average separation (for $r=2, 3$) between successive transfers $\overline{W}_{eff} = 80$ eV nm $\times \rho$	$\overline{W}_{eff} = 200$ eV nm $\times \rho$	A (μm^2)	A_{O_2}/A_{N_2}
Inactivation of bacteria							
Aerobic conditions							
E. coli WP2$_{her+}$	1	1	76	1·0		0·32	
(Munson *et al.* 1967)	2	1	20	4·0		0·12	
Shigella flexneri Y6R	1	1	76	1·0			
(Alper *et al.* 1967)	2	1	20	4·0			
Induction of mutations in Saccharomyces cerevisiae (Mortimer *et al.* 1965)							
Tr_F^+—aerobic	1	1	16	5·0		$3{\cdot}6 \times 10^{-8}$	4·4
	2	1	98	0·82		$1{\cdot}15 \times 10^{-6}$	1·6
Tr_F^+—anoxic	1	1	5·9	14		$8{\cdot}2 \times 10^{-9}$	
	2	1	108	0·74		$7{\cdot}3 \times 10^{-7}$	
Tr_{NF}^+—aerobic	2	1	2·1	38		$3{\cdot}04 \times 10^{-8}$	3·4
	2	1	96	0·8		$2{\cdot}65 \times 10^{-6}$	1·4
Tr_{NF}^+—anoxic	2	1	2·0	40		$9{\cdot}0 \times 10^{-9}$	
	2	1	113	0·7		$1{\cdot}95 \times 10^{-6}$	
Chromatid aberrations in Tradescantia bracteata (Neary and Savage 1966)							
Chromatid intrachanges aerobic	1	2	71		2·8		
Chromatid interchanges aerobic	1	2	72		2·8		
Isochromatid breaks aerobic	1	2	69		2·9		$\simeq 1$
Isochromatid breaks anoxic	1	2	104		1·9		
Single gaps anoxic	1	2	86		2·3		
Inactivation of human cells of kidney origin (Barendsen *et al.* 1966)							
Low dose region, aerobic	1	1	1·2	67		$5{\cdot}1 \times 10^{-2}$	3·2
	1	2	21	3·8	10	2·3	3·0
	1	3	90		2·2	32·7	$\simeq 1$
Low dose region, anoxic	1	1	0·8	100		$1{\cdot}6 \times 10^{-2}$	
	1	2	21	3·8	10	0·76	
	1	3	115		1·7	36·2	
High dose region, aerobic	1	1	10	8		0·88	3·4
	1	2	21	3·8	10	2·1	2·1
	1	3	90		2·2	32·7	$\simeq 1$
High dose region, anoxic	1	1	6	13		0·26	
	1	2	21	3·8	10	1·0	
	1	3	115		1·7	36·0	

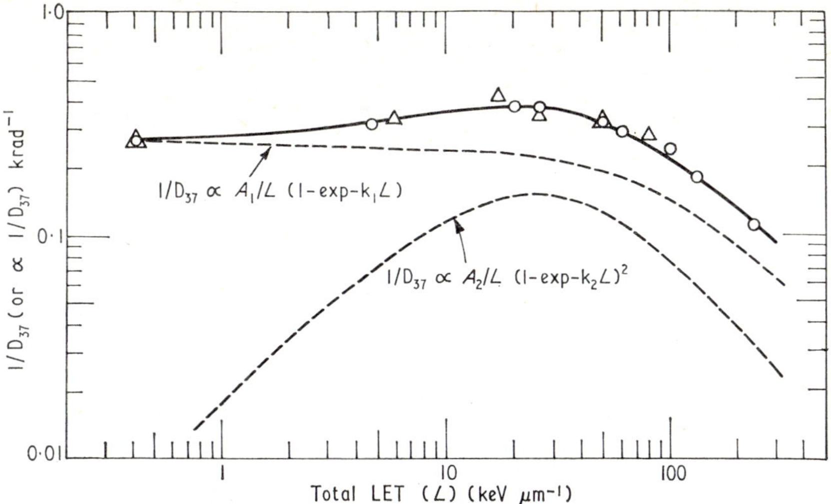

Figure 1. ○—Inactivation of bacteria, strain *E. coli* WP2$_{hcr}$+ (Munson *et al.* 1967). △—Inactivation of *Shigella flexneri* Y6R (Alper, Moore and Bewley 1967). Aerobic conditions. Ordinate: $1/D_{37}$ (which is proportional to A_L/L) for the data of Munson *et al.* (1967); the data of Alper *et al.* (1967) are normalized to those of Munson *et al.* (1967) at low L. Abscissa: total LET, (L). See table 1 for analysis.

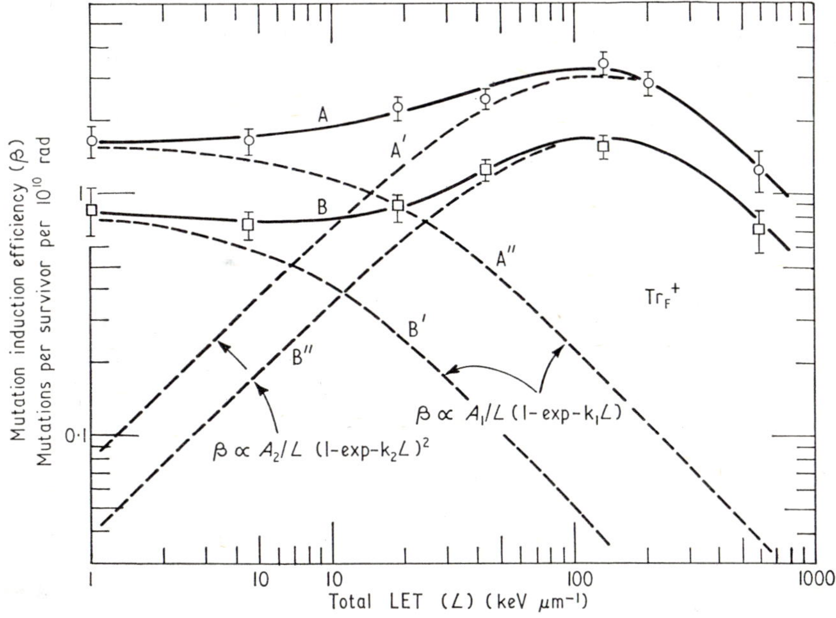

Figure 2. Yield of Tr_F^+ mutations per survivor, per 10^{10} rad (proportional to A_L/L) in the diploid strain X841 of the yeast *Saccharomyces cerevisiae*, as a function of total LET (Mortimer, Brustad and Cormack 1965). Curves, A, A′ and A″ refer to irradiation in air; B, B′ and B″ refer to irradiation in nitrogen. See table 1 for analysis.

3.2. *Mutations in yeast*

Figure 2 shows the results obtained by Mortimer, Brustad and Cormack (1965) for the induction of Tr_F^+ mutants among survivors of the irradiated diploid strain X841 of *Saccharomyces cerevisiae*. The general form of the curves for irradiation under

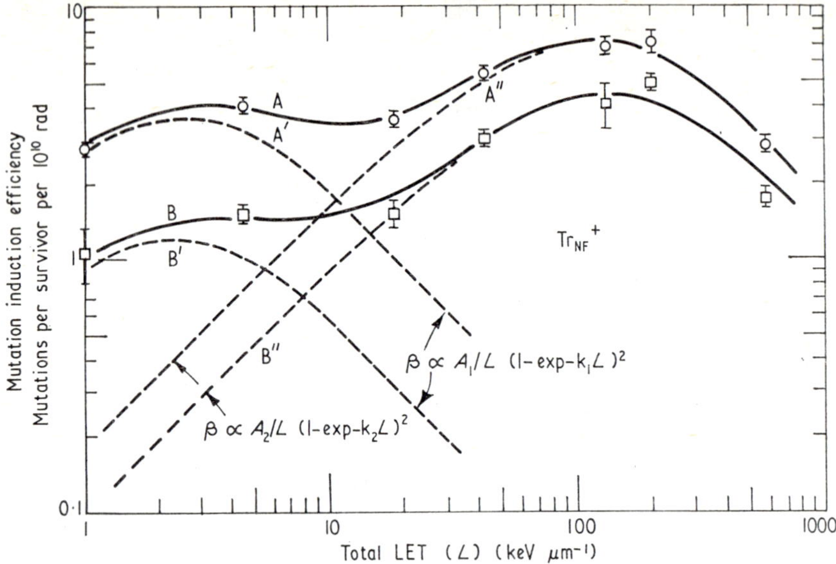

Figure 3. As for figure 2. Induction of Tr_{NF}^{+} mutations in *Saccharomyces cerevisiae* (Mortimer *et al.* 1965).

aerobic and anoxic conditions resembles that in figure 1. Especially interesting is the conspicuous oxygen effect at high values of LET, from 20 to 600 keV μm^{-1}. Parameters characterizing the theoretical curves with $n=1$, $r=1$, and $n=2$, $r=1$ are listed in table 1. Similar results for the induction of Tr_{NF}^{+} mutants are illustrated in figure 3 and summarized in table 1. Here both modes appear to be of the form $n=2$, $r=1$.

3.3. *Chromatid aberrations in Tradescantia Bracteata*

Neary and Savage's (1966) data for the induction of chromatid interchanges (figure 4) give a slightly better fit to $A_L = A(1 - \exp - kL^2)$, than to $A_L = A(1 - \exp - kL)^2$. Further data beyond $L = 100$ keV μm^{-1} would resolve the ambiguity. The value of L making kL^2 unity in the equation $A_L = A(1 - \exp - kL^2)$, is about 72 keV μm^{-1}. Data for interchanges, single gaps (anoxic), and isochromatid breaks (anoxic and aerobic) also show a similar dependence on L (see summary in table 1).

Dose-response relations for these various chromatid aberrations did not differ significantly from linearity (Neary and Savage 1966) and hence each type of aberration was generally produced by a single primary particle.

3.4. *Inhibition of clone formation by cells of human origin*

Experimental results (Deering and Rice 1962, Barendsen *et al.* 1966) for the inhibition of clone formation by cells of human origin, can be represented by the sum of three versions of equation (6) (see figures 5 to 8). Although the values of n and r for each mode of inactivation do not change with certain conditions of irradiation ('low' and 'high' dose regions; aerobic, anoxic) values of A and k can differ appreciably for two of the three modes (see table 1), but especially for the mode $n=1$, $r=1$ which predominates at low L (around 2 keV μm^{-1}).

The analysis of results obtained by Barendsen *et al.* (1966) at low L, in the 'low dose region'—corresponding to survival between 1·0 and 0·5—depends on the significance

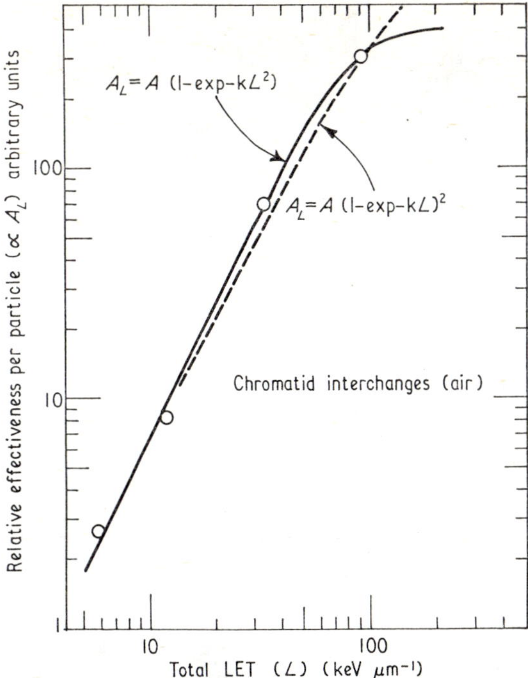

Figure 4. Relative effectiveness per particle (proportional to A_L) for the induction of chromatid interchanges (in air) as a function of total LET (Neary and Savage 1966). Chromatid aberrations were scored in the generative nucleus of isolated mature pollen grains of *Tradescantia bracteata* at the pollen tube mitosis. Results obtained with α-particles at 91 keV μm^{-1} in Neary and Savage's (1966) second experiment, have been normalized to those with protons in their first experiment, through the findings with 11·8 keV μm^{-1} protons in both experiments.

of the shoulder on type C survival curves. If such curves imply traditional multi-hit or multi-target models, then the analysis at low L in figures 5 to 8 would be invalid. However, if we adopt Haynes's (1966) view that one mode of inactivation is reversible, and that the shoulder arises from a dose-dependent 'repair term', then the interpretation shown in figures 5 to 8 and table 1 can be sustained. Haynes's (1966) survival-curve equation for homogeneous cells is

$$S_D = \exp\{-kD + \alpha(1 - \exp - \beta d)\} \tag{8}$$

Parameter k characterizes the intrinsic inactivation probability per unit dose. The 'repair term', $\alpha(1 - \exp - \beta D)$ relates to a reversible mode of injury, where the extent of repair declines with increasing dose. The magnitude of β depends on factors such as dose-rate and cell phase. In both modes of injury a single primary particle is responsible for the lethal event. If repair processes cause the delay in cell division, then because the extent of delay is directly proportional to dose, repair may well involve a sequential mechanism.

The data analysed in figures 5 to 8 are consistent with the basic assumptions underlying Haynes's (1966) theory. Values of k and A for the modes $n=1$, $r=2$ and $n=1$, $r=3$ stay approximately constant at 'low' and 'high' doses—indicating the absence of repair and recovery—but they differ markedly for the mode $n=1$, $r=1$. Under aerobic conditions—see table 1—the value of A for this latter mode is about 17 times larger at 'high' than at 'low' dose; under anoxic conditions the corresponding ratio is very

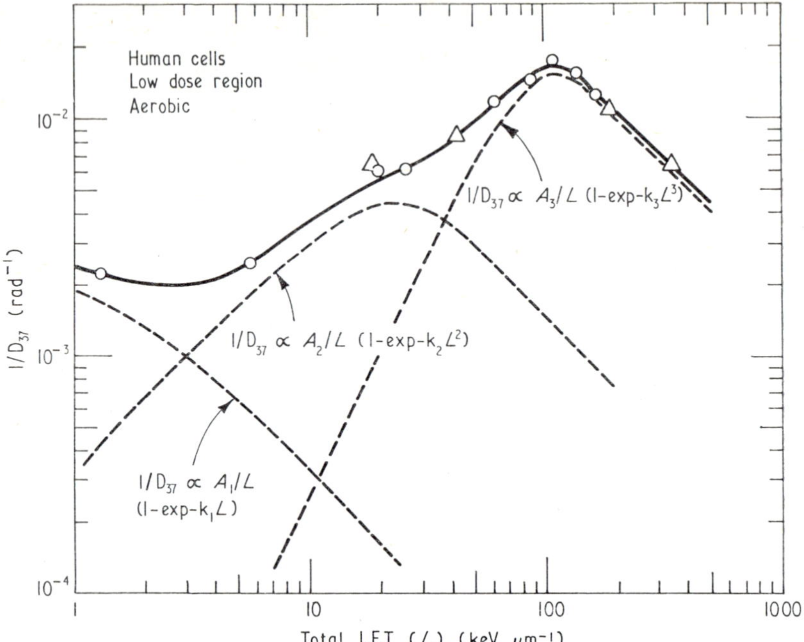

Figure 5. Inhibition ($1/D_{37}$) of clone-formation by cells derived from man, as a function of total LET. Aerobic conditions. ○—Results of Barendsen *et al.* (1966) for cells of kidney origin, obtained in the 'low dose region' for the surviving fraction 1·0 to 0·5. △—Results of Deering and Rice (1962) for HeLa cells, normalized to the results of Barendsen *et al.* (1966) at high L.

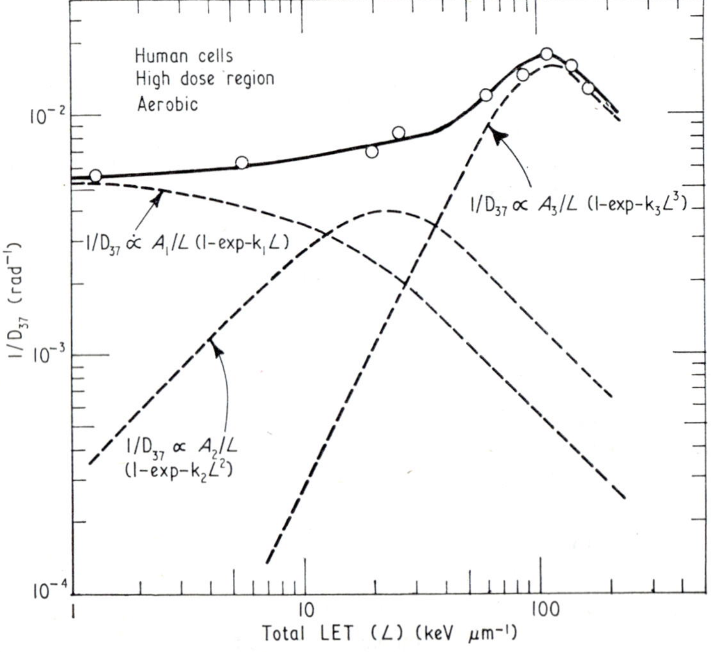

Figure 6. As for figure 5. Results of Barendsen *et al.* (1966) for aerobic conditions, and the 'high dose region', corresponding to the surviving fraction 0·1 to 0·01.

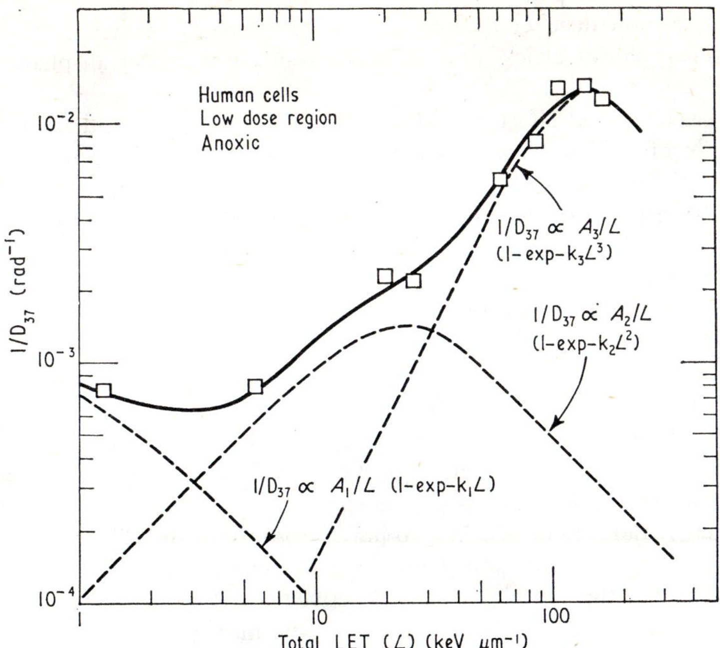

Figure 7. As for figure 5. Results of Barendsen *et al.* (1966) for the 'low dose region' and anoxic conditions.

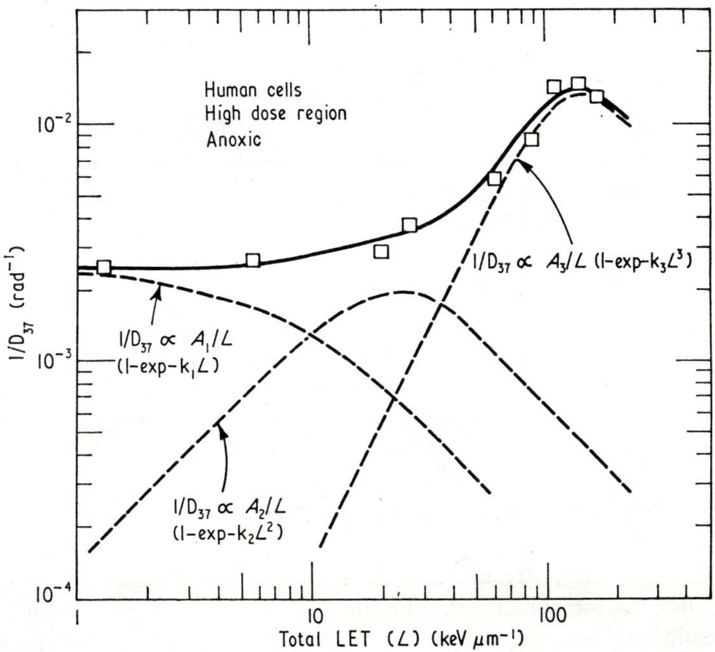

Figure 8. As for figure 5. Results of Barendsen *et al.* (1966) for the 'high dose region' and anoxic conditions.

similar—about 16. Repair and/or recovery in the mode $n=1$, $r=1$ is therefore extensive under 'low dose' conditions. Because these cells were unsynchronized, the degree of repair shown at low dose represents an average over all phases of the cell cycle.

Oxygen has a marked effect on A for the modes $n=1$, $r=1$ and $n=1$, $r=2$; but a negligible effect for the mode $n=1$, $r=3$. Nevertheless, oxygen seems to exert a small effect on k in the mode $n=1$, $r=3$. Changes in the physical dimensions of a complex target, involving the separation of distinctive elements, would affect the value of k.

4. Discussion

More detailed experimental findings, at narrow intervals of LET, are required to validate the provisional interpretations shown in figures 2 to 8 and table 1. These tentative analyses represent the simplest way of describing the available experimental data in terms of models 1 to 3. They reveal no obvious conflict between theory and experiment but more stringent tests are clearly needed.

We must now inquire whether the neglect of δ-tracks seriously distorts the interpretation. The answer to this query depends, to a large extent, on the nature of the effective energy transfers or 'hits'. To judge from the wide differences in the values of L making kL or kL^r unity, and the several values of the oxygen enhancement ratio, the energy transfers that are biologically effective often differ from one mode of change to another. When LET–RBE relations satisfy model 3, certain limitations are imposed on the nature of effective energy transfers, because the product $c_i l_i L$ must be appreciably <1 for all $c_i l_i L$ of interest. From figures 5 to 8, we see that $c_i l_i L \ll 1$ at $L \simeq 100$ keV μm^{-1}, for the mode $n=1$, $r=3$. Suppose effective transfers in this mode can be *any* form of primary ionization. For energetic heavy particles, the average energy per primary ionization is about 80 eV (Lea 1946). Hence, $c_i \simeq (1/80)$ eV^{-1}. It follows that l_i should be much less than $1/\rho$ nm, where ρ is the average density of the target medium. In other words, *if* the hit corresponds to *any* form of primary ionization, it must occur within a region of atomic or subatomic dimensions.

The following argument shows, however, that the hit cannot be any form of primary ionization. For suppose an ionization produced at any point along a δ-track can be biologically effective. Then for δ-tracks with an energy of several hundred eV, ionizations (and therefore hits) could occur at any range and direction up to about 10 nm from the origin of the δ-track in the path of the primary particle. This is inconsistent with the inequality requirement: $l_i \ll 1/\rho$ nm.

The dilemma can be satisfactorily resolved if the hits of model 3 correspond to relatively high energy events such as any one or more of the following: (i) an ion cluster or δ-track in a distinctive and narrowly localized geometrical pattern; (ii) removal of an L-shell electron (128 eV) from a phosphorus atom; (iii) removal of a K-shell electron (284 eV) from a carbon atom. For events such as these, $\overline{W}_{\text{eff}}$, the average energy of those hits that result in biological change will be considerably in excess of 80 eV. The criterion: $c_i l_i L \ll 1$, can then be readily met for L-values in the region of 100 keV μm^{-1}, if l_i is of atomic dimensions ($\sim 0{\cdot}1$ nm).

The results of Neary, Preston and Savage (1967) with low energy electrons are consistent with this interpretation. Figure 3 shows that the induction of chromatid interchanges in *Tradescantia* can probably be explained in terms of the model $n=1$, $r=2$. Neary *et al.* (1967) found that 2·54 keV photoelectrons together with 0·50 keV Auger electrons induce such aberrations as efficiently as protons of the same track-average LET. However, 0·96 keV photoelectrons together with 0·5 keV Auger electrons are about 0·28 times, and 0·73 keV photoelectrons with 0·5 keV Auger electrons are only one-tenth as effective as protons of the same track-average LET. The form of their figure 1 suggests that the efficiency of electrons with energy below about 0·5 keV should fall close to zero for one-track mechanisms of interaction.

If we can assume that the complex modes $n=1$, $r=2$ and $n=1$, $r=3$ associated with the inactivation of mammalian cells often result in double-strand scissions in DNA, then further support for this hypothesis is given by the findings of Corry and Cole (1968). These authors find that a double-strand scission of the DNA of mammalian metaphase chromosomes requires an average energy dissipation in the DNA itself of 600 eV, or 300 eV per strand. However, it seems unlikely that all high energy transfers to DNA, especially those to bases, will break the sugar–phosphate backbone, and hence the average energy of hits actually leading to a double-strand scission is probably less than 300 eV per strand: a value in the region of 200 eV may be more reasonable.

I suggest, therefore, that where the modes $n=1$, $r=2$ or $n=1$, $r=3$ are involved (figures 4 to 8), most δ-tracks in these experiments were much less efficient than the heavy primary at inducing change: neglecting them will not introduce serious interpretational error.

Putting $W_{\text{eff}} \simeq 200$ eV, the average spacing between successive hits at

$$L = 100 \text{ keV } \mu\text{m}^{-1}$$

is about $2/\rho$ nm, and at $L=20$ keV μm^{-1} it is about $10/\rho$ nm. These spacings should be compared with the diameter of the Watson–Crick double-helix (2 nm for purified DNA) and the outside diameter—about 17 nm—of chromatin in the cells of higher organisms (Davies and Small 1968). From X-ray diffraction studies it is suspected that 3 nm diameter DNA in nucleohistone may form a coil (super-helix) of diameter 10 nm between the axial centre of the threads, and of pitch 12 nm (Pardon, Wilkins and Richards 1967). Hence the typical spacing between effective transfers in modes $n=1$, $r=2$ and $n=1$, $r=3$ is consistent with the widely held and supported view that nuclear deoxyribonucleoprotein constitutes the main target where cytotoxic events are concerned.

Additional support for this view comes from the estimated total value of A for human cells (about 36 μm^2—see table 1) which approximates to the area presented by the nucleus of a cell (Barendsen 1964). This identification of A with the approximate cross-sectional area of the nucleus of the mammalian cell should be submitted to further experimental test, using cells with different nuclear sizes, and radiations of high LET.

The nature of hits involved in the mode $n=1$, $r=1$ cannot be readily assessed. This mode often predominates at low L (below ~10 keV μm^{-1}) (see figures 1 and 2 and 5 to 8) and in this region RBE tends to be almost independent of L. Because the LET of most of the tracks associated with conventional X- and γ-irradiations lies in this range, this property of near-independence of RBE is of great practical convenience. However, theoretical estimates of A in this LET region probably fail to correspond with the physical target area. The high values of l that are sometimes associated with the mode $n=1$, $r=1$ (see table 1) point to indirect action. If DNA, for example, is damaged by diffusing H^0 and/or OH^0 radicals, then estimates of l and A will include a substantial contribution from the mean effective diffusion path. Although survival curve statistics would correspond to those of 'direct' action, the mechanism of change is that of 'indirect' action. The neglect of energetic δ-tracks and the diffusion of radicals, will cause estimates of A to exceed the physical cross section. Similarly, l will exceed the 'thickness' of the target depending on diffusion distances and the range of δ-tracks.

4.1. *Contrasts with earlier models*

To explain a rising RBE with LET, models have been invoked in which multiple ionizations are deposited in the target zone (Lea 1946, Howard-Flanders 1958, Brustad 1962, Barendsen 1964, Randolph 1964), or in which interaction occurs between two parts of the same ionizing track (Tobias and Manney 1964, Todd 1964).

The essential distinction between all these earlier models and the present one concerns the *manner* in which multiple energy depositions occur. The target zone of the earlier models has extension along the direction of the track, but no pertinent internal structure. Calculations by Lea (1946), Howard-Flanders (1958), Brustad (1962), and Barendsen (1964) have assessed the minimum number of ions needed to produce a specific biological injury, and the length of track (t) within which they must be formed. These authors assumed that the *distribution* of ions within t has no relevance.

In my model, a rising RBE with LET signifies a complex target, with n, r, or nr discrete elements, each one of which receives an independent transfer of energy from the primary particle. That is to say, radiobiologists may have to consider the internal structure of complex targets such as genes and chromosomes—as well as the structure of the track. At high LET, we may be unable to regard deoxyribonucleoprotein as a homogeneous target volume.

Fortunately, these two approaches lead to some contrasting predictions. In favourable situations, where only one mode of change such as $n=2$, $r=1$; $n=1$, $r=2$; or $n=1$, $r=3$ might be involved, sufficiently detailed and accurate experimental relations between A_L and L could alone discriminate fairly reliably between the two models. However, when the inactivation of the proliferative capacity of mammalian cells constitutes the biological endpoint, the complexity of the relation between A_L and L presents difficulties. Barendsen's (1964) analysis, based on a 'featureless-target' model, fails to distinguish between oxygen-dependent and oxygen-independent modes of injury; furthermore, it makes no provision for the changes in the slope of survival curves between 'low-', and 'high-' doses. Nevertheless, by postulating multiple modes of inactivation, it might be possible to reconcile such a model with the overall experimental findings.

Certain restrictions on the statistics of my complex-target model should be emphasized. Consider the induction of a specific point mutation, of a specific gene. Here, the number of targets per cell will be small, and it will usually be determined by the ploidy of the cell. Consequently, models 1 and 2 may be relevant to this situation, whereas model 3 cannot be. For the induction of specific mutations in yeast (figures 2 and 3), we find the relations between A_L and L can in fact be described by modes $n=1$, $r=1$ and $n=2$, $r=1$. Model 3 can be applied only when the biological endpoint may be brought about through multiple independent hits, on any *one* of a *very large number* of complex targets. Thus, while model 3 may be relevant to the induction of a general class of chromosomal aberration in, say, mammalian cells, it could not be applied if we considered the induction of a specific aberration of a specific chromosome. All the data analysed are consistent with these restrictions of the complex-target model.

In certain situations, the response of a system to low energy electrons should distinguish between 'complex-' and 'featureless-' target models. Consider the mode

$n=1$, $r=2$, figures 5 to 8, in which the average energy per effective transfer (complex model) is, perhaps, in the region of 200 eV. For electrons with energy falling below 2 keV, the effectiveness compared with protons of equivalent LET should fall off rapidly to become almost zero at 500 eV. On the other hand, the featureless-target model would probably predict a high efficiency for electrons of energy in the region of 1 keV: such electrons have an effective LET close to the optimum value for this mode.

Another potentially useful test concerns the statistical fluctuations in biological response, in relation to corresponding fluctuations in energy deposition. Generally, the 'featureless-target' model requires more discrete energy transfers per lesion at high LET than the 'complex-target' model 3, and hence the variance of the biological response for the former model will be less than for the latter.

A much more difficult, but ultimately essential test in connexion with the inactivation of mammalian cells, calls for the independent identification of each hypothetical mode of injury using cytological, cytogenetic, biochemical or other suitable techniques. Once a specific mode of inactivation has been defined, its variation with L can be studied and checked against the appropriate theoretical model. Furthermore, the 'complex-' and 'featureless-' target models will often make different predictions concerning the average energy per lesion: direct measurements of this quantity should therefore provide another experimental means for distinguishing between the models.

To conclude, my main purpose in this paper has been to suggest that in concentrating on the structure of the track we should not overlook the possible importance of the structure of the target.

References

ALPER, T., MOORE, J. L., and BEWLEY, D. K., 1967, *Radiat. Res.*, **32**, 277.

BARENDSEN, G. W., 1964, *Int. J. Radiat. Biol.*, **8**, 453.

BARENDSEN, G. W., KOOT, C. J., VAN KERSEN, G. R., BEWLEY, D. K., FIELD, S. B., and PARNELL, C. J., 1966, *Int. J. Radiat. Biol.*, **10**, 317.

BRUSTAD, T., 1962, *Adv. biol. med. Phys.*, **8**, 161.

BURCH, P. R. J., 1968, *An Inquiry Concerning Growth, Disease and Ageing* (Edinburgh: Oliver and Boyd).

CORRY, P. M., and COLE, A., 1968, *Radiat. Res.*, **36**, 528.

DAVIES, H. G., and SMALL, J. V., 1968, *Nature*, **217**, 1122.

DEERING, R. A., and RICE, R., 1962, *Radiat. Res.*, **17**, 774.

HAYNES, R. H., 1966, *Radiat. Res. Suppl.*, **6**, 1.

HOWARD-FLANDERS, P., 1958, *Adv. biol. med. Phys.*, **6**, 553.

LEA, D. E., 1946, *Actions of Radiations on Living Cells* (London: Cambridge University Press).

MORTIMER, R., BRUSTAD, T., and CORMACK, D. V., 1965, *Radiat. Res.*, **26**, 465.

MUNSON, R. J., NEARY, G. J., BRIDGES, B. A., and PRESTON, R. J., 1967, *Int. J. Radiat. Biol.*, **13**, 205.

NEARY, G. J., PRESTON, R. J., and SAVAGE, J. R. K., 1967, *Int. J. Radiat. Biol.*, **12**, 317.

NEARY, G. J., and SAVAGE, J. R. K., 1966, *Int. J. Radiat. Biol.*, **11**, 209.

PARDON, J. F., WILKINS, M. H. F., and RICHARDS, B. M., 1967, *Nature*, **215**, 508.

RANDOLPH, M. L., 1964, *Ann. N.Y. Acad. Sci.*, **114**, Art. 1, 85.

TOBIAS, C. A., and MANNEY, T., 1964, *Ann. N.Y. Acad. Sci.*, **114**, Art. 1, 16.

TODD, P. W., 1964, *Reversible and Irreversible effects of Ionizing Radiations on the Reproductive Integrity of Mammalian Cells Cultured in vitro.*, Ph.D. Thesis, University of California.

The effect of track structure on OER at high LET

S. B. CURTIS

Lawrence Radiation Laboratory, Berkeley, California

Abstract. It is suggested that the difference in measured OER of human kidney cells *in vitro* obtained with alpha-particle beams and with high energy heavy-ion beams at the same high LET is due to the different energy density distribution in the tracks of the particles involved. In particular, the data from both experiments appear consistent when plotted against the parameter Z^{*2}/β^2, where Z^* is the effective charge on the incoming ion and β is its velocity relative to that of light. This parameter is a measure of the energy density from 'close-collision' delta rays at a point near, but not necessarily in, the track core.

It has been noted (Bewley 1968) that there is an apparent difference in the experimental values of the Oxygen Enhancement Ratio (OER) for the proliferative capacity of human kidney cells (T1) *in vitro* in the same high range (100–400 keV μm^{-1}) of LET_∞ (total dE/dx) depending on whether low energy helium ions (α-particles)

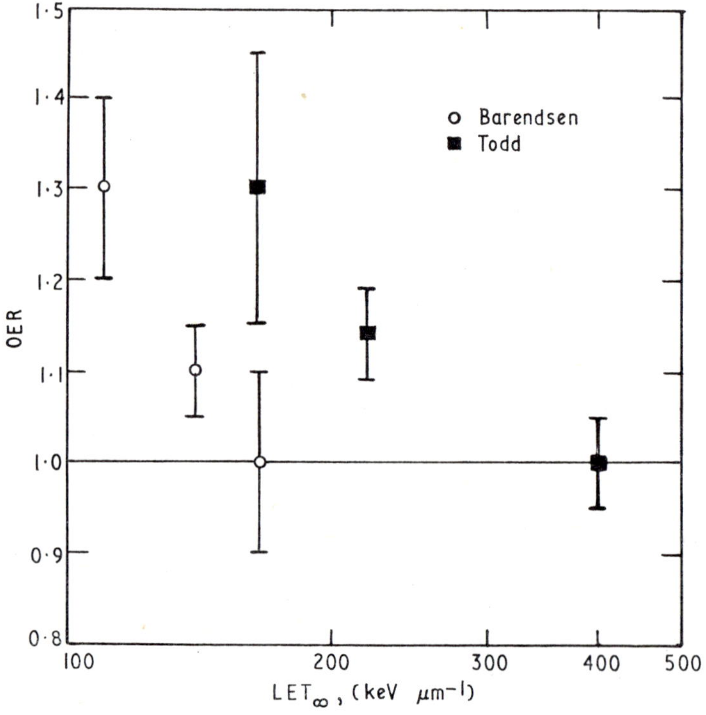

Figure 1. The Oxygen Enhancement Ratio for the proliferative capacity of human kidney cells (T1) *in vitro* plotted as a function of LET_∞ in the range 100 to 400 keV μm^{-1}. The open circles are low energy alpha-particle data (Barendsen *et al.* 1966) and the solid squares are high energy, heavy-ion data (Todd 1966).

(Barendsen *et al.* 1966) or high energy heavy ions (Todd 1966) are used. The experimental results are shown in figure 1. Open circles denote the low-energy α-particle data of Barendsen and the solid squares denote the high energy, heavy ion data of Todd. It has been suggested (Barendsen 1968) in connection with the measured inactivation cross sections that the different δ-ray distributions from the two kinds of ions at the same LET_∞ cause differences in experimental results at very high LET values. This paper is concerned specifically with the OER results and their interpretation by examination of the track structure of the α-particles and heavy ions involved.

The average radial distribution of energy density in a charged particle track can be assumed in a restricted region to decrease as $1/r^2$, where r is the distance from the centre of the track. This dependence is roughly correct only in the region where 'close' collisions play a dominant role in the energy deposition (i.e., not in the 'track core' itself) and where r is small compared with the range of an electron receiving the maximum transferrable energy. It is easily shown (Butts and Katz 1967) that in this region, the energy density is proportional to Z^{*2}/β^2 of the incoming ion, where Z^* is its effective charge (taking into account charge pick-up) and β is its velocity relative to the velocity of light. It is of interest to see if this parameter, which is a measure of the height of the 'shoulder' of the energy density curve, might be more relevant to plot against the OER data. This has been done for the data of figure 1, and the results are shown in figure 2. It is seen that the data from the two investigators overlap to a much greater extent and are consistent with lying on the same curve.

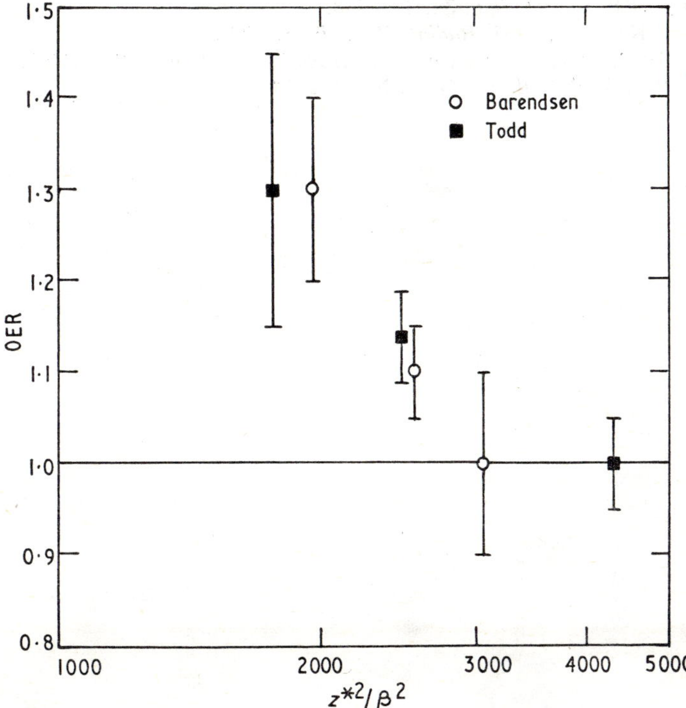

Figure 2. The same data as in figure 1 plotted as a function of the parameter Z^{*2}/β^2, where Z^* is the effective charge of the particle and β is the velocity of the particle relative to that of light.

It is tempting to speculate that the reason for the agreement in figure 2 is that the 'close in' shoulder of the energy density of a particle track determines whether the presence of oxygen in the cell is important in its inactivation. When the density reaches a critical value, the probability for inactivation becomes unity, irrespective of the initial presence or absence of oxygen. After this density is reached, the remaining *tail* of the δ-ray distribution has no effect in the inactivation process. In this view, it is the energy density caused by the δ-ray distribution near, but not necessarily in, the core of the track that determines the importance of oxygen to inactivation of kidney cells *in vitro*. It should be added, however, that this does not explain the continued increase in *inactivation cross section* measured by Todd at even higher LET (>400 keV μm^{-1}).

We note that Fowler (1968) suggested in the discussion of a paper by Barendsen (1968) at the Ispra Microdosimetry Conference (1967) that it might be of interest to plot the RBE results as a function of 'track core' LET. That suggestion has much the same physical content as the one made here, but the 'track core' LET has a logarithmic dependence on β^2 which does not appear in the present treatment.

It is a pleasure to acknowledge discussions with J. F. Fowler and G. W. Barendsen on the points discussed in this paper.

References

BARENDSEN, G. W., KOOT, C. J., VAN KERSEN, G. R., BEWLEY, D. K., FIELD, S. B., and PARNELL, C. J., 1966, *Int. J. Radiat. Biol.*, **10**, 317–327.

BARENDSEN, G. W., 1968, *Proc. of the Sym. on Microdosimetry*, (Ispra-1967), Ed. H. G. Ebert, 249–263.

BEWLEY, D. K., 1968, *Radiat. Res.*, **34**, 446–458.

BUTTS, J. J., and KATZ, R., 1967, *Radiat. Res.*, **30**, 855–871.

FOWLER, J. F., 1968, *Proc. of the Sym. on Microdosimetry* (Ispra-1967), Ed. H. G. Ebert, 264.

TODD, P. W., 1966, *Med. Coll. Va. Q.*, **1**, No. 4, 2–14.

1970 CH. PART. TR. SOL. LIQ.

The influence of track structure in radiation chemistry

W. G. BURNS

Chemistry Division, AERE, Harwell, Berkshire

Abstract. Recent developments in models for aqueous radiation chemistry, which depend on the distribution in energy of primary ionizing events, and therefore on the electron degradation spectrum, are considered first. The following phenomena, which appear to depend quantitatively on initial track structure, are then reviewed:

(i) the conductivity of irradiated organic liquids and the yield of free ions;
(ii) the effect of ionic scavengers on radiation chemical yields;
(iii) some results of kinetic spectroscopy, following pulse radiolysis;
(iv) some optical, ESR and thermoluminescent properties of irradiated cooled glasses;
(v) the observation of high LET of differing product G values for radiation of the same LET
   (a) for particles of differing mass
   (b) for the same particle on either side of the Bragg maximum.

The review contains a comparison of current models of ionization in irradiated liquids, and concludes with suggestions for further work.

One of the aims of the radiation chemist is an understanding of the events leading from the absorption of primary radiation to the formation of stable products, and in principle the details of radial track structure should form an important part of this understanding. Many measurements, however, are the result of the inseparable grand total of a multitude of events of dissimilar energies, and the chemist has in the past found a description of events in terms of effective average values of track parameters to be adequate.

For instance early models for the radiation chemistry of aqueous systems (Samuel and Magee 1953, Ganguly and Magee 1956) postulate that radicals (R·) are formed in spherical groups in spurs with a Gaussian (or other) spatial distribution with radius r_0

$$M \rightsquigarrow 2\mathrm{R}\cdot \quad G_1.$$

They may then undergo a first order reaction with the scavenger present (S)

$$\mathrm{R}\cdot + \mathrm{S} \longrightarrow \mathrm{RS} \quad k_1$$

to give the observed yield of oxidizing (OH) or reducing (H and e_{aq}) radicals. A second order reaction may also occur,

$$\mathrm{R}\cdot + \mathrm{R}\cdot \longrightarrow \mathrm{R}_2 \quad k_2$$

to give molecular products H_2, H_2O_2. In the course of these reactions diffusion of the radicals (diffusion constant D), and sometimes of the scavenger is assumed. The LET of the radiation enters the calculation as the reciprocal of the mean spur spacing —at high LET reaction of radicals from neighbouring spurs is predominant and the

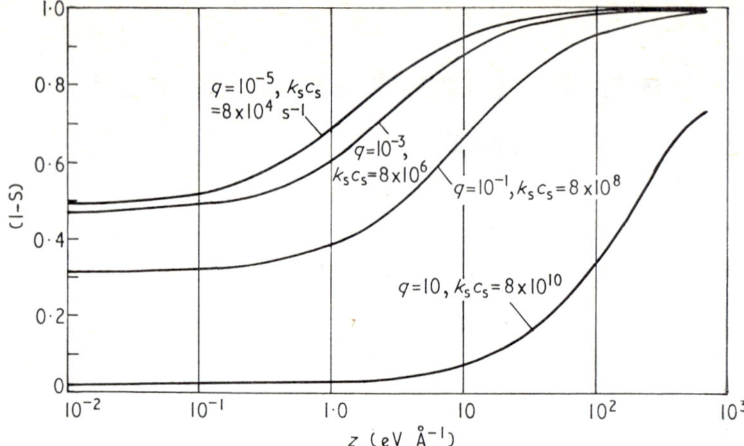

Figure 1. (1 − S), fraction of radicals which combine, against mean LET, z, for various values of $q=k_S c_S t_0$. No. of eV per spur = 100, No. of radicals per spur = 6, $t_0=1{\cdot}25\times10^{-10}$ s, $D=2\times10^{-3}\ \text{mm}^2\ \text{s}^{-1}$, $r_0=10$ Å, $k=3\times10^9\ \text{M}^{-1}\ \text{s}^{-1}$.

geometry becomes cylindrical resulting in high molecular yields. The distribution in energy of secondary electrons enters by way of the energy per spur, and r_0, which is related to the effective mean range of the secondary electrons. The main experimental features of aqueous systems, i.e. the change in G value of molecular products with scavenger concentration (spur erosion), and the change in G values of radical and molecular products with LET are both explained by this simple model. Figure 1 and figure 2 show theoretical predictions of the dependence of (1 − S), the fraction of radicals which combine, on the mean LET

$$z=\frac{1}{E}\int_0^E\left(-\frac{dE}{dx}\right)dE$$

and on $q=k_S c_S t_0$, where $t_0=r_0^2/4D$, which can give reasonable overall agreement with experiment.

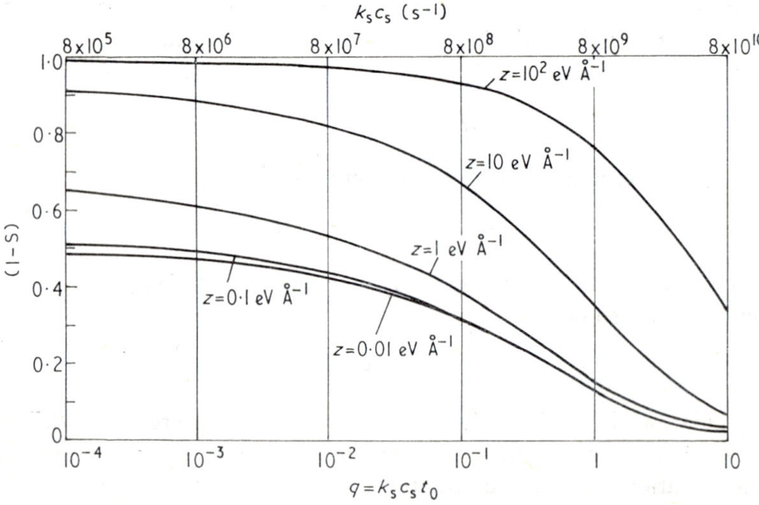

Figure 2. (1 − S) against $q=k_S c_S t_0$ at various values of z. Constants as for figure 1.

Although for many years authors, including Platzman (1958) and Burch (1957) have been urging the importance of the distribution in energy of secondary electrons and of the electron degradation spectrum, it is only recently that these considerations have been used widely in the interpretation of radiation chemical phenomena. Some examples of experimental phenomena which seem to require quantitative explanation in terms of radial track structure are as follows:

(i) the conductivity of irradiated liquids and the yield of free ions;
(ii) the effect of ionic scavengers on radiation chemical yields;
(iii) some results of kinetic spectroscopy, both emission and absorption, following pulse radiolysis;
(iv) some optical, ESR and thermoluminescent properties of irradiated cooled organic and inorganic glasses containing suitable solutes
(v) in the field of high LET radiation chemistry the observation of differing product G values for radiation of the same LET
 (a) for particles of differing mass,
 (b) for the same particle in different parts of track on either side of the Bragg maximum in LET.

I propose to treat these topics in turn briefly and to conclude with an opinion of what new experiments and theoretical explorations would be valuable. Before doing this I wish to consider the nature of the energy degradation spectrum, and an improvement of the early diffusion model for the aqueous system, made by Mozumder and Magee (1966a and b).

Figures 3, 4, 5 show graphs in which the area under the curve between abscissa values E_1 and E_2 represents the contribution to energy loss or ionization by electrons of energy between E_1 and E_2. The construction of such curves involves a knowledge of appropriate cross sections, e.g. those for generating electrons in one energy group by electrons in another, and a moderate amount of book-keeping, since it is important not

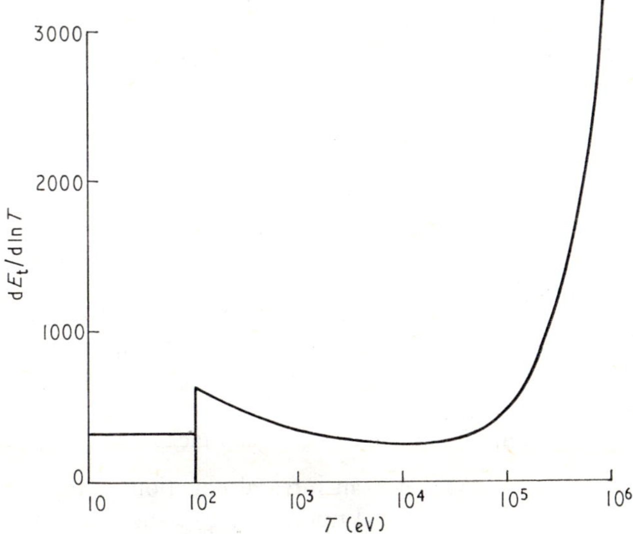

Figure 3. Contribution of different part of the degradation spectrum to energy loss in events of <100 eV for a 1 MeV electron in water. Burch, 1957a. $T\,dE_t/dT$ is plotted against $\log_{10} T$, where E_t is 6243, so that the total area is $6243/\log_e 10$.

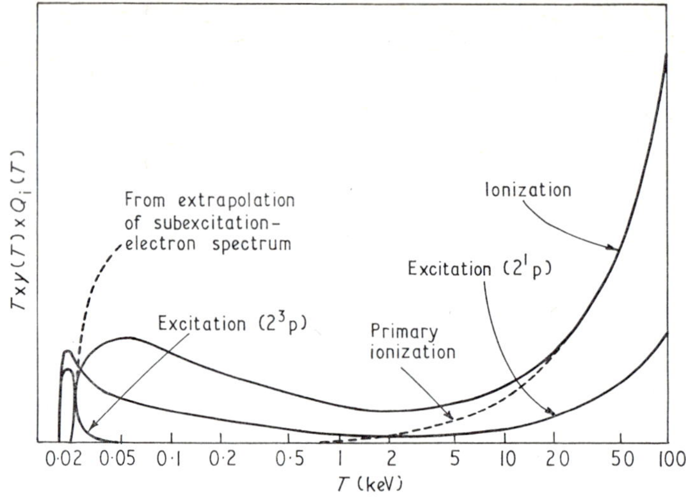

Figure 4. Contribution of different portions of the electron degradation spectrum to the total ionization and excitation: 100 keV electrons in helium gas, Platzman (1961).

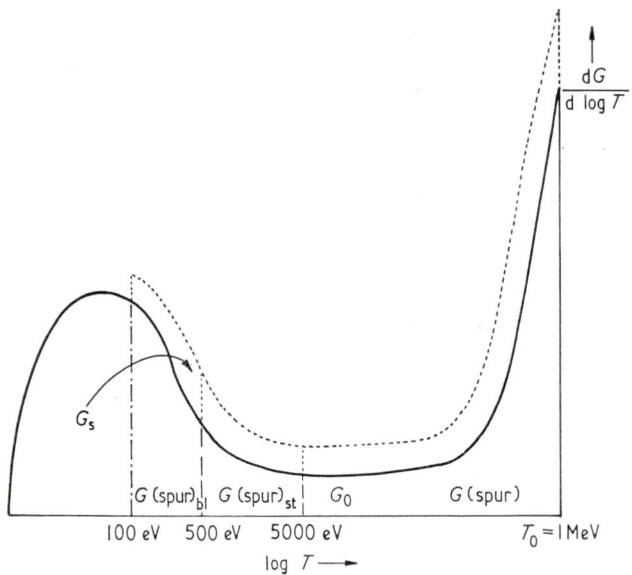

Figure 5. Subdivision of the degradation spectrum, and contributions of its various regions to the total yield of primary activations. Santar and Bednář, (1968, 1969).

to count energy losses more than once. The primary electron produces a first generation of secondaries, each of which produces a second generation, and so on, and for each energy loss the electron considered is itself degraded in energy and so enters another group. The book-keeping problems have been described by Burch (1957), and by Platzman (1961), and the problem has been tackled in three different ways.

(a) Spencer and Fano (1954) used an integral equation method, presenting their results in the form of differential track length versus electron energy.
(b) Burch (1957) used an algorithmic approach based on probabilities and energy groups.
(c) Magee and Mozumder (1966c) used a Monte Carlo method.

The results of the different methods give reasonable agreement and all show the bimodal nature of the energy depositions.

As part of their treatment Mozumder and Magee (1966a and b) have presented approximate methods for dividing energy loss events into 'spurs', 'blobs', and 'short tracks'. The spurs comprise all events of energy up to 100 eV not generated in blobs and short tracks; the blobs comprise all events of energy 100 to 500 eV not generated in short tracks; the short tracks comprise events of energy 500 to 5000 with their arising secondaries. One argument for these separations is that the model for chemical reactions has different symmetries in the different entities, the spurs and blobs being spherical, and, the short tracks, which are high LET components, being cylindrical. Figure 6 shows the variation in the distribution between these entities with primary

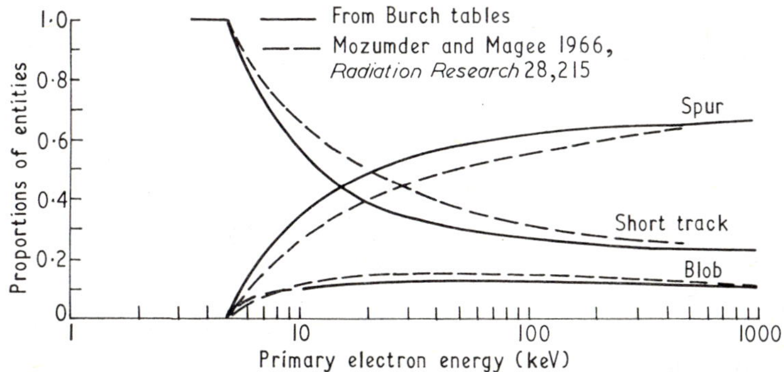

Figure 6. Dependence of the proportions of the entities, spurs, blobs, and short tracks on primary electron energy. Mozumder and Magee (1966a), and as calculated by summing appropriate parts of preliminary tables in Burch (1957a).

electron energy, as calculated by Mozumder and Magee, and also as can be obtained from energy loss tables according to Burch, showing reasonable agreement. In applying these distributions to the diffusion kinetics problem Mozumder and Magee found it necessary to use the more detailed distribution in 'spur' and 'blob' sizes which were the result of the Monte Carlo method (1966b). This application to the aqueous case was an important demonstration of the use of a more refined model of track structure, but in some respects the final outcome for lightly ionizing radiation was not very different from that predicted by less elaborate models, see figure 7. Figure 8

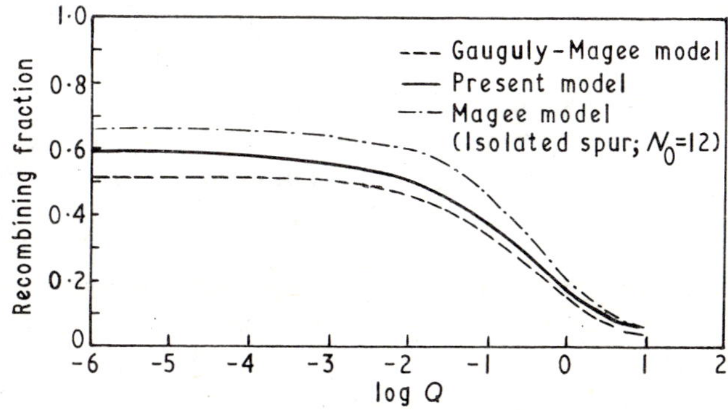

Figure 7. $(1-S)$, fraction of radicals which combine, against $q = k_S c_S t_0$, for various track models. Mozumder and Magee (1966a).

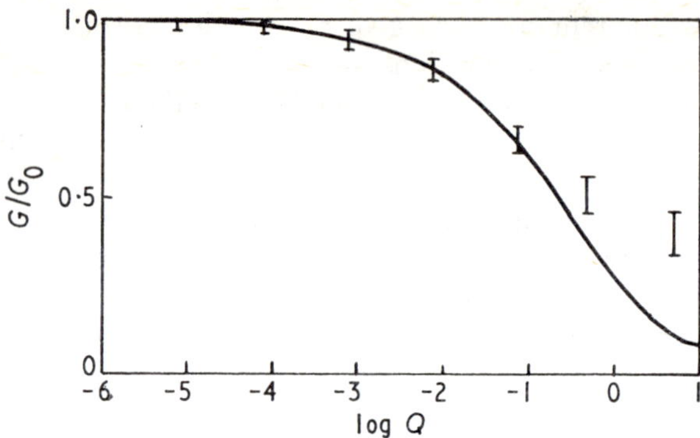

Figure 8. Comparison of theory and experiment in respect of lowering of molecular yield with scavenger concentration. Full line, calculated on present theory for a 500 keV electron; vertical bars are experimental points according to H. A. Schwarz, with height denoting approximate scatter (A. O. Allen 1961). Some experimental points for aerated solutions (outside the general Schwartz curve) have been ignored. (Mozumder and Magee 1966a.)

shows that the ratio of G values of molecular product, G/G_0, with and without scavenger present, as a function of $q = k_s c_s t_0$ predicted by the new model, agrees well with experiment over a wide concentration range; however, this dependence is not very sensitive to the spur size distribution, or even to the model used. Kupperman (1967) has given a detailed diffusion model description of the aqueous case in which energy depositions by various parts of the electron energy spectrum were calculated by the Burch method, but it is not clear that this latter elaboration was essential to give agreement between calculation and experiment. It is clear, however, that the dependence of G values on scavenger concentration and on LET require non-homogeneous kinetics.

The field of ionization of liquids is one in which track structure plays an important role. The experiments of Allen and Hummel (1966) and of Freeman (1965), on the conductivity of organic liquids under irradiation, and on the mobility of the charge-carrying ions, showed that for a liquid such as n-hexane, G (free ions) is 0·1, with a marked temperature coefficient.

If we accept that the probability of mutual escape for an ion pair is that given by Onsager, $e^{-r_c/r}$ (where r = the ion pair separation, $r_c = e^2/(\epsilon kT)$, where e is the electronic charge, ϵ the dielectric constant, and k Boltzmann's constant, giving $r_c = 300$ Å at 293 °C for hexane) and that this probability is applicable to the furthest separated pair of a spur, then only those spurs with separation above ~300 Å will contribute substantially to ionization. Using the normal range/energy relation to deduce the energy corresponding to this 300 Å range, we arrive at an energy of ~1000 eV, above which electrons can contribute sensibly to the ion-pair yield. Table 1 shows the number of energy loss events in various energy intervals, calculated from the degradation spectrum, produced by a 1 MeV electron. The number of events above 1000 eV is only ~67, i.e. they have a G value of $67 \times 100/10^6$ which is 0·0067, about an order of magnitude too low. Allen and Hummel concluded that the large number of events at lower energies must also contribute to the ion-pair yield, and that low energy electrons must therefore travel further than their normal range before the escape probability formula can be applied.

This concept was elaborated by Mozumder and Magee (1967) who considered the thermalization of a subvibrational electron losing energy to the surrounding medium in

Table 1. The number of energy-loss events for some intervals of W for a 1 MeV electron during slowing down

Energy interval ΔW in eV	Number of events, $n(W)\Delta W$ This work	Magee *et al.**
> 5000	11	
1000–5000	56	44
500–1000	76	55
300–500	95	88
100–300	548	457
80–100		347
30–80		4876
10–30		27991

* MAGEE, J. L., FUNABASHI, K., and MOZUMDER, A., 1964, *A.E.C. Rep.* COO-38-378.

a random walk process in the force field of its sibling positive ion. They concluded that, in a non-polar medium, a low energy electron (10–40 eV) can be expected to travel on average a distance of ~80 Å from its conjugate positive ion before thermalization. After adjustment of the relevant parameters of the random-walk process for agreement with experiment at one temperature and for lightly ionizing radiation, the theory of Mozumder and Magee gave good agreement with the observed temperature effect for lightly ionizing radiation, and also agreed well with the observed yield of free ions for ^{37}Ar radiation, see table 2.

Table 2. Ion-pair yield in n-hexane at different temperatures. Experimental values from Allen and Hummel (1966). Predicted values according to Mozumder and Magee (1967)

Temperature K	Onsager critical distance r_c in Å	Experimental value of G(ip)	$G(ip)$ according to theory	Effective thermalization length for a small spur
293	312	0·095	0·087	80·5
283	320	0·089	0·083	81·6
273	330	0·084	0·078	82·7
263	339	0·078	0·074	83·8
253	351	0·071	0·070	85·0
243	363	0·066	0·065	86·1
233	375	0·060	0·060	87·4
223	389	0·054	0·056	88·6
213	405	0·047	0·050	89·9
203	422	0·042	0·046	91·2

Mozumder and Magee's treatment involves the integration over all relevant energy loss events of the product of the initial G value (obtained from the energy degradation spectrum) and the escape probability p, which for each event depends on a distribution of distances due to the random walk process, each with its own escape probability. If each energy loss event had given rise to a single thermalization distance l, then p would be given by $p = e^{-r_c/l}$, and for each event, l, the effective thermalization length, can be calculated from p. Table 2 gives values of l for the numerous low energy events (10–40 eV). The treatment is an extension of Hummel's (1967), in which the distribution of thermalization distances for each event was assumed to be Gaussian, and agreement with experiment was obtained by adding $4{\cdot}0 \times 10^{-4}$ mg cm^{-2} ($\equiv 60$ Å in n-hexane at 20 °C) to the ranges in hexane below 80 eV to obtain the most probable thermalization separations. At higher energies the normal range is used, and the treatment gives reasonable agreement with experiment for the effect of temperature.

The concept that low energy electrons travel farther than their normal range in liquids like n-hexane is not admitted by Freeman (1965, 1967), who uses normal ranges applied to equation (1)

$$G = \sum G(r)\, e^{-r_c/r} \tag{1}$$

but achieves agreement with experiment by severely underestimating the G values of electrons in the low energy region (Hummel 1966, 1967 and Burns 1968).

Figures 9 and 10 show graphs in which the areas under the graph are proportional to the G value of energy loss events in the relevant energy interval. Figure 9 is for the initial yield of ions and figure 10 for the escaped, or free ions. The total values of G(initial ions) ~ 3 and of G(free ions) $\sim 0{\cdot}1$, are similar for the treatments of Freeman,

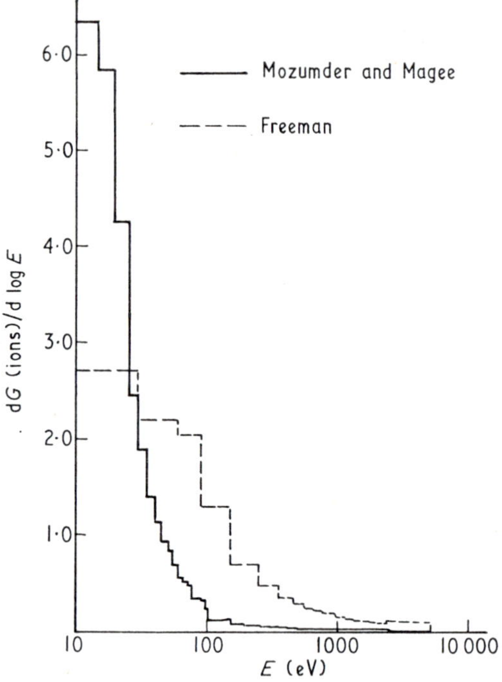

Figure 9. Distribution in energy of the initial yield of ions: unbroken line: Mozumder and Magee (1967); broken line: Freeman (1965, 1967).

and of Mozumder and Magee, but it can be seen that the energy distributions of the initial and escaped ion pairs are very different. The initial and the free ion yield distribution of Mozumder and Magee is peaked in the low energy region 10–40 eV, and although Freeman's initial yield is greatest at low energies, it is more spread out in energy, producing a free ion yield which is important at 1000 eV. The distributions of the yields in separation at thermalization show further differences since the initial yields are peaked at ~ 17 Å for Freeman, and ~ 80 Å for Mozumder and Magee, reflecting the different energy/separation criteria, and the free ion distribution of

Mozumder and Magee is peaked at around 100 Å, whereas Freeman's is spread out over the range 100 to 10 000 Å.

Further experiments by Schmidt and Allen (1968) on the measurement of *G*(free ions), this time using a clearing field, have revealed substantial differences in this quantity for a variety of non-polar liquids (table 3) so that although, as Freeman (1965) has pointed out, the dielectric constant has an important influence, other considerations, possibly to do with the energy loss of subexcitational electrons, also play their part. Schmidt and Allen represent their results in terms of *b*, the most probable ion-pair separation in a Gaussian distribution for the whole electron energy spectrum, so no

Table 3. Free ion yields in pure liquids at 23 ° (Schmidt and Allen 1968b)

Compound	No. of detns[a]	*G*, ion pair/ 100 eV	Av devn, %	ε^b	r^c, Å	b/r^c	*b*, Å	d,[b] g cm^{-3}	*bd*, g cm^{-2} $\times 10^8$
n-Pentane	3	0·145	3·4	1·842	306	0·234	71·5	0·623	44·6
Isopentane	4	0·170	0·9	1·838	307	0·248	76·0	0·617	46·9
Neopentane	5	0·857	3·4	1·777[c]	318	0·561	178·4	0·588	104·9
Cyclopentane	1	0·155	—	1·960	288	0·239	68·9	0·742	51·1
n-Hexane	6	0·131	3·8	1·885	299	0·226	67·4	0·656	44·2
3-Methylpentane	2	0·146	4·1	1·901	297	0·235	69·6	0·662	46·0
2,3-Dimethylbutane	2	0·192	1·3	1·953	289	0·259	74·9	0·659	49·4
2,2-Dimethylbutane	2	0·304	3·0	1·926	293	0·314	92·0	0·646	59·5
Hexene-1	2	0·062	3·2	2·046	276	0·178	49·1	0·670	32·9
Cyclohexane	9	0·148	5·5	2·022	279	0·237	66·1	0·776	51·3
Cyclohexene	2	0·150	2·0	2·222	254	0·237	60·1	0·806	48·5
Benzene	3	0·053	1·9	2·278	248	0·170	42·1	0·876	36·9
n-Octane	2	0·124	0·8	1·944	290	0·221	64·2	0·700	44·9
2,2,4-Trimethylpentane	4	0·332	1·1	1·936	291	0·326	95·0	0·689	65·4
1,4-Dioxane	3	0·046	5·0	2·212	255	0·164	41·7	1·030	42·9
Diethyl ether	2	0·350	1·1	4·280	132	0·335	44·2	0·710	31·4
Carbon tetrachloride	2	0·096	1·3	2·232	253	0·203	51·4	1·590	81·7
Germanium tetrachloride	1	0·127	—	2·435	232	0·223	51·7	1·870	96·7
Carbon disulfide	2	0·314	1·3	2·633	214	0·336	71·8	1·259	90·4

[a] Separate fillings of the measurement cell; often different cells with different electrode spacings were used. [b] Data from 'Landolt-Börnstein Zahlenwerte und Funktionen,' Berlin: Springer. 'Handbook of Chemistry and Physics,' Chemical Rubber Co., Cleveland, Ohio; 'Organic Solvents', (New York: Interscience Publishers). [c] Calculated from the molar refraction; no data available.

assumptions are made about the energy spectrum apart from the value of *G*(initial ions). Values of *b* are similar to Mozumder and Magee's value of *l* for low energy events, and the distance required to be added to the range at low energies by Hummel, since the energy spectrum is dominated by low energy events. They find that *bd* where *d* is the density, for n-hexane, n-pentane, and cyclohexane, is independent of temperature. This is consistent with the approximate agreement between predicted and experimental temperature effects found in Hummel's treatment.

I wish to consider now the subject of ion scavenging in organic liquids. T. F. Williams (1964) first correlated the yield of free ions, as measured by conductivity, with the yield of scavengeable ions. He dissolved deuterated ammonia in cyclohexane

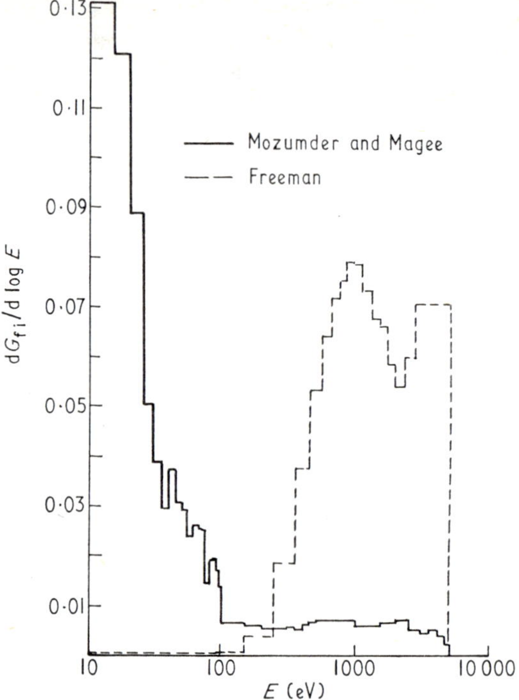

Figure 10. Distribution in initial energy of the yield of free ions: unbroken line: Mozumder and Magee (1967); broken line: Freeman (1965, 1967).

and measured *G*(HD), which was supposed to arise as follows:

$$RH^+ + ND_3 \xrightarrow{k_s} ND_3H^+ + R \qquad \text{proton transfer, } k_s \sim 10^9 \text{ M}^{-1}\text{ s}^{-1}$$

$$(ND_3H^+ + ND_3 \longleftrightarrow ND_4^+ + NH_3D) \qquad \text{exchange}$$

$$ND_4^+ + e \longrightarrow D + ND_3 \qquad \text{neutralization}$$

$$D + C_6H_{12} \longrightarrow HD + C_6H_{11} \qquad \text{abstraction}$$

The yield of HD is therefore a direct measure of the yield of available (i.e. not geminately recombined) positive ions. At low concentrations, $[ND_3] < \sim 10^{-2}$ M, *G*(HD) was 0·08, agreeing with *G*(free ions), and higher yields at higher concentrations were attributed to the scavenging of ions which otherwise would recombine geminately. On the basis of the Nernst–Einstein relationship,

$$\frac{\partial r/\partial t}{\partial E(r)/\partial r} = \frac{D}{kT}$$

Williams argued that the recombination time for ions separated by a distance r should be given by

$$\tau = \frac{kT}{3De^2} r^3$$

and that the onset of ion scavenging occurs for $r_c = \sim 300$ Å, whence $\tau = 10^{-7}$ s which can be equated with the half-life of the ion with respect to proton transfer at

$[ND_3] \sim 10^{-2}$ M, viz. $1/k_s[ND_3] = 10^{-7}$ s. In this, and another study using ethanol (Buchanan and Williams 1966) as scavenger, he estimated the median initial ion separation to be ~ 50 Å. The importance of Williams's considerations was that the escaped electron appeared to move with mobility near that of a molecular entity rather than that of a free electron. Other important experimental work on electron scavenging has been done by the Newcastle school (Scholes and Simic 1964) and others (Sagert and Blair 1967, Sato *et al.* 1967, Sherman 1966) using N_2O which produces nitrogen gas,

$$N_2O + e \longrightarrow N_2 + O^-$$

and on alkyl halides by workers at the Mellon Institute (Warman, Asmus, and Schuler 1968).

$$CH_3Br + e \longrightarrow CH_3\cdot + Br^-$$

The latter school has shown that the following empirical relation has a wide application

$$G(S^-) = G_{fi} + \frac{G_{gi}\alpha[S]^{1/2}}{1 + \alpha[S]^{1/2}}$$

where G_{fi} is the yield of free ions and G_{gi} the yield of ions which recombine geminately. An interesting recent development is the determination of a variety of values of G_{fi} for non-polar liquids by extrapolation of the above scavenging formula (Rzad and Warman 1968) and by other chemical methods (Capellos and Allen 1968), which agree with Schmidt and Allen's conductivity determinations (Schmidt and Allen 1968a and b).

Theoretical explanations (Freeman 1967, Buchanan and Williams 1966, Sato *et al.*

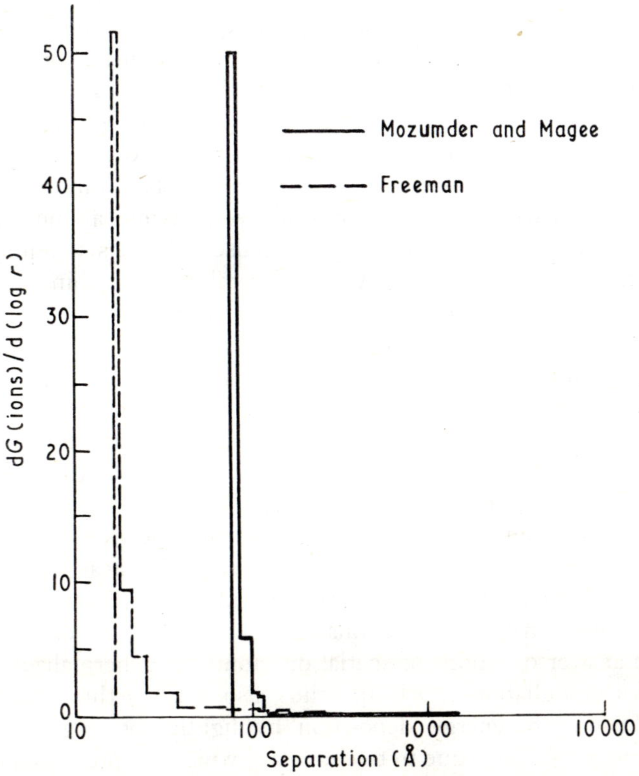

Figure 11. Distribution in thermalization separation of the initial yield of ions: unbroken line: Mozumder and Magee (1967); broken line: Freeman (1965, 1967).

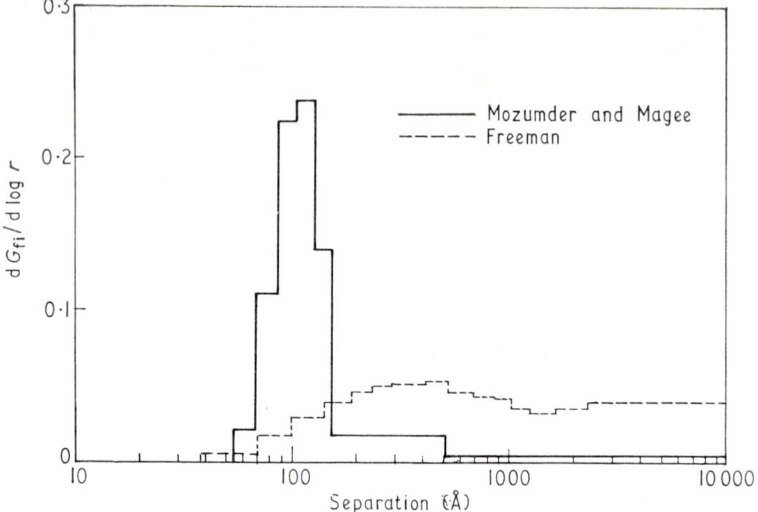

Figure 12. Distribution in thermalization separation of the yield of free ions: unbroken line: Mozumder and Magee (1967); broken line: Freeman (1965, 1967).

1967) have normally followed the line of taking an initial distribution in separations, giving rise to a distribution of return times proportional to r^3. From these times the number of diffusive jumps for return is obtained, and using statistical considerations the probability of encountering a solute molecule is calculated. The use of the r^3 law has been criticized by Mozumder (1968), who has questioned the applicability of the Nernst–Einstein relationship to these situations. In another approach not explicitly involving the time scale of the events, Hummel (1968a) has applied a probability of reaction with scavenger, calculated by Monchik (1956), to the initial distribution of separations which gave agreement with his conductivity experiments. The resulting relation, Figure 13, seems so far the best founded correlation in this field.

A type of study in which the time scale of events is of paramount importance is kinetic absorption spectroscopy following the pulse radiolysis of liquids. The times involved need to be very short, $\gtrsim 1$ ns, to follow geminate recombination. Cooper and Thomas (1968) observed the absorption of a positive ion of CCl_4 which had a rapid first order decay, with $t_{1/2} = 15$ ns, followed by a span of recombination times up to microseconds, as might be expected from a broad distribution of separations. The situation has been treated theoretically by Freeman (1968), using the same electron range distribution and relation between separation and return time, as were used to explain the yield of free ions and the effects of ion scavengers. More recently Hummel (1968b) has given a derivation of ion lifetimes which depends ultimately on experiment, and which includes the effect on the ion lifetimes when the identity of the negative ion is changed by scavenging. The form of the decay of luminescence of dilute organic scintillators in aliphatic hydrocarbon solutions (Ludwig and Burton 1968, Ludwig and Huque 1968) is also attributed to geminate recombination and observed decay times should help to answer questions of spatial distribution of thermalized electrons.

Some studies of radiation effects in solids also help to throw light on the non-homogeneous nature of energy deposition by lightly ionizing radiations. These include observation of effects due to free radicals, which are presumed to be formed in that part of the track where ionization and excitation occur, and of effects due to trapped electrons.

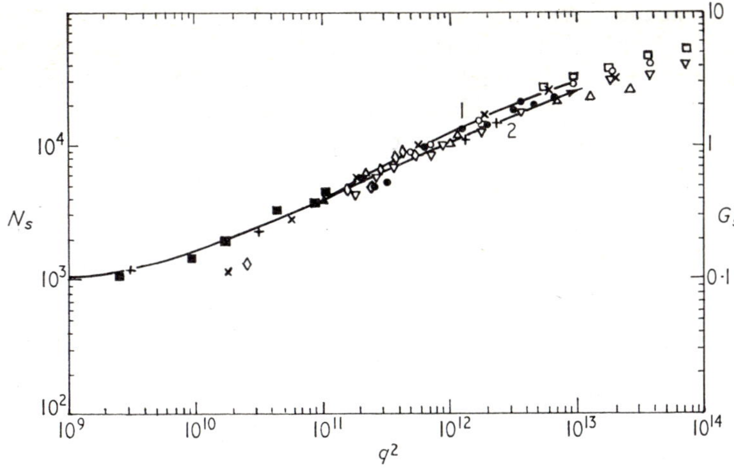

Figure 13. The yield of scavenged ions in irradiated liquids. Solid lines: N_s, the calculated number of ions originating during the thermalization of a 1 MeV electron in hexane reacting with a scavenger as a function of $q^2 = k_S c_S/D$ (D is the sum of the diffusion constants of the positive and negative ions). Experimental results are plotted against αc_S, where c_S is the scavenger concentration in cm^{-3}. ○, □, ▽-N_2 yield from N_2O in cyclohexane (Scholes and Simic 1964, Sato *et al.* 1967, W. V. Sherman 1966), $\alpha = 3{\cdot}0 \times 10^{-7}$ cm; ●—trans-stilbene isomerization in cylcohexane (R. R. Hentz *et al.* 1966), $\alpha = 1{\cdot}1 \times 10^{-7}$cm; ■—HD yield from ND_3 in cyclohexane (F. Williams 1964) (the yields are multiplied by 4/3 to account for the isotopic distribution), $\alpha = 4{\cdot}9 \times 10^{-9}$ cm; X—HD yield from C_2H_5OD in cyclohexane (Buchanan and Williams 1966) (the yields are multiplied by 2), $\alpha = 9{\cdot}8 \times 10^{-9}$ cm; ◇—2-butene isomerization in dodecane (Cundall and Griffith 1963), $\alpha = 4{\cdot}3 \times 10^{-9}$ cm; + —$C_3D_6H_2$ yield from C_3D_6 in isopentane (195 K), (Scala Lias and Ausloos 1966), △—benzyl radical yield from α-chlorotoluene in cyclohexane (Hagemann and Schwarz 1967), $\alpha = 1{\cdot}7 \times 10^{-7}$ cm.

The results of ESR measurements, including distinguished work done in the Soviet Union, have already been summarized in this conference by Dr Wyard, and I merely wish to mention some typical findings. By measuring the broadening of ESR lines in irradiated polymethylmethacrylate, Bullock, Griffiths, and Sutcliffe estimated the local free radical concentration and compared it with the overall concentration (Bullock, Griffiths, and Sutcliffe 1967). Whereas the overall concentration increased with irradiation time from $\sim 10^{17}$ to 10^{18} spins cm^{-3}, the local concentration stayed constant at $1{\cdot}8 \times 10^{19}$ spins cm^{-3}. Similar differences in the local and average concentration of trapped electrons in 2-methyltetrahydrofuran irradiated at 77 K had been found by Smith and Pieroni (1965). Such experiments lead to *average* local concentrations in spurs. It should be pointed out that observations of increases in local radical concentrations are not universal, and in irradiated calcite Cunningham (1967) found no evidence for spur formation. In another type of experiment on irradiated cooled organic and inorganic glasses, concerning the decay of free radical (Willard 1968) and electron trap (Dyne and Miller 1965) concentration with optical irradiation or heating, results have been interpreted on the basis of a distribution of separation distances. Observations of thermally or optically stimulated luminescence also come into this category (Deroulede, Kieffer and Magat 1968, Brocklehurst and Russel 1967). A large initial fraction of mutual neutralization occurs with a greater efficiency than the remainder, and this is often attributed to the lower separation distances of the majority of ion-pairs or free radicals. Schulte-Frohlinde (Vacek and Schulte-Frohlinde 1968) and others have warned that the observation of such kinetics does not always imply an inhomogeneous spatial distribution. In some circumstances the mutual neutralization of nearest neighbours in a homogeneous distribution leads to greater separations for the remainder and hence increased difficulty of further reaction.

I wish now to deal with the question of differences in G values for heavy particles which have the same LET. This situation can occur along the track of a single particle if we select two energies on either side of the Bragg maximum such that they give rise to the same LET. The instantaneous G value, measured by differentiating the curve of $G \times E$ against E, is found to have different values for the same LET on either side of the Bragg maximum for the only cases where such measurements have been made, i.e. $G(H_2)$ from benzene (Burns and Marsh 1968) and $G(Fe^{+++})$ from aqueous Fricke dosimeter solutions (Collinson *et al.* 1961) for ^4He ions, figure 14. We can postulate that these differences are due to the differences in associated electron spectra. The observed instantaneous G value, g_k, is due to the effect of local energy losses, of energy less than a value η, which form the dense central track, together with the effect of more energetic secondary electrons which can be considered separately

$$g_{\mathrm{k}} = H_{\mathrm{k}} h_{\mathrm{k}} + \sum_{\mathrm{m}} L_{\mathrm{m}} l_{\mathrm{k,m}}$$

Here H_{k} is the instantaneous local G value of the heavy particle, and h_{k} the fraction of its energy deposited in local events, L_{m} is the G value for electrons in a particular energy group m, and $l_{\mathrm{k,m}}$ is the fraction of energy deposited in electrons of energy group m by heavy particles in group k. In order to form the value of H_{k}, the local instantaneous

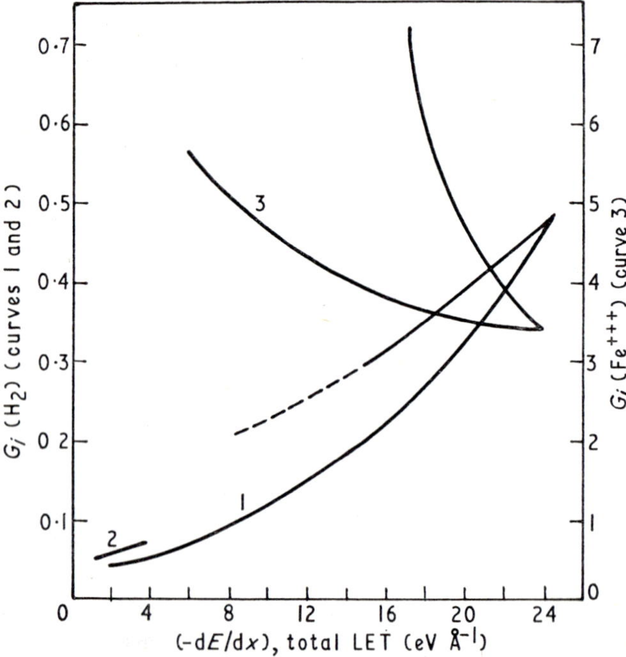

Figure 14. Instantaneous G values against total LET in eV Å^{-1}. Curve 1, $G(H_2)$ from benzene for ^4He ions (Burns and Marsh 1968); Curve 2, $G(H_2)$ from benzene for protons (Burns and Marsh 1968); Curve 3, $G(Fe^{+++})$ from aqueous Fricke dosimeter solutions for ^{210}Po α-particles (Collinson *et al.* 1961).

G value, given by

$$H_k = (g_k - \sum_m L_m l_{k,m})/h_k$$

we need the values of L_m. As a first approximation we assume that instantaneous *G* values for electrons L_m are the same as those for ^4He ions at the same LET, and that for these electrons, whose LET does not extend beyond 10 eV Å^{-1}, local and total instantaneous *G* values are equal. For low energy electrons this approximation is, in any case, a good one. Using values of $l_{k,m}$ and h_k from tables like those of Burch (1957a and b), we can obtain a new corrected set of values of H_k, which when plotted against local $(-dE/dx)$ give rise to a corrected set of values of L_m. This procedure, applied iteratively, was found to yield a single curve of *H* against local LET for low energy ^4He ions with $\eta = 300$ eV both for $G(H_2)$ from benzene (Burns and Marsh 1968) and $G(Fe^{+++})$ from aqueous ferrous solutions (see figure 15). With η set to 200 or 400 eV, curves giving rise to two *G* values at the same LET were obtained.

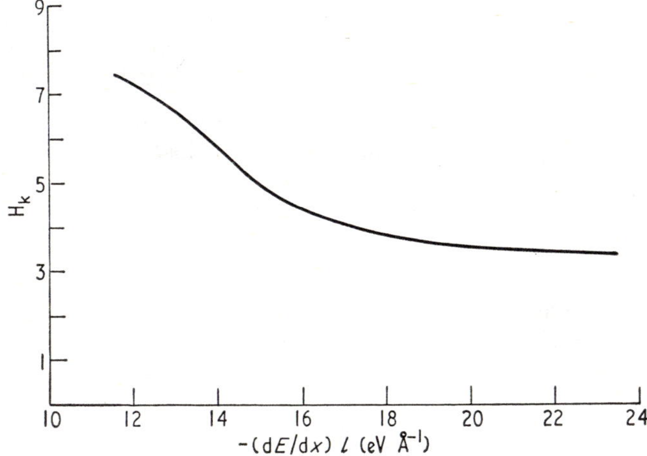

Figure 15. Local instantaneous $G(Fe^{+++})$ from aqueous Fricke dosimeter solutions against local LET. A single curve is obtained for both sides of the Bragg maximum when local events are defined as having energy up to 300 eV.

Heavy particles of differing mass can also have the same LET (figure 14), e.g. 1·4 MeV protons and 32·5 MeV ^4He ions, and are observed to give rise to different *G* values. The instantaneous $G(H_2)$ from benzene for 0·5 to 4 MeV protons is ~20% higher than for ^4He ions of the same LET, and although the absolute difference is diminished if we calculate local instantaneous values, H_k, the relative difference is increased since H_k is lower than the corresponding g_k. A similar difficulty occurs with proton and ^4He ion $G(Fe^{+++})$ values for aqueous ferrous solutions (Schuler and Allen 1957). Other examples are the *G* values of products from cyclohexane (Burns *et al.* 1968), and of hydrogen from benzene (Burns and Marsh) for 114 MeV ^{14}N ions. The qualitative effect is always that the higher energy heavier ions yield products with a *G* value closer to that for fast electrons than the lighter, less energetic ion, and the effect is larger than can be explained on the basis of secondary electron spectra. A possible cause of the effect is another radial track phenomenon pointed out by Mozumder *et al.* (1968), that the maximum impact parameter depends on the particle velocity, and above ~3 MeV amu^{-1} a broadening of the track core begins to occur so that a core radius of ~15 Å broadens to ~30 Å for a particle of 10 MeV amu^{-1}. The differences in concentration of active species at the track core are quite sufficient to account quantitatively for differences in product *G* values for the cases for which we have made calculations.

I wish to conclude with an opinion of what further experiments and theoretical investigations would be most helpful in elucidating the problem of track structure in liquids and solids. All the types of experiment mentioned above can be used to help our understanding, and further work on nanosecond and even picosecond pulse radiolysis would seem the most direct means of following the effects of track structure in liquids. In order to test current theories, measurements of ion yields and of charge scavenging, using different kinds of radiation including heavy particles, are important. In the latter field we have made a start in investigations at Harwell (W. G. Burns and C. R. V. Reed, unpublished), and these show, as might be expected, that both positive and negative ion scavenging is less effective at the same scavenger concentration with densely than with lightly ionizing radiation. To correlate experimental results information is required on the expected distribution in energy, and, more important, in separation, of electrons and positive ions, the mechanism of thermalization, the time scale of return, and the probability of reacting with scavengers during return.

Work on the degradation spectrum, going to low energies, using the sum rule and such approximations as the optical approximation, has been done by Magee and Platzman (1967), and by Santar and Bednář (1968 and 1969), and the last authors' prediction of a large effect of molecular structure on the proportion of 'spurs', seems worth further investigation. Magee's first approach to the problems of predicting thermalization separations has been partially successful, but even here the calculated value of *G* (free ions) seems to be quite sensitive to the values used for the electronic range of electrons of energies in the groups 10–15, 15–20, 20–25 eV, where our knowledge is inadequate, and further work in this field would be valuable. The fact that Freeman, on the one hand, and Mozumder and Magee and Hummel on the other, have been able to obtain agreement with experiment using different energy and separation spectra shows that the effects of separation distribution and of probability of reaction during return, are interwoven to an extent that the compensation of errors is a possibility. In any event the prediction of electron behaviour before and after thermalization are equally important in correlating experimental results in the field of ionic conductivity and scavenging.

I wish to thank Mr C. R. V. Reed for help with calculations, Drs A. O. Allen, J. Bednar, A. Hummel, Professor J. L. Magee and Professor R. L. Platzman for permission to reproduce tables and figures.

References

Allen, A. O., 1961, *The Radiation Chemistry of Water and Aqueous Solutions*, (London: D. van Nostrand), p. 64, figures 5, 6.

Brocklehurst, B., and Russell, R. D., 1967, *Nature*, **213**, 65.

Buchanan, J. W., and Williams, T. F., 1966, *J. chem. Phys.*, **44**, 4377.

Bullock, A. T., Griffiths, W. E., and Sutcliffe, L. H., 1967, *Trans. Faraday Soc.*, **63**, 1846.

Burch, P. R. J., 1957a, *Radiation Research*, **6**, 289.

—— 1957b, *Br. J. Radiol.*, **30**, 524.

Burns, W. G., and Marsh, W. R., 1968, *Trans. Faraday Soc.*, **64**, 2375.

Burns, W. G., Marsh, W. T., and Reed, C. R. V., 1968, *Nature*, **218**, 867.

BURNS, W. G., 1968, *J. chem. Phys.*, **48**, 1876.
CAPELLOS, C., and ALLEN, A. O., 1968, *Science*, **160**, 302.
COLLINSON, E., DAINTON, F. S., and KROH, J., 1961, *Proc. R. Soc.* A, **265**, 407.
COOPER, R., and THOMAS, J. K., 1968, *Advances in Chemistry* (Series 82), *Radiation Chemistry* II (Washington: American Chemical Society), p. 351.
CUNDALL, R. B., and GRIFFITH, P. A., 1963, *Discuss. Faraday Soc.*, **36**, 111.
CUNNINGHAM, J., 1967, *J. phys. Chem.*, **71**, 1967.
DEROULEDE, A., KIEFFER, F., and MAGAT, M., 1968, *Advances in Chemistry* (Series 82), *Radiation Chemistry* II (Washington: American Chemical Society), p. 401.
DYNE, P. J., and MILLER, O. A., 1965, *Can. J. Chem.*, **43**, 2696.
FREEMAN, G. R., and FAYADH, J. M., 1965, *J. chem. Phys.*, **43**, 86.
FREEMAN, G. R., 1967, *J. chem. Phys.*, **46**, 2822.
—— 1968, *Advances in Chemistry* (Series 82), *Radiation Chemistry* II (Washington: American Chemical Society), p. 339.
GANGULY, A. K., and MAGEE, J. L., 1956, *J. chem. Phys.*, **25**, 129.
HAGEMANN, R. J., and SCHWARZ, H. A., 1967, *J. phys. Chem.*, **71**, 2694.
HENTZ, R. R., PETERSON, D. B., SRIVASTAVA, S. B., BARZYNSKI, H. F., and BURTON, M., 1966, *J. phys. Chem.*, **70**, 2362.
HUMMEL, A., and ALLEN, A. O., 1966, *J. chem. Phys.*, **44**, 3426.
HUMMEL, A., ALLEN, A. O., and WATSON, F. H., 1966, *J. chem. Phys.*, **44**, 3431
HUMMEL, A., 1967, Thesis, Free University, Amsterdam.
—— 1968a, *J. chem. Phys.*, **48**, 3268.
—— 1968b, *J. chem. Phys.*, **49**, 4840.
KUPPERMAN, A., 1967, *Radiation Research*, Ed. G. Silini (Amsterdam: North Holland), p. 212.
LUDWIG, P. K., and BURTON, M., 1968, *Advances in Chemistry* (Series 82), *Radiation Chemistry* (Washington: American Chemical Society), p. 542.
LUDWIG, P. K., and HUQUE, M. M., 1968, *J. chem. Phys.*, **49**, 805.
MONCHICK, L., 1956, *J. chem. Phys.*, **24**, 381.
MOZUMDER, A., 1968, *J. chem. Phys.*, **48**, 1659.
MOZUMDER, A., CHATTERJEE, A., and MAGEE, J. L., 1968, *Advances in Chemistry* (Series 81), *Radiation Chemistry* I (Washington: American Chemical Society), p. 27.
MOZUMDER, A., and MAGEE, J. L., 1966a, *Radiat. Res.*, **28**, 215 and **28**, 203.
—— 1966b, *J. chem. Phys.*, **45**, 3332.
—— 1967, *J. chem. Phys.*, **47**, 939.
PLATZMAN, R. L., 1958, *Radiation Biology and Medicine*, Ed. W. D. Claus (Reading, Mass: Addison-Wesley), p. 15.
—— 1961, *Int. J. App. Rad. and Isotopes*, **10**, 116.
—— 1967, *Radiation Research* Ed. G. Silini (Amsterdam: North Holland), p. 20.
RZAD, S. J., and WARMAN, J. M., 1968, *J. chem. Phys.*, **49**, 2861.
SAGERT, N. H., and BLAIR, A. S., 1967, *Can. J. Chem.*, **45**, 1351.
SAMUEL, A. H., and MAGEE, J. L., 1953, *J. chem. Phys.*, **21,** 1080.
SANTAR, J., and BEDNAR, J., 1968, *Advances in Chemistry* (Series 81), *Radiation Chemistry* I (Washington: American Chemical Society), p. 523.
—— 1969, *Int. J. Radiat. phys. Chem.*, **1**, 133.
SATO, S., TERAO, T., KONO, M., and SHIDA, S., 1967, *Bull. Chem. Soc. Japan*, **40**, 1818.
SCALA, A. A., LIAS, S. G., and AUSLOOS, P., 1966, *J. Amer. Chem. Soc.*, **88**, 5701.
SCHMIDT, W. F., and ALLEN, A. O., 1968a, *Science*, **160**, 300.
—— 1968b, *J. phys. Chem.*, **11**, 3730.
SCHOLES, G., and SIMIC, M., 1964, *Nature*, **202**, 895.
SCHULER, R. H., and ALLEN, A. O., 1957, *J. Amer. Chem. Soc.*, **79**, 1565.
SHERMAN, W. V., 1966, *J. Chem. Soc. A*, 599.
SMITH, D. R., and PIERONI, J. J., 1965, *Can. J. Chem.*, **43**, 876
SPENCER, L. V., and FANO, U., 1954, *Phys. Rev.*, **93**, 1172.
VACEK, K., and SCHULTE-FROHLINDE, D., 1968, *J. chem. Phys.*, **72**, 2686.
WARMAN, J. M., ASMUS, K.-D., and SCHULER, R. H., 1968, *Advances in Chemistry* (Series 82), *Radiation Chemistry* II (Washington: American Chemical Society), p. 25.
WILLARD, J. E., 1968, *Advances in Chemistry* (Series 82), *Radiation Chemistry* II (Washington: American Chemical Society), p. 540.
WILLIAMS, T. F., 1964, *J. Amer. Chem. Soc.*, **86**, 3954.

Similarity of secondary electron tracks

D. HARDER

Physikalisches Institut der Universität Würzburg, Germany

Abstract. The scaling law or similarity rule for multiple scattering and slowing down of electrons considerably simplifies the physics of secondary electron tracks. The Lewis–Spencer transport theory shows that the directional and spectral distribution of the electron fluence behaves as a universal function if the space-coordinates and the residual path length are scaled with reference to the average path length. The way the universal function changes from high to medium energy electrons is in agreement with experiments. Screening effects reducing the curvature of electron tracks disturb the scaling behaviour at low energies. For very low energies another rise of the curvature is expected.

1. Introduction

One feature of the geometrical track structure of any fast particle is the behaviour of the secondary electrons. Relevant physical information is the expected value of the directional and spectral distribution of the fluence of secondary electrons, $f(\boldsymbol{x}, T, \boldsymbol{\theta})$, at any point $\boldsymbol{x}$ on or off the geometrical track line. The quantities in bold type are the space vector and the unit vector of direction, while T is the kinetic energy.

This information may be obtained by an integration over the starting points, initial energies and initial directions of all the secondary electrons provided that the solution of the penetration problem of electrons (multiple scattering and slowing down) is sufficiently known in the region of interest.

The purpose of this communication is not to present results of an integration of this kind, but to discuss a general regularity governing the combined scattering and slowing down phenomenon which is known as the 'scaling law'. This regularity, well known from medium energy electron beam physics, has already been discussed from the theoretical point of view by Blanchard and Fano (1951) and by Spencer (1955, 1959) and was extensively used by the latter as an aid to calculation.

As an illustrative example of the scaling law we take the statement that the shape of the transmission curve should not depend on electron energy if plotted as a function of the fractional range (ratio of target thickness to average path length or true range). This behaviour is clearly illustrated in figure 1. On the other hand, the region of validity of this rule is limited, as figure 2 shows for the high energy region. Limitations may also occur in the low energy region.

The derivation of the scaling law, its modifications in different regions of initial energy, and its range of validity will be discussed here.

2. Derivation of the scaling law

The scaling properties may be derived theoretically from the transport theory of electron scattering and slowing down. Lewis (1950) finds that, in the 'continuous slowing-down' approximation, the transport equation may be shaped into a system

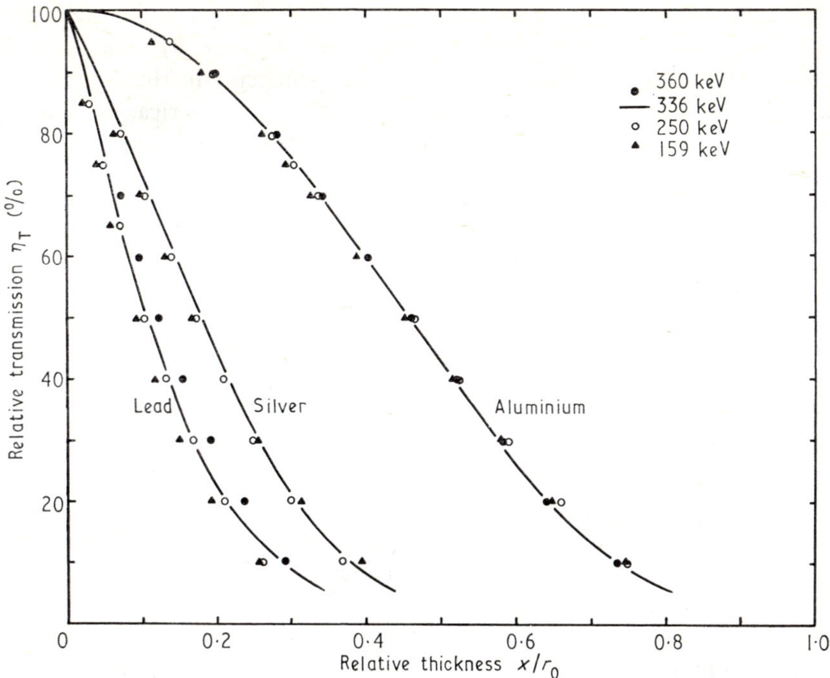

Figure 1. Relative transmission curves for medium energy electrons in aluminium, silver, and lead as a function of the relative target thickness. After Seliger (1955). Values of r_0 were taken from Spencer (1955).

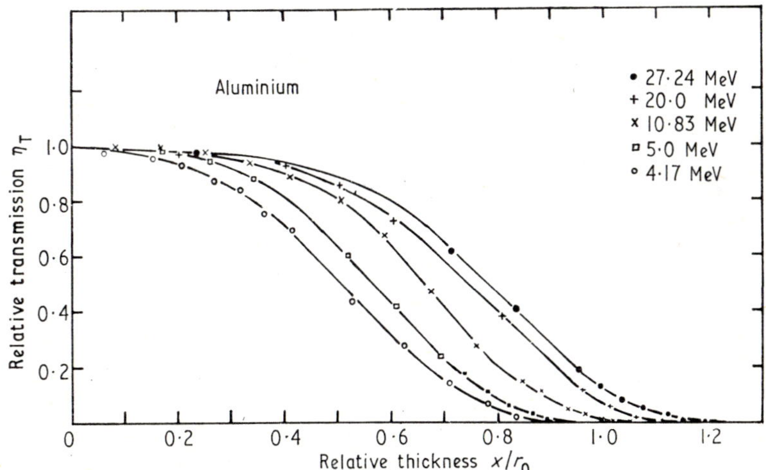

Figure 2. Relative transmission curves for high energy electrons in aluminium as a function of the relative target thickness. After Harder and Poschet (1967). Values of r_0 were taken from Berger and Seltzer (1964).

of coupled differential equations of the following type

$$D_1\{f_{\lambda\mu}(r, x)\} = N\sigma_l(r) f_{lm}(r, x) \tag{1}$$

Here r is the residual path length, which serves as a measure of the kinetic energy in the continuous slowing-down approximation, and D_1 is a linear differential operator of the first order in r and in the space-coordinates. The $f_{\lambda\mu}$ and f_{lm} are the coefficients in the expansion of the unknown distribution function $f(\boldsymbol{x}, r, \boldsymbol{\theta})$ with respect to the normalized surface harmonics, and $\sigma_l(r)$ are the coefficients in the expansion of the nuclear elastic scattering cross section with respect to the spherical harmonics.

If we introduce scaled coordinates $\boldsymbol{\xi}=\boldsymbol{x}/r_0$ and $t=r/r_0$ with r_0=average true range or average path length, equation (1) may be rewritten as

$$D_1\{f_{\lambda\mu}(\boldsymbol{\xi}, t)\}=r_0 N\sigma_l(t) f_{lm}(\boldsymbol{\xi}, t) \tag{2}$$

due to the linearity and first order of the differential operator. It is the quantity $r_0 N\sigma_l(t)$ which contains the whole physical input information except the boundary conditions. N is the number of atoms per unit of volume.

We may therefore immediately give the general formulation of the scaling law: For a set of given scattering media and given initial electron energies, for which the term $r_0 N\sigma_l(t)$ takes the same values irrespective of l and t, the solutions of the transport equation will be the same in scaled coordinates, provided that the boundary conditions also agree in scaled coordinates. The actual coordinates will then be in the proportion of the average path lengths. The cases where the provision concerning $r_0 N\sigma_l(t)$ holds, will be shown up by a closer discussion of this term.

First, it is obvious that N drops out since r_0 is inversely proportional to N. Moreover, an illustrative presentation of the whole term has been found by Blanchard (viz. Spencer (1955))

$$r_0 N\sigma_l(t)\sim\frac{\alpha d_l}{t(t+\alpha)} \tag{3}$$

where

$$\alpha=\left(\frac{3N_A\phi_0 ZB}{4A}\right)^{-1}\frac{4}{r_0} \tag{4}$$

$$d_l=\frac{1}{4}(Z+1)\frac{1}{B}C_l(Z, T) \tag{5}$$

$$\phi_0=\frac{8\pi e^4}{3(m_e c^2)^2} \tag{6}$$

T is the kinetic energy divided by $m_e c^2$. The meaning of B (stopping number) and C_l follows from*

$$r_0=\left(\frac{3N_A\phi_0 ZB}{4A}\right)^{-1}\frac{T_0^2}{(T_0+1)} \tag{7}$$

$$N\sigma_l(T)=\frac{3}{4}(Z+1)\left(N_A\phi_0\frac{Z}{A}\right)C_l\frac{(T+1)^2}{T^2(T+2)^2} \tag{8}$$

The behaviour of equation (3) may easily be discussed because it varies mainly with α, whereas d_l, as the quotient of two logarithmic functions of the electron energy, remains nearly constant.

* The nomenclature follows the original publication by Spencer (1955). A modified presentation, where B is not treated as a constant in calculating r_0, is given by Harder (1967).

2.1 *Case* $\alpha \gg 1$

For initial energies below 0·9 $m_e c^2$, t may be neglected compared with α, so that equation (3) becomes

$$r_0 N \sigma_l(t) \sim \frac{d_l}{t} = \frac{(Z+1)\, C_l}{4Bt} \tag{9}$$

Identical solutions of the transport equation will therefore be obtained for media of the same atomic number irrespective of the initial energy of the electrons. This is the well-known scaling of the penetration phenomena in the medium energy region, for which figure 1 was the example.

2.2 *Case* $\alpha \ll 1$

For initial energies higher than 40 $m_e c^2$, α may be neglected compared with t, so that equation (3) becomes

$$r_0 N \sigma_l(t) \sim \frac{\alpha d_l}{t^2} = \frac{(Z+1)\, C_l}{Bt^2 T_0} \tag{10}$$

Identical solutions of the transport equation will now be obtained for cases which agree in the quotient $(Z+1)/T_0$. Due to the t^2 in the denominator, the type of solution should be different from the solutions of the transport equation in case $\alpha \gg 1$.

This explains the absence of scaling in figure 2. We have tried to prove the predictions of equation (10), and the result is shown in figure 3. This example and

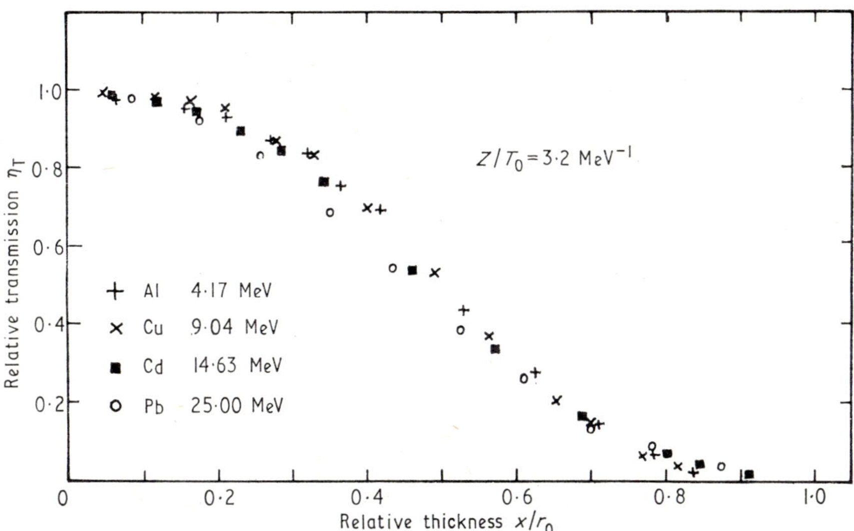

Figure 3. Relative transmission curves for high energy electrons in aluminium, copper, cadmium, and lead as a function of the relative target thickness. After Harder and Poschet (1967). With r_0 from Berger and Seltzer (1964).

a number of others (Harder 1967, 1969) have shown that in the high energy region the scaling of penetration phenomena occurs for sets of experiments which agree in $(Z+1)/T_0$ or Z/T_0. (It is a matter of the C_l whether we write Z or $Z+1$.) The first to predict the high energy behaviour of a scaled quantity—the backscattering coefficient—was Frank (1959).

3. Limits of validity

It will be of practical as well as of theoretical interest to investigate the width of the energy regions in which scaling of the two types indicated does occur. Since the shape of the transmission curve should depend solely on Z in the case of energies below $0{\cdot}9\ m_e c^2$ and solely on $(Z+1)/T_0$ or Z/T_0 in the high energy case, we can take one parameter of this shape, viz. the 'detour factor', as a criterion for this general examination.

Figure 4 shows the *high energy* behaviour of the detour factor. In the region investigated (Harder and Poschet 1967) (from 4 to 30 MeV and from $Z=6$ to $Z=82$) there is a fair agreement of the detour factor with a single function of Z/T_0. This was found for the two detour factors which can be calculated from the practical range and from the average range.

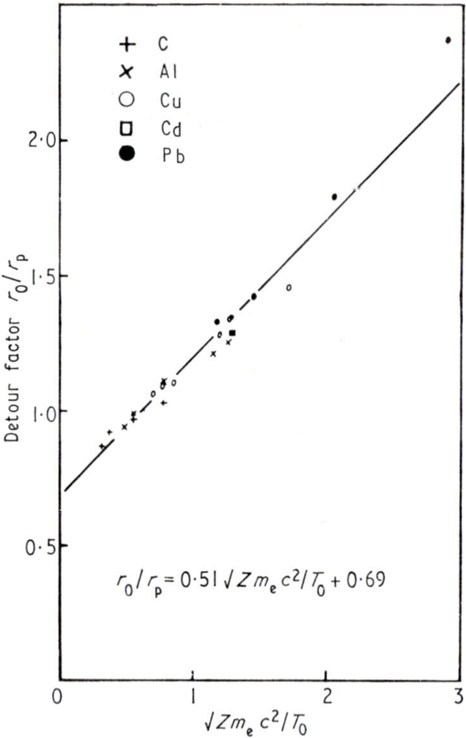

Figure 4. Detour factor (average path length divided by practical range) for high energy electrons as function of $Zm_e c^2/T_0$. (T_0 = Initial kinetic energy.) After Harder and Poschet (1967).

Figure 5 summarizes experimental results over four decades of the initial energy. The behaviour predicted by equation (9) is actually exhibited below $0{\cdot}9\ m_e c^2$. But at low energies an unexpected fall-off of the detour factor sets in. Therefore a lower limit to the region where equation (9) is valid does exist, so that the well-known type of scaling for media of the same atomic number irrespective of electron energy turns out to be the type of scaling which is characteristic of the *medium energy* region.

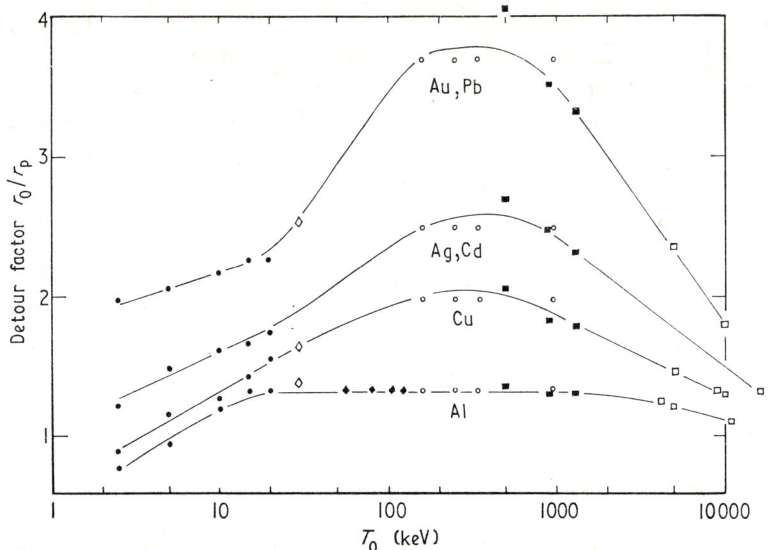

Figure 5. Detour factor (average path length divided by practical range) for electrons as function of the initial energy and the atomic number of the scatterer. ● Coslett and Thomas (1964), ○ Seliger (1955), ◇ Schonland (1923), ■ Fleeman (1954), ♦ Huffman *et al.* (1957), □ Harder and Poschet (1967).

In order to understand the behaviour of the detour factor in the *low energy* region, we refer to the fact that the detour factor may be approximated by

$$r_0/\lambda_1 = r_0 N \sigma_1 \qquad (t=1) \tag{11}$$

where λ_1 is the transport mean free path at the initial energy. Together with equation (9) this immediately shows that an increase of the stopping number B or a decrease of the scattering factor C_l would produce a decrease of the detour factor. Since B will decrease rather than increase with falling energies, a pronounced decrease of the C_l must be the reason for the falling detour factor.

This decrease of the C_l is well known from scattering experiments as a consequence of the screening of the nuclear Coulomb potential by the atomic electrons. Screening becomes important when the electron wavelength becomes comparable to the Thomas–Fermi atom radius.

For the track structure, decreasing detour factors indicate that the tracks become more and more straight with decreasing energy. Detour factors in the medium energy region, if applied to the low energy region, may lead to errors in the interpretation of track structure. The density of the medium also affects the amount of screening (Fink and Kessler 1966). At energies below 100 eV the picture may change again due to the sharp fall-off of the stopping number B and to resonances in elastic scattering. The detour factor will then increase again.

It would be an interesting task to investigate the type of scaling law which holds for the low energy region. Machov (1960) has proposed an empirical formula summarizing low energy transmission data (viz. Schumacher 1965). The shapes of the transmission curves measured by Cole (1969) indicate that for collodion as scattering medium the detour factor is a constant in the region 0·5 to 18 keV initial energy, whereas it rises again for energies below 200 eV.

4. Summary

The scaling law for electron penetration phenomena was derived from the Lewis–Spencer transport theory. The conditions pertaining to the scattering media and the range of electron energies, which form a set characterized by scaled transmission curves, are different in the high, medium and low energy regions. The reduction of scattering effects at low energies due to the screening of the nuclear Coulomb potential has a remarkable effect on the track structure of electrons. There is an indication that at very low energies another increase of the track detours occurs.

References

Berger, M. J., and Seltzer, S. M., 1964, *NASA Rep.* SP-3012.
Blanchard, C. H., and Fano, U., 1951, *Phys. Rev.* **82**, 767.
Cole, A., 1969, *Radiat. Res.*, **38**, 7.
Coslett, V. E., and Thomas, R. N., 1964, *J. appl. Phys.*, **15**, 128.
Fleeman, J., 1954, *Natn. Bur. Stand. Circ.*, **527**, 91.
Fink, M., and Kessler, J., 1966, *Z. Phys.*, **196**, 1.
Frank, H., 1959, *Z. Naturf.*, **4A**, 247.
Harder, D., 1967, *Argonne National Laboratory*, ANL-TRANS-608.
—— 1970, *Proc. Second Symp. on Microdosimetry*, 1969, *Stresa, Italy*, Ed. H. G. Ebert, Euratom, Brussels.
Harder, D., and Poschett, G., 1967, *Phys. Lett.*, **24B**, 519.
Huffman, F. N., Cheka, J. S., Saunders, B. G., Ritchie, R. H., and Birkhoff, R. D., 1957, *Phys. Rev.*, **106**, 435.
Lewis, H. W., 1950, *Phys. Rev.*, **78**, 526.
Machov, A. F., 1960, *Fiz. Tver. Tela*, **2**, 2161.
Schonland, F. J., 1923, *Proc. R. Soc. Lond.*, **104**, 235
—— 1925, *Proc. R. Soc. Lond.*, **108**, 187.
Schumacher, B. W., 1965, *Proc. Int. Conf. on Electron and Ion Beam Science and Technology*, 1964 (New York: Wiley & Sons).
Seliger, H. H., 1955, *Phys. Rev.*, **100**, 1029.
Spencer, L. V., 1955, *Phys. Rev.*, **98**, 1597.
—— 1959, *Natn. Bur. Stand. Monogr.*, 1.

1970 CH. PART. TR. SOL. LIQ.

Charged particle tracks in crystalline solids

D. V. MORGAN

Nuclear Physics Division, AERE, Harwell, Berks.

Abstract. When energetic charged particles pass through some crystalline solids, a permanent trail of damage delineates their path. In this communication two problems are considered, firstly the track storing property of a crystal is related to its electrical conductivity, and secondly, the intermittent nature of tracks observed in MoS_2 is related to changes in the energy loss of the charged particles.

1. Introduction

The tracks produced by heavy charged particles in crystalline matter, which are observable in the electron microscope, fall into two broad classes (Bowden and Chadderton 1962, Chadderton 1964, Chadderton *et al.* 1966). In the thinner crystals ($\lesssim 100$ Å) each track is generally revealed as a light area on a dark background and is a line of enhanced electron transmission owing to the removal of some of the crystal (figure 1). For thicker crystals, however, the tracks are dark on a light background. This is due to a rearrangement of the crystal lattice leading to the strain observed by the electron microscope (figure 2). This communication will be concerned with the latter. Experiments are described which consider two problems, firstly, an investigation into the relationship between electrical conductivity of a crystal and its track storing properties; secondly, an attempt has been made to see if the intermittency of the particle tracks observed in some solids (figure 2) can be related to local changes in the rate of energy loss of the charged particle.

2. Experimental

Molybdenum disulphide is one of a very large family of semiconductors which crystallize in a layer structure. They each contain basal planes with strong bonding and high atomic density, separated by relatively large interplanar channels, where the bonding is mainly due to the weak Van der Waals forces, figure 3. The structure and physical properties of these materials do not differ appreciably, whereas their conductivities do, ranging from 10^{-2} $(\Omega\ \mathrm{cm})^{-1}$ in MoS_2 to $3{\cdot}84 \times 10^3$ $(\Omega\ \mathrm{cm})^{-1}$ in NbT_2. The materials are therefore suitable for an investigation of the role of conductivity in track registration. Alternatively, the conductivity of any one of these semiconductors may be varied, by the addition of a small number of impurity atoms which do not disturb the parent lattice. This provides an additional and independent approach to the problem.

In the first experiment fission fragment irradiation was carried out on a selection of twelve crystals from the MoS_2 family (table). The conductivity varied over a range of 10^5 $(\Omega\ \mathrm{cm})^{-1}$ with MoS_2 at one extreme and $NbTe_2$ at the other. In the second experiment WSe_2 and WTe_2 were doped with between (0–7)% weight niobium. By the addition of such small quantities of Nb impurity the conductivity of these crystals was increased. Consequently crystals were grown whose conductivity varied from 10^{-1}–10^3 $(\Omega\ \mathrm{cm})^{-1}$.

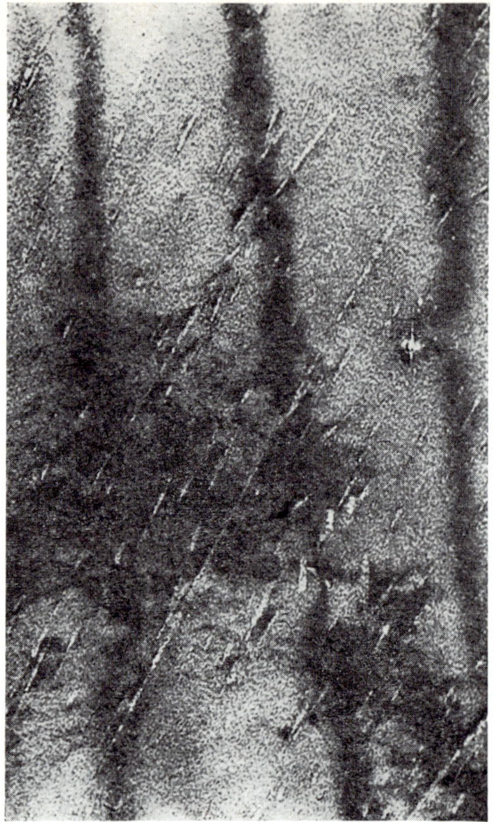

Figure 1. Tracks in a crystal of molybdenum trioxide formed by ^{12}C ions. These tracks are 50–80 Å in width (after Torrens, unpublished).

Table

Compound	Distance between layers (Å)	Electrical conductivity at 25 °C $(\Omega\ cm)^{-1}$	Damage observed by the electron microscope
MoS_2	6·13	10^{-2}	Clearly defined track
$\alpha MoTe_2$	6·97	$1 \cdot 23 \times 10^{-1}$	Clearly defined track
WSe_2	6·47	2×10^{-1}	Tracks
$MoSe_2$	6·45	$3 \cdot 33 \times 10^{-1}$	Tracks
WS_2	6·18	1	Tracks
WTe_2	6·27	$3 \cdot 70 \times 10^{2}$	Tracks
$TiSe_2$	6·004	5×10^{2}	No visible tracks
$\beta MoTe_2$	6·43	1×10^{3}	Tracks
$TaSe_2$	6·37	$2 \cdot 5 \times 10^{3}$	Occasional ill-defined surface track
$NbSe_2$	6·50	$2 \cdot 86 \times 10^{3}$	No visible tracks
$TaTe_2$	6·69	$2 \cdot 94 \times 10^{3}$	Occasional ill-defined surface track
$NbTe_2$	6·26	$3 \cdot 84 \times 10^{3}$	No visible tracks

Thin films suitable for transmission electron microscopy were prepared by successively cleaving crystals along their basal planes using cellulose adhesive tape until they were reduced to a suitable thickness. They were then mounted on electron microscope grids and irradiated with fission fragments by placing the grids in contact with natural uranium foil $\sim 10\ \mu m$ thick and placing them for a specified time in the BEPO reactor

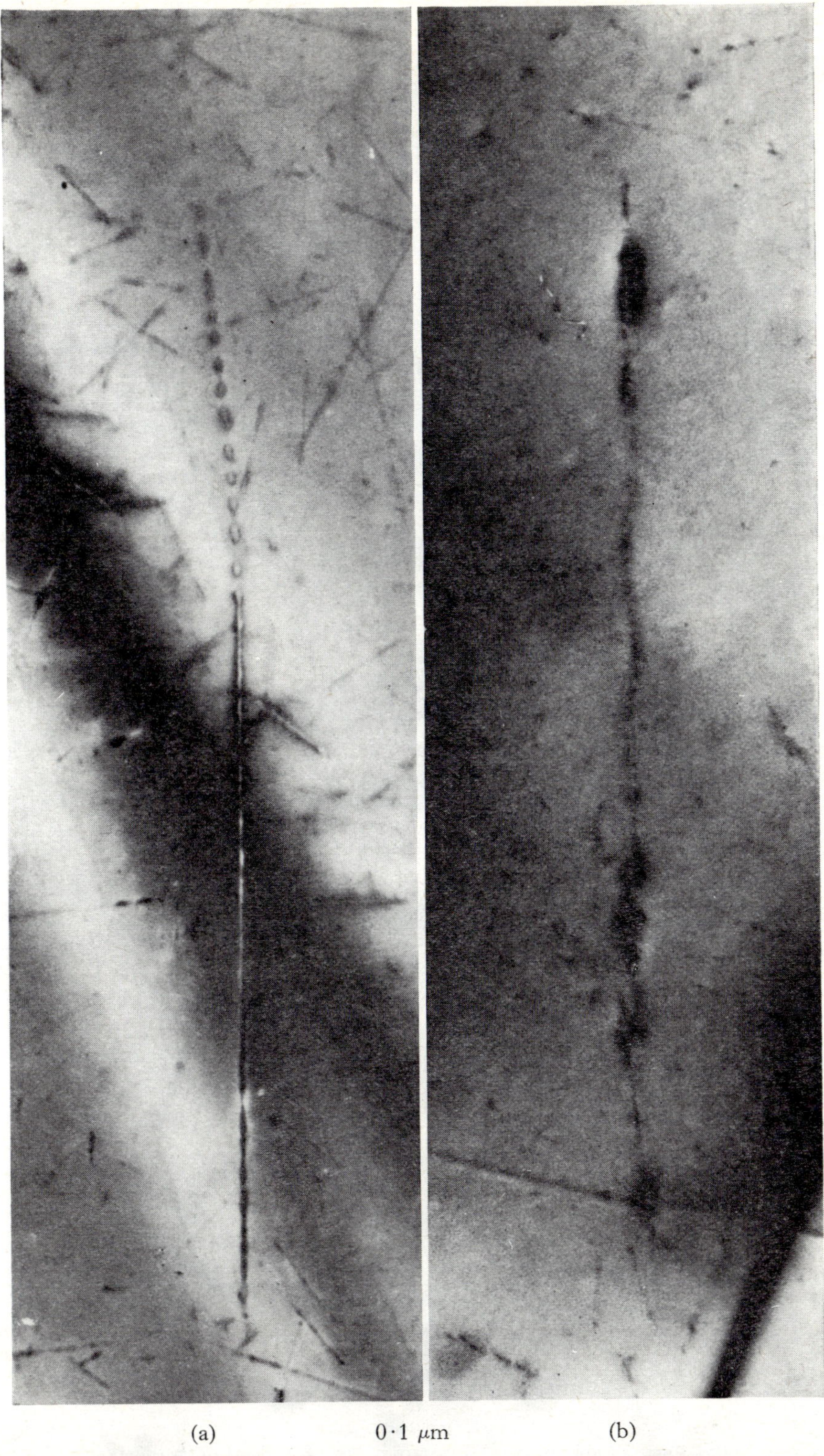

(a) 0·1 μm (b)

Figure 2. Electron micrograph of dark tracks: (a) a large angle collision in MoS_2 showing the change in intermittency after the collision, (b) an intermittent track where the spots are just resolved.

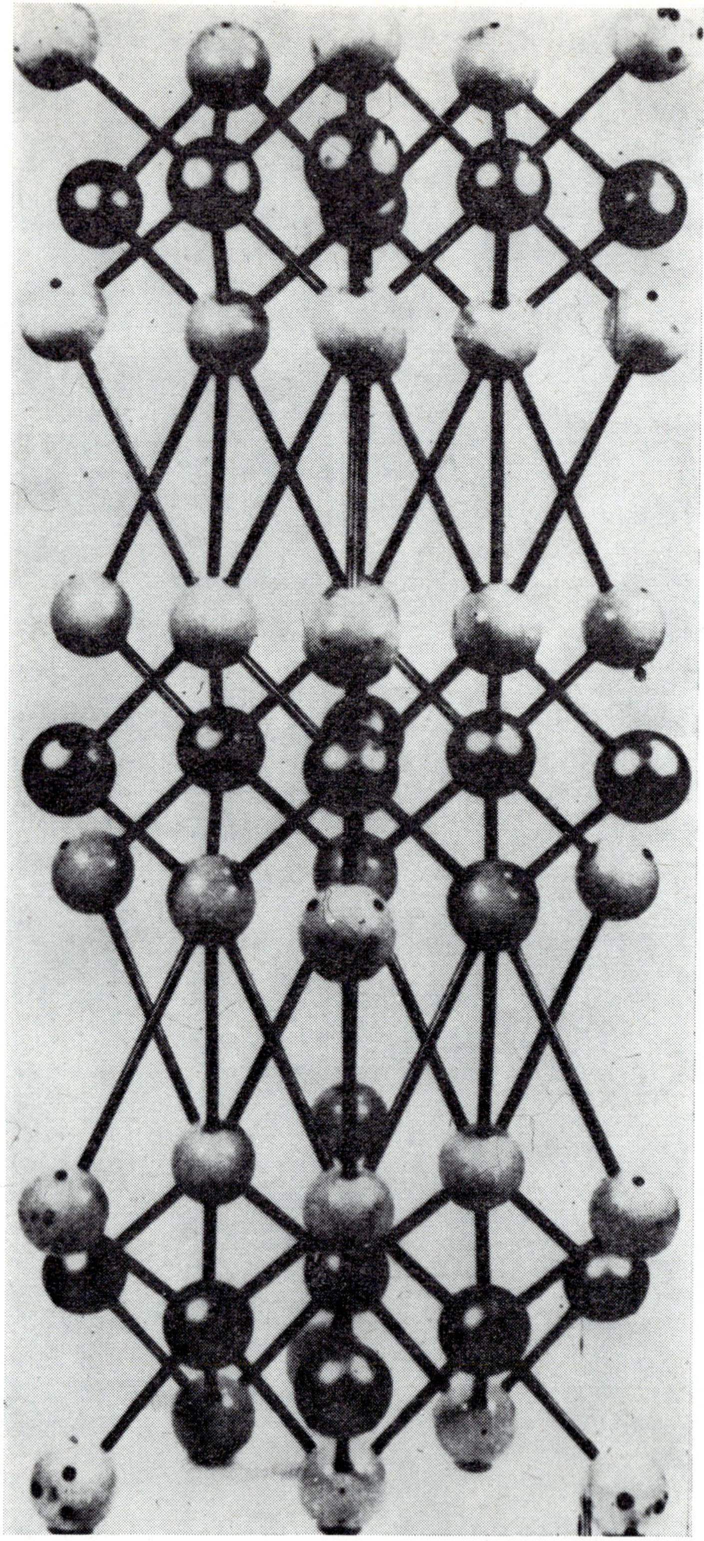

Figure 3. A model of the crystal structure of MoS_2. The dark balls are molybdenum and the light ones the sulphur.

at Harwell (Chadderton 1962). Post irradiation observations were carried out in an AEI EM6 electron microscope working at 100 keV.

3. Results

The results of the investigation are illustrated in the table. In the first experiment tracks were formed only in the materials with the smaller conductivities. Between tungsten ditelluride $3{\cdot}7 \times 10^2\ (\Omega\ \text{cm})^{-1}$ and tantalum diselenide $2{\cdot}5 \times 10^3\ (\Omega\ \text{cm})^{-1}$ track registration presumably ceased—the only visible damage was the occasional surface track-like feature (i.e. absorption contrast). For this family of semiconducting compounds, tracks are evidently only formed when the conductivity is less than $2 \times 10^3\ (\Omega\ \text{cm})^{-1}$.

In the second experiment, using selectively doped crystals, the same general picture emerged. Fission tracks were produced in all the crystals except the WTe_2–7%Nb. However, in these crystals, tracks were not completely absent—a very small density of ill-defined tracks was observed in some of the crystals analysed. From the low efficiency of track registration and the relatively weak contrast from these tracks, it seemed clear that they had been converted from track storing to non-storing material, by the change in electrical conductivity—all other factors being environmentally similar. This result is in close agreement with the first experiment since it is only in the case of the 7% doped crystals that the conductivity was increased to above $10^3\ (\Omega\ \text{cm})^{-1}$. It was not possible to determine the conductivity accurately and the figure $10^3\ (\Omega\ \text{cm})^{-1}$ must only be regarded as an order of magnitude estimate. Further, it was also observed that in all these materials in which fission tracks were registered, the latter were of the same intermittent character as those found in MoS_2.

4. Discussion

A fission fragment passing through a crystal loses its energy in collisions with both the nuclei and the electrons of the crystal. The majority of the energy lost is given to the electrons, and this has given rise to two postulated mechanisms of track formation: firstly, the modified thermal spike (Chadderton *et al.* 1966, Morgan 1966) and secondly, the ion explosion spike (Noggle and Stiegler 1961, Fleischer Price and Walker 1965, Chadderton Morgan and Torrens 1966). Both these models predict that fission tracks are formed only if the conductivity is less than some critical value, which is characteristic of the material considered, thereby establishing a dependence of track registration upon conductivity of the type we have reported here. Since both models predict a similar dependence upon conductivity, these results alone do not allow more specific conclusions to be drawn which favour either model.

4.1. *Intermittent tracks*

When a fission fragment passes through a crystal of MoS_2 at some angle α to the basal plane, it will cross successive basal sandwiches of high atomic density (A–B, figure 4*a*) and through channel regions of low atomic density (B–C, figure 4*a*). Goland (1965) suggests that the rate of energy loss $(-\mathrm{d}E/\mathrm{d}x)$ will be greater along A–B than it will be along B–C; this is essentially a crystallographic energy loss effect similar to channelling between the layers. A cursory estimate of the energy loss to electrons $(-\mathrm{d}E/\mathrm{d}x)$ for the region AB and BC for a heavy fission fragment (Mass 140 and energy 60 MeV) travelling through MoS_2 yield $(-\mathrm{d}E/\mathrm{d}x)\ \text{AB} \simeq 4000$ eV Å^{-1} and $(-\mathrm{d}E/\mathrm{d}x)\text{BC} \simeq 0$. Intermittent tracks may be formed by either, the thermal spike or, the ion explosion spike provided that

$$(-\mathrm{d}E/\mathrm{d}x)\mathrm{AB} > (-\mathrm{d}E/\mathrm{d}x)_{\mathrm{crit}} > (-\mathrm{d}E/\mathrm{d}x)\mathrm{BC}$$

where $(\mathrm{d}E/\mathrm{d}x)_{\mathrm{crit}}$ is the critical rate of energy loss required for track formation. This condition can be satisfied for MoS_2. Clearly, the angle α must be small enough for the defect regions to be well separated and must be greater than a certain critical value, so that more than one defect region is formed. As α increases the defect regions will merge to form a continuous track.

Consider an intermittent track making an angle α to the basal plane. Provided that the crystal is perfectly flat, the 'intermittency' d will be a constant, and the product of the total number of spots N with the interlayer distance l will equal the specimen thickness $t(N \gg 1)$ given by

$$Nl \simeq t \tag{1}$$

Thus if a permanent visible defect is produced every time the fragment passes through a dense layer, then the measurement of the crystal thickness, using (1), should agree with the thickness measured by some other independent method. Most crystals, however, are not planar. They suffer considerable and random bending when they are deposited on the microscope grids. Hence, for a particular track the intermittency may vary along its length, because α is now a function of position in the crystal. This may account for the gradual changes in track intermittency observed in bent crystals. An extreme example of this occurs when a fragment undergoes a close encounter with a lattice atom, where α changes abruptly, and when d will do the same. This was again found to be the case in the analysis of a large number of individual tracks. Figure 2*a* provides an example of this.

Consider a number of fission particle tracks in a relatively flat region of crystal. From similar triangles (figures 4*a* and 4*b*))

$$R = (t/l)d \tag{2}$$

where R = length of track; d = average spacing between spots (d is only constant if α is a constant). If α is small (i.e. R is large), then d is large and defective regions are well separated and easily resolved. However, as R decreases, d decreases and the individual regions get closer together. Their associated elastic strain fields begin to interact

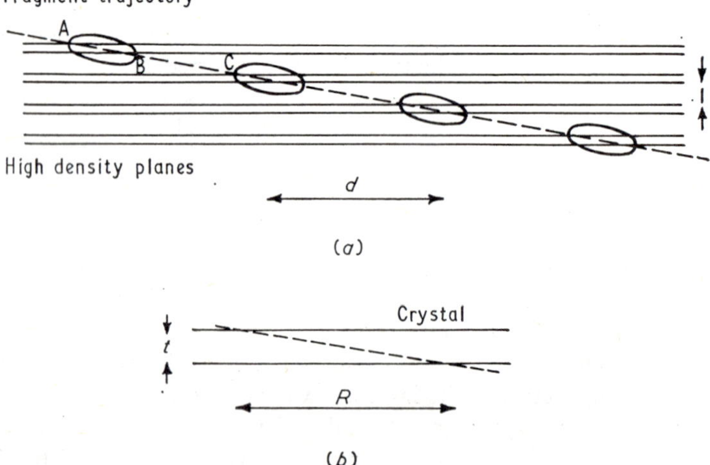

Figure 4. Model for the formation of intermittent tracks in layer structures.

giving rise to a continuous line of diffraction contrast. After careful analysis of individual small tracks this fine structure has been observed even for tracks which initially seemed continuous (figure 2*b*). However, provided R is sufficiently large for individual dots to be resolved equation (2) is obeyed.

In order to prove equation (1) a value for N is required. A thin crystal of molybdenum disulphide, which contained a large number of intermittent tracks suitable for analysis, was therefore selected. If such a crystal is of constant thickness (as might reasonably be expected for this type of cleaved crystal), then the independent values of N calculated from each track should be consistent with each other, and from these it is possible to calculate an average value of t. A value of 420 ± 30 Å was obtained for t in this way, and this compared very favourably with the value of 450 Å calculated using the bent foil technique and observations of the diffraction pattern (Siems Delavignette and Amelinckx 1962). The value is approximately 30 Å too small, but to some extent this smaller value was expected since, as the track approaches or leaves the specimen surface, the damage mechanism is predominantly affected by the close proximity of the surface. The intermittent character is replaced by a short continuous white line and some uncertainty is introduced into the value of N. A similar consistency was achieved in the four specimens analysed, three molybdenum disulphide crystals and one tungsten diselenide.

5. Conclusion

On the basis of these observations the following conclusions may be drawn.

1. Fission fragment irradiation of these semiconducting compounds has established a dependence of track registration on the conductivity of the crystals, when other factors such as structure, point defect mobility and melting temperature do not differ appreciably.

2. The intermittent tracks observed in these crystals are ascribed to the reduced energy loss when the fission fragment travels in the interplanar channel. Analysis of the fine structure of the tracks confirm the presence of one defect aggregate per sandwich plane. The changes in the separation of defect aggregates along the length of a track arise from a localized bending of the irradiated specimens, or from changes of particle direction due to nuclear collision events.

Many different kinds of crystal have now been irradiated with energetic ions and examined in the electron microscope. It is clear that we are moving towards some kind of overall picture of the mechanism involved. However, the major question is still unanswered; namely, what is the relative importance of the ion explosion spike and the modified thermal spike in track formation? This problem is of fundamental importance and it is in this direction that future work should be directed.

References

Bowden, F. P., and Chadderton, L. T., 1962, *Proc. R. Soc.* A, **269**, 143.

Chadderton, L. T., 1962, *Ph.D. Thesis*, University of Cambridge.

Chadderton, L. T., 1964, *Proc. R. Soc.* A, **280**, 110.

Chadderton, L. T., Morgan, D. V., and Torrens, I. McC., 1966, *Solid St. Comm.*, **4**, 391.

Chadderton, L. T., Morgan, D. V., Torrens, I. McC., and Van Vliet, D., 1966, *Phil. Mag.* **13**, 185.

Fleischer, R. L., Price, P. B., and Walker, R. M., 1965, *J. appl. Phys.*, **36**, 3645.

Goland, A. N., 1965, *Brookhaven National Laboratory Rep.*, BNL 9169.

Morgan, D. V., 1966, *Ph.D. Thesis*, University of Cambridge.

Noggle, T. S., and Stiegler, J. O., 1961, *Symp. on Radiation Effects*, American Society for Testing Materials.

Siems, R., Delavignette, P., and Amelinckx, S., 1962, *Phys. St. Solida*, **4**, 421.

The nature and reactions of initial species in radiolysis†

J. K. THOMAS

Argonne National Laboratory, Argonne, Illinois 60439

Abstract. The fundamental processes occurring in the radiolysis of liquids are discussed in terms of the reactions of ions and excited states. Experimental evidence to collaborate these reactions are provided from nanosecond pulse radiolysis studies of water, ethanol, aniline, cyclohexane and benzene. The bearing that the experimental evidence has on the 'spurs' or short tracks is also discussed.

1. Introduction

From cloud chamber experiments and from radiation chemical studies, it is generally assumed that energy is lost heterogeneously to an irradiated medium. At low LET the region of high energy loss is called a spur, while at high LET a string of spurs closely spaced together is called a track. The purpose of the present talk is to present data which has a bearing on spurs at low LET.

Figure 1 outlines the radiation-induced processes which may occur in the radiolysis of liquids. Both excited species and ions may be produced in the primary radiation chemical event. The extent of these processes has been investigated theoretically by the so-called optical approximation (Platzman 1966), which has a bearing on the later discussion. On experimental grounds it is found that ion chemistry predominates in

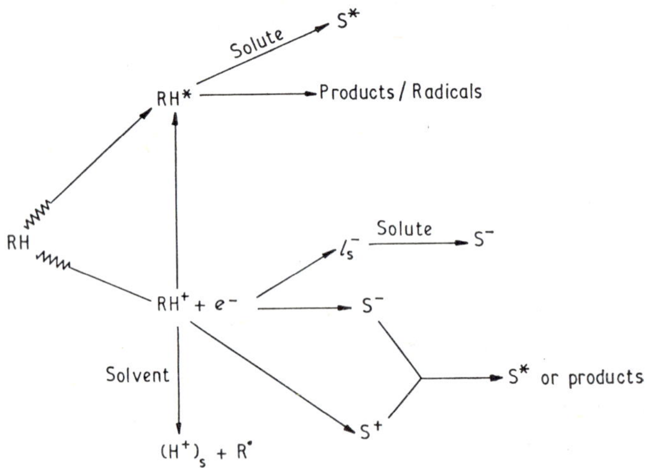

Figure 1. Primary processes in the radiolysis of liquids.

† Work sponsored by the US Atomic Energy Commission.

the radiolysis of polar liquids like water and alcohols, while excited states predominate in the radiolysis of less polar liquids like benzene and cyclohexane. There are exceptions to this broad ruling; for example, the radiolysis of some nonpolar liquids such as tetrahydrofuran produces significant yields of ions and low yields of excited states. The radiolysis of other liquids, aniline and dioxane, produces significant yields of both excited states and ions.

Of course, the mere production of an excited state in radiolysis does not necessitate that a molecule has been directly excited by secondary electrons to this state, as ion-neutralization processes may also lead to excited molecules, as well as to stable products or free radicals. In polar liquids such as water and alcohols, the parent positive ion is rapidly destroyed by an ion–molecule reaction involving the solvent. In water it is estimated that the lifetime of H_2O^+ is less than 10^{-14} s. The fate of the positive ion is therefore forever sealed in these liquids. However, in less polar liquids the solvent cation may have a significant lifetime and produce a rich array of chemical processes, one of which is the formation of solute cations.

The electron which is produced along with the cation, may react with a solute molecule producing an anion, it may recombine with the cation, or it may be solvated by the solvent forming a solvated electron. It appears that solvated electrons are formed in all solvents, and so will be formed in the radiolysis of liquids. The solvated electron again has a wide span of chemical reactions producing solute anions, etc. In the absence of suitable scavengers, some solvated electrons are stable, e.g. in ammonia. To the best of our knowledge, in water and alcohols they slowly decay to give H atoms (Hart Gordon and Fielden 1966).

It is of interest to inquire about the spatial distribution of these species, i.e., the possibility of spurs which lead to a rapid initial decay of the primary species after the initial energy loss, followed by a much slower reaction of the species which have diffused away into the bulk of the liquid. The separation of the electron from the positive ion is of prime importance as ion-pairs are formed in the radiolysis of most liquids leading to heterogeneous ion-kinetics, a process not to be confused with spurs, where more than one ion-pair exists in a discrete and small volume of the solvent.

It may be argued that the above discussion does not fully describe the radiolysis of liquids as the roles of free radicals and other species have not been emphasized. However, the present work is an attempt to describe the primary processes in the radiolysis of liquids, and in many instances the primary yield of free radicals is small compared to the yields of ions and excited states.

2. Experimental

In this work the experimental technique used to investigate the nature and reactions of the primary species in the radiolysis of liquids is that of nanosecond pulse radiolysis. This technique has been described already (Hunt and Thomas 1967, Thomas *et al.* 1968) and only a brief outline will be given here. Pulses of 3 MeV electrons are used to bombard liquids contained in a suprasil cell. The duration of the electron pulse can be as short as 1 ns or as long as 1 μs; a one nanosecond pulse has sufficient intensity to produce 2 μM of hydrated electrons, e_{aq}^-, in degassed water. This is equivalent to a dose of 4×10^{19} eV l^{-1} pulse. The short-lived primary species are observed by fast absorption or emission spectroscopy. This type of apparatus has a response time of 2 ns in the present experiments but has been recently developed to a response as short as 0·1 ns by suitable adaptation (J. K. Thomas, to be published).

2.1. *Water*

The decay of e_{aq}^- in the nanosecond pulse radiolysis of pure water is shown in figure 2. A 3 ns pulse of 3 MeV electrons is used in this experiment and the e_{aq}^- is monitored by the absorption at 600 mμ. The portion XY of the trace represents the I_0 of the analysing light before the radiation pulse appears at Y. An absorption is produced and the trace moves to A corresponding to a concentration of e_{aq}^- of about

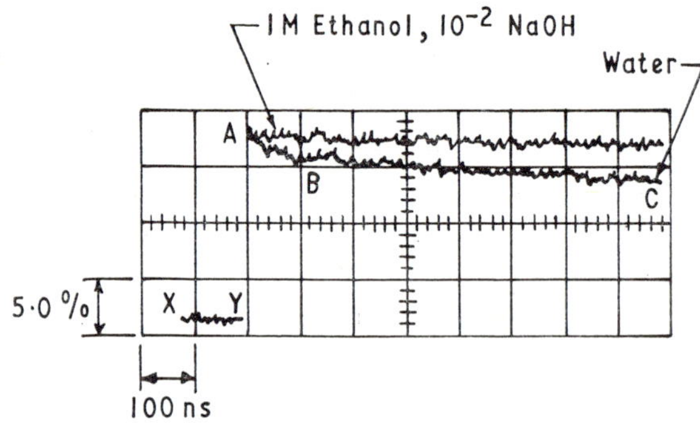

Figure 2. Decay of e_{aq}^- in the pulse radiolysis of water and 10^{-2}M NaOH 1M C_2H_5OH.

6 μM l^{-1}. The decay of e_{aq}^- exhibits an initial rapid portion which is independent of radiation intensity AB followed by a slower decay over the region BC which is intensity-dependent. The decay BC is similar to that observed in μs pulse radiolysis and corresponds to the decay of e_{aq}^- with the other products of the radiolysis reactions (1)–(4).

$$e_{aq}^- + e_{aq}^- \rightarrow H_2 \tag{1}$$

$$e_{aq}^- + H^+ \rightarrow H \tag{2}$$

$$e_{aq}^- + OH^- \rightarrow OH^- \tag{3}$$

$$e_{aq}^- + H_2O_2 \rightarrow H_2O_2. \tag{4}$$

Dosimetry shows that the initial yield of e_{aq}^-, $G(e_{aq}^-)$ is 3·4 immediately after the pulse and that the rapid decay AB corresponds to a $G = 0{\cdot}6$.

Addition of ethyl alcohol up to 1 M and alkali up to 10^{-2} M, causes the rapid portion AB to decrease by about 50%, while the initial yield of $G(e_{aq}^-) = 3{\cdot}4$ is unchanged. The addition of both alcohol and alkali removes the rapid decay while $G(e_{aq}^-)$ still remains at 3·4. This is illustrated in figure 2. The effect of the two scavengers, alkali which removes OH^- via (5), and alcohol which removes OH radical via (6), suggests

$$H^+ + OH^- \rightarrow H_2O \tag{5}$$

$$OH + C_2H_5OH \rightarrow H_2O + C_2H_4OH \tag{6}$$

that the rapid decay AB corresponds to a reaction of e_{aq}^- with H^+ and OH, reactions (2) and (3) and that these reactions correspond to yields of H and OH^- of about 0·3.

If we now expand a small time region such as A to a much more rapid time sweep, we obtain an oscilloscope trace similar to that shown in figure 3. The bottom smooth trace represents a condition with no analysing light in the irradiation cell. Under

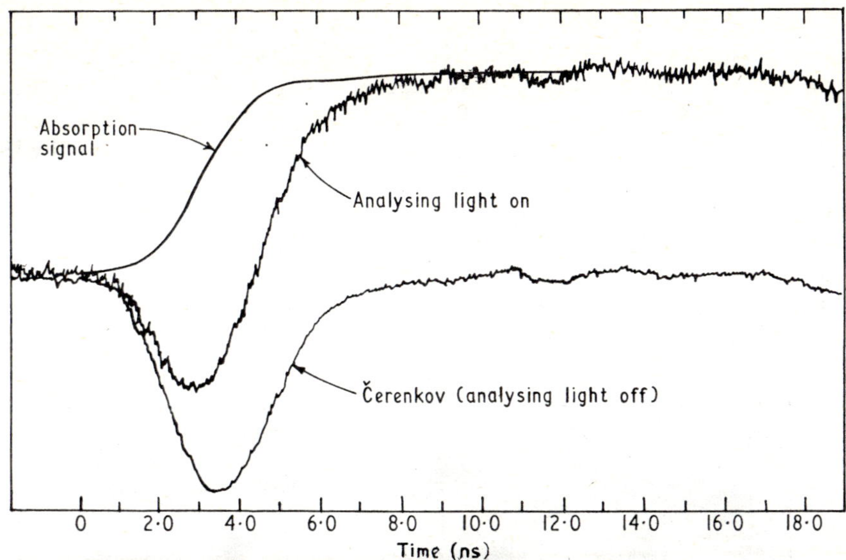

Figure 3. Nanosecond pulse radiolysis of water. 3 ns pulse.

these conditions only the Čerenkov radiation is observed. With the analysing light passing through the cell, a composite signal of Čerenkov radiation and e_{aq}^- absorption is observed. The difference in the curves gives the absorption alone. It can be seen that no rapid initial decay occurs in 1 or 2 ns in contrast to the sharp decay over 100 ns.

Instead of increasing the time response of the apparatus in order to observe 'spur' or rapid reactions, it is possible to slow up the events by cooling the irradiated solution. The reaction rates, diffusion and mobility of the primary species are reduced by this technique. In order to avoid the formation of a crystalline solid on cooling water, 6M sodium hydroxide was used. This forms a transparent glass at low temperatures and provides a convenient method of studying the radiolysis of a system similar to water down to 77 K. At room temperature the irradiation of 6M sodium hydroxide produced kinetic traces similar to those shown in figure 2. However, reducing the temperature to $-80\,°C$ did not affect appreciably the initial yield of electrons or affect the initial rapid decay of e_{aq}^-. It may still be argued that spur reactions involving a $G(e_{aq}^-) > 3{\cdot}4$ are still too rapid for observation in the present apparatus, even at low temperatures.

The rapid decay of e_{aq}^- with a $G(e_{aq}^-) = 0{\cdot}6$ may represent the decay of the tail end of a spur. Spur calculations (Samuel and Magee 1953, Kupperman 1961, H. A. Schwartz, private communication) predict that a $G(e_{aq}^-)$ of 5 to 6 initially exists in the radiolysis of water and that the e_{aq}^- decay to a $G(e_{aq}^-) = 2{\cdot}8$ over about 20 ns. Indeed Schwarz has been able to fit the present data with calculations from his modified spur model. The present data may be interpreted in terms of a spur model where a rapid decay of e_{aq}^- occurs over 0·1 ns followed by a longer tail over tens of ns. On the other hand, the data can be explained in terms of isolated ion-pairs which exhibit rapid neutralization over tens of ns. It is then necessary to explain the molecular products such as H_2 and H_2O_2 in terms of other unknown processes. The subsequent data on the decay of e_{aq}^- in organic liquids provides some explanation for the above discussion of water radiolysis.

2.2. *Alcohols*

The decay kinetics of the solvated electron in ethanol following a 12 ns pulse is shown in figure 4 (Thomas and Bensasson 1967). Again an initial sharp decay is observed followed by a slower decay. The initial $G(e_S^-) = 1{\cdot}5$ and the rapidly decaying portion corresponds to a $G(e_S^-) = 0{\cdot}5$. The addition of alkali, either sodium hydroxide or sodium ethoxide, results in an increase in $G(e_S^-)$ beyond 1·5, while the rapidly decaying portion is removed. This data is similar to that in water, but in the initial rapid decay the e_S^- only decays by reaction with a positive ion. The increase in $G(e_S^-)$ with alkali suggests that a larger yield of electrons may be present at times short compared with 1 ns; the alkali removes positive ions, thereby inhibiting the back reaction of electrons to form neutral species.

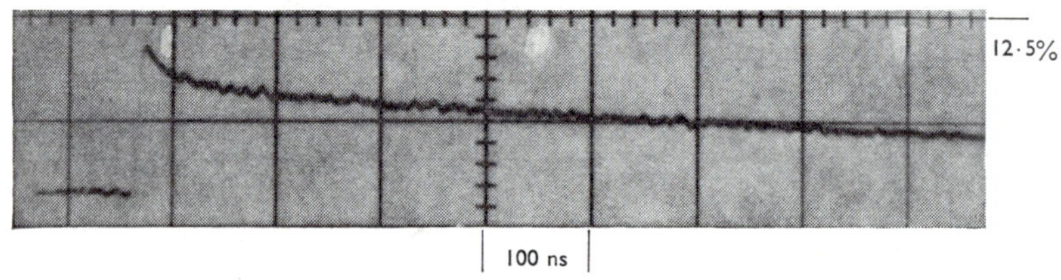

Figure 4. Decay of the solvated electron in the pulse radiolysis of pure ethanol.

If the ethanol is cooled down to −150 °C the yield of e_S^- is unchanged, and the general kinetic behaviour observed at room temperature still prevails. If the heterogeneous decay or spur extends beyond $G(e_S^-) = 1{\cdot}5$ to larger yields at times short compared to 1 ns, then cooling the alcohol should result in slower reaction rates, leading to a larger initial $G(e_S^-)$ than that observed at room temperature. However, the increased yield of e_S^- in the presence of alkali shows that $G(e^-)$ is greater than 1·5 although $G(e_S^-) = 1{\cdot}5$. It may be concluded that some of the electrons are not solvated prior to recapture by the positive ion, whereas destruction of the positive ion leads to an increased lifetime which permits solvation.

2.3. *Aniline*

In the previous liquids significant yields of ions are observed while excited molecules are not observed. In aniline both ions and excited states are observed.

Figure 5 shows the transient spectrum produced in the pulse radiolysis of pure deaerated aniline (Cooper and Thomas 1968). The sharp peak at 390 mμ is attributed to the cation of aniline, as it is scavenged by solutes of lower ionization potential such as TMPD (N, N, N′N′-tetramethyl paraphenylene diamine), while the characteristic cation of TMPD is formed. The absorption increasing into the red is attributed to the solvated electron in aniline as the absorption is removed by typical e_S^- scavengers such as nitrous oxide and oxygen. Indeed aniline is the first liquid in which positive and negative ions have been observed. The decay of the positive ion is shown in figure 6. An initial rapid decay is observed followed by a much slower decay. The solvated electron shows identical kinetics and in figure 6, the decay of the negative ion of biphenyl in aniline is shown. This is formed by reaction of e_S^- with biphenyl, the kinetics of decay are identical to those of e_S^-. Using the known extinction coefficient of the biphenylide $G(e_S^-) \equiv G(An^+) \equiv G(\phi_2^-) = 0{\cdot}95$ and 65% of the ions exhibit an initial rapid decay.

Excited states of biphenyl, anthracene, naphthalene, and 1 : 2 benzanthracene are

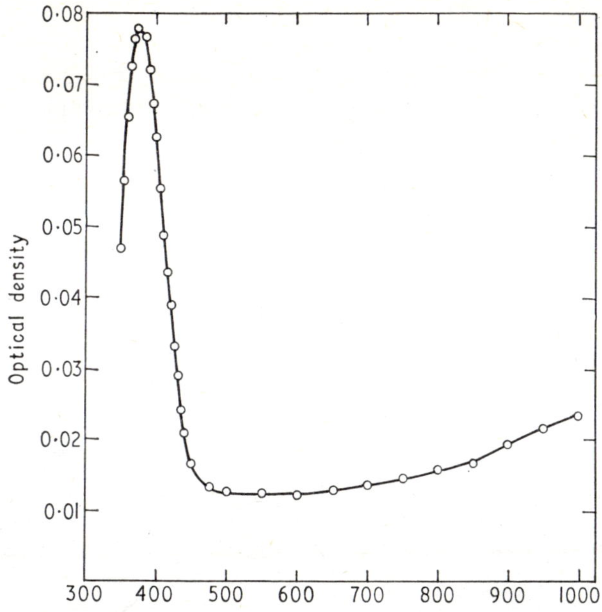

Figure 5. Spectrum of transients formed in the pulse radiolysis of pure aniline.

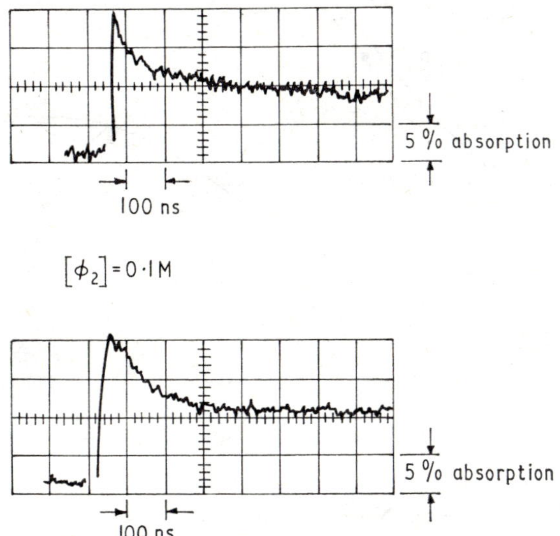

Figure 6. Decay of the cation of aniline and of the biphenylide ion in the pulse radiolysis of pure aniline and 0·1M biphenyl in aniline, respectively.

formed when solutions of these molecules are pulse-irradiated in aniline. It can be shown that at least a part of the excited state yield is derived from the recombination of the solute ions. This will be dealt with in more detail in cyclohexane.

2.4. *Cyclohexane and 3-methylpentane*

Only one absorbing species is observed in the pulse radiolysis of pure degassed cyclohexane at room temperature. This is attributed to the cyclohexyl radical which has a weak absorption spectrum below 3000 Å (Sauer and Mani 1968). In the nanosecond time region no formation or decay of this radical is observed within 2 ns following a 3 ns pulse. This shows that the rate constant of reaction (7) is greater than 4×10^7 s^{-1} and that any 'spur' reactions of this radical are complete within 2 ns.

$$H + C_6H_{12} \rightarrow H_2 + C_6H_{11}. \quad (7)$$

At room temperature no absorption spectra of solvated electrons or solvent cations can be observed. At lower temperature approaching 77 K, solvated electrons can be readily detected in 3-methylpentane in γ-radiolysis (Skelly and Hamill 1966) and in nanosecond pulse radiolysis (J. Richards and J. K. Thomas, unpublished).

It is possible to observe the ions produced in the radiolysis of C_6H_{12} at room temperature, by the addition of suitable aromatic solutes. Figure 7 shows the

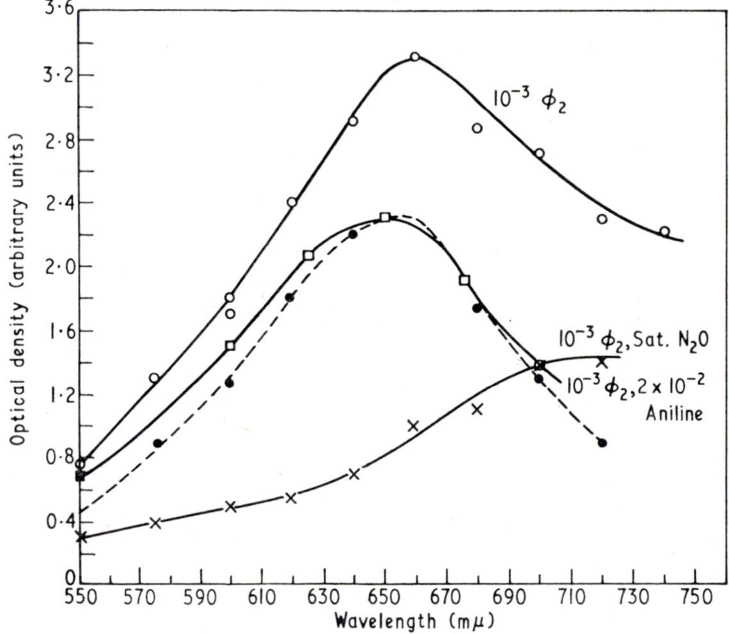

Figure 7. Spectra of the transients formed in the pulse radiolysis of 10^{-3}M biphenyl/C_6H_{12}.

○ 10^{-3}Mϕ_2. □ 10^{-3}Mϕ_2, 2×10^{-2}M aniline.
× 10^{-3}Mϕ_2, saturated N_2O. ● difference of 1 and 2.

transient spectra observed in the pulse radiolysis of 10^{-3} M biphenyl (ϕ_2) in C_6H_{12}. A peak is observed at 650 nm and also another stronger peak at 410. This is similar to the spectrum of the biphenylide ion and this ion is formed in radiolysis by electron capture by ϕ_2. The addition of nitrous oxide causes a decrease in the intensity of the absorption together with a movement of the peak to longer wavelengths. This shows that the peak at 650 nm is composed of at least two separate components. The addition of aniline to 10^{-3} ϕ_2/C_6H_{12} causes a decrease in the intensity of the 650 nm absorption and a blue shift of the maximum. Aniline removes the cations in the solution and hence leaves only the anion of biphenyl, ϕ_2^-, while N_2O removes electrons and leaves only the cation of biphenyl, ϕ_2^+. These experiments show that solute cations as well as anions may be produced in radiolysis provided the solvent cation does not decay very rapidly as in water and ethanol. The proposed reactions for the formation of the ions are (8) and (9).

$$e^- + \phi_2 \rightarrow \phi_2^- \quad (8)$$

$$C_6H_{12}^+ + \phi_2 \rightarrow C_6H_{12} + \phi_2^+ \quad (9)$$

Figure 8 illustrates the decay of ϕ_2^- in a pulse-irradiated solution of 0·1M and 10^{-3} M biphenyl in C_6H_{12}. In 0·1M biphenyl the initial $G(\phi_2^-)$ is 1·03 and about 90% of these ions exhibit a rapid decay which is independent of radiation intensity. A

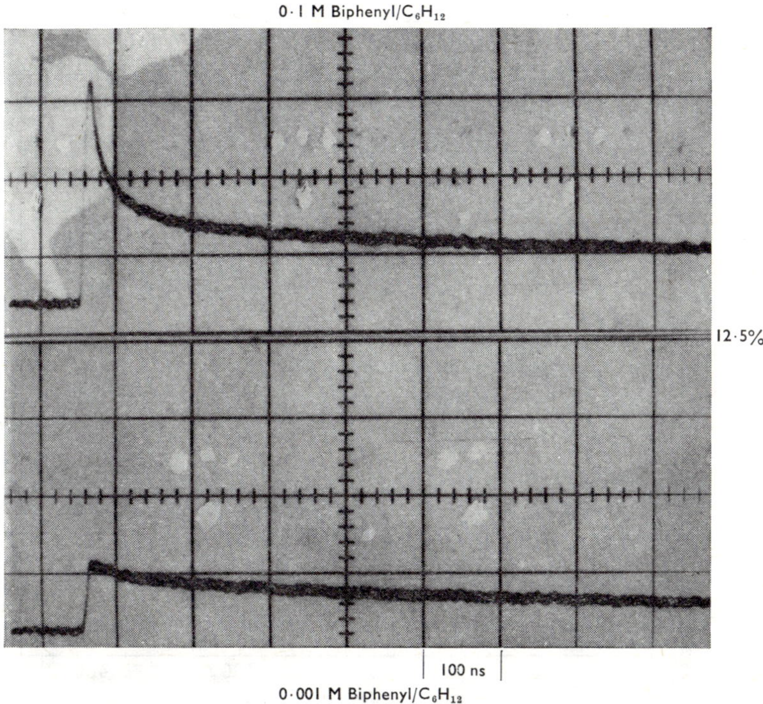

Figure 8. Decay of biphenylide ion in the pulse radiolysis of 0·1M biphenyl and 10^{-3}M biphenyl in C_6H_{12}.

$G(\phi_2^-) \sim 0{\cdot}1$ exhibits a slower decay over many μs. In 10^{-3}M biphenyl in C_6H_{12} the initial $G(\phi_2^-)$ is much smaller than in 0·1 M, while the yields at 1 μs after the pulse are comparable. An initial rapid decay is observed in 10^{-3}M biphenyl but it is a smaller effect than at higher concentrations. These results may be explained in terms of geminate ion recombination in C_6H_{12}. The addition of biphenyl captures some of the ions prior to recombination, resulting in the observed biphenyl ions which also undergo geminate recombination. The higher the concentration of biphenyl, the larger the number of solvent ions captured. Solutions of other aromatic compounds in C_6H_{12} show similar ionic chemistry. In the case of anthracene in C_6H_{12} the decay of the anion of anthracene may be observed at 700 nm while the triplet excited state may be observed at 425 nm (Thomas *et al.* 1968). Figure 9 shows the kinetic of formation of the triplet states of anthracene and naphthalene in pulse-irradiated C_6H_{12}. Both solutes show that about 70% of the triplet state is produced very rapidly with the pulse, while 30% grows in subsequently. Figure 10 shows a plot of the rate of decay of the

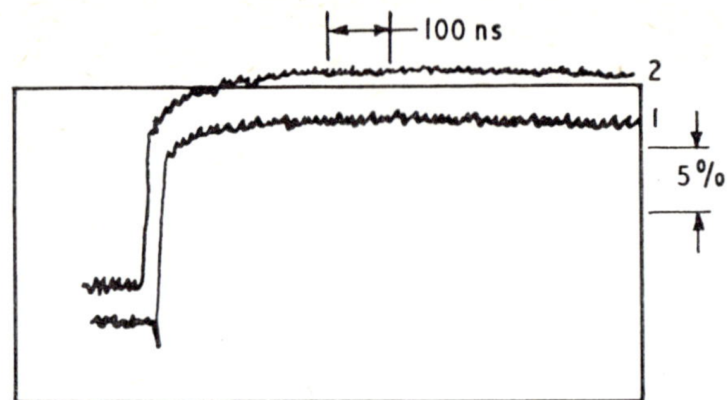

Figure 9. Formation of triplets of anthracene and naphthalene in the pulse radiolysis of C_6H_{12} solutions of these solutes.

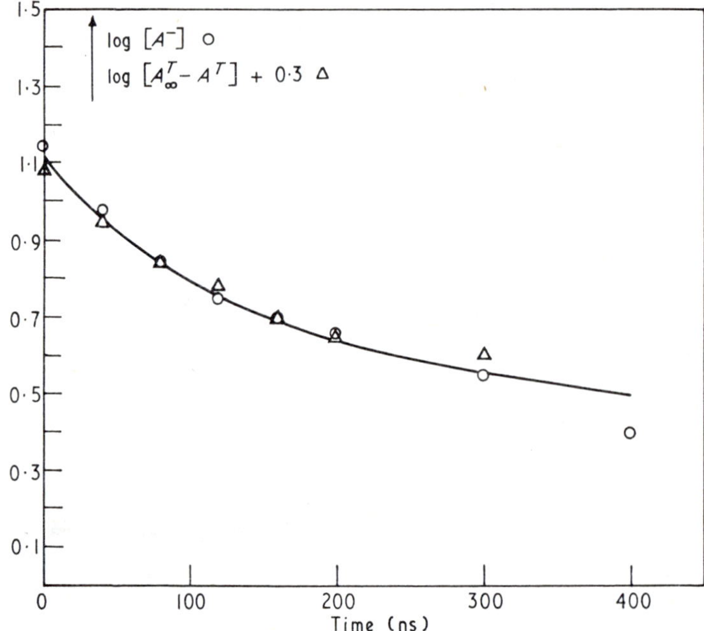

Figure 10. Decay of anthracene anion (○) and growth of anthracene triplet state (△) in pulse radiolysis of 10^{-2} M anthracene in C_6H_{12}.

anion and the rate of growth of the triplet state of anthracene. The fact that both growth and decay match suggests that the ion-neutralization leads to excited states. Indeed the fact that ion scavengers such as NH_3 and SF_6 decrease both the yield of aromatic ions and the yield of the growing portion of the excited state, confirms this conclusion.

It is pertinent to inquire about the nature of the electron in irradiated hydrocarbons. As solvated electrons are only observed at low temperature, the ion recombination reaction may be too rapid for observation at room temperature. The significant lifetime of the biphenyl ions argue against this point. It is more probable that the electrons in the hydrocarbons are not solvated at room temperature. Two pieces of experimental evidence support this conclusion.

In liquids such as alcohols where electrons are solvated, the rate constant for reaction of e_s^- with ϕ_2 is $4{\cdot}3 \times 10^9$ s^{-1} (Anbar and Neta 1965). This means that the half-life of formation of ϕ_2^-, $t_{1/2}$ is $1{\cdot}6 \times 10^{-6}$ s in 10^{-4}M biphenyl. The pulse radiolysis of

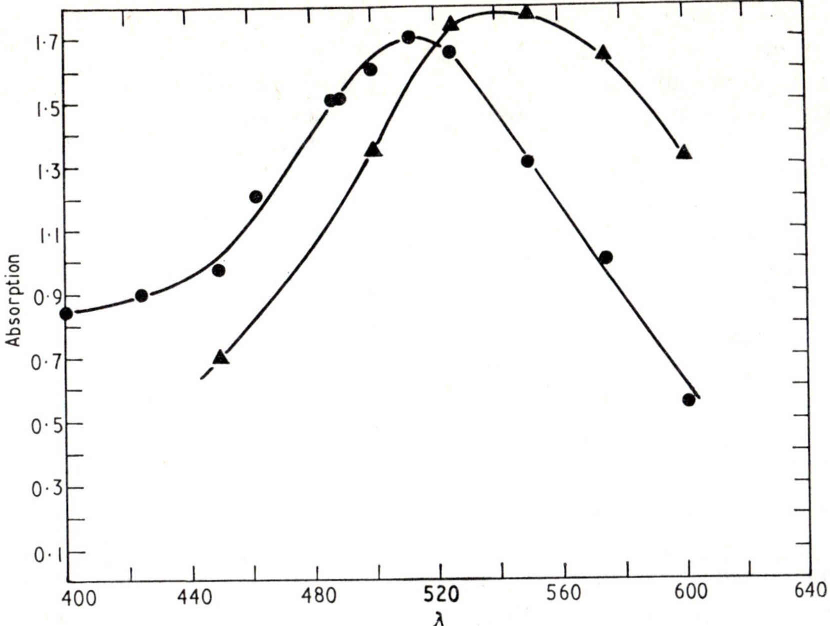

Figure 11. Transition spectra of excited excimer states in the pulse radiolysis of benzene (●) and toluene (▲).

10^{-4} biphenyl in C_6H_{12} shows that ϕ_2^- is formed within 2 ns following a 3 ns pulse, i.e.,

$$k_{e^- + \phi_2} > 10^{12}$$

At lower temperatures where solvated electrons are observed in 3-methyl pentane (3MP), a slow growth of ϕ_2^- is observed concurrent with a decay of e_s^-. However, even here most of the ϕ_2^- is formed at a rate which is much faster than the time resolution of the apparatus.

It is possible to study electron transfer rates in C_6H_{12} from ϕ_2^- to several secondary solutes such as O_2, benzyl chloride and SF_6 (J. Richards and J. K. Thomas, unpublished). Diffusion controlled rate constants are obtained with $k \sim 10^{10}$ s^{-1}. In the case of benzyl chloride and SF_6 it is found that these solutes compete with biphenyl for electrons hence reducing the initial yield of ϕ_2^-. However, O_2 does not compete with biphenyl for electrons but does accept electrons from ϕ_2^-. Hence O_2 does not react with the primary state of the electron in the hydrocarbons, a condition which is contrary to its properties in solvents where e^- is solvated.

Both series of experiments suggest that electrons are not solvated in hydrocarbons at room temperature, but have high mobility and react rapidly with suitable solutes forming solute anions. Many other liquids have been studied by the technique of nanosecond pulsed radiolysis, but time does not permit their discussion here. However, the broad outlines of the radiolysis of benzene will be considered due to its importance in scintillation studies and because, in contrast to the other liquids considered, benzene has a long-lived excited state.

2.5. *Benzene*

The nanosecond pulse radiolysis of benzene and toluene (Cooper and Thomas 1968b) shows that although the initial yield of ions is large, the yield observed at 1 ns is only 0·3 due to a rapid neutralization reaction. This reaction leads to excited states of benzene, the singlet $^1B_{2u}$, the excimer singlet $^1B_{1g}$ and the triplet $^3B_{1u}$. The spectra of the excimer singlets are shown in figure 10; in benzene a peak is observed at 515 nm. This corresponds to a transition $^1E_{1u} \leftarrow {}^1B_{1g}$ with ΔE of 2·4 V agreeing with the observed maximum. The nanosecond flash photolysis of 10% benzene/C_6H_{12} with the 265 nm line of a laser shows the same short-lived species (J. K. Thomas, unpublished). In the presence of a scintillator such as anthracene, naphthalene, etc., the singlet and triplet benzene energy is transferred to the solute producing excited singlet and triplet states. The higher excited singlet state of benzene $^1E_{1u}$ is also formed directly by secondary electrons and this cascades inefficiently to the first excited singlet (S. Lipsky, private communication). In other aromatic molecules, e.g., p-xylene, the cascade process is efficient leading to a larger yield of excited singlet states and an enhanced scintillation efficiency.

3. Conclusions

The experiments illustrate most of the processes shown in figure 1. Heterogeneous kinetics are observed for the decay of ions, but not neutral species, in all solvents from water to cyclohexane. In hydrocarbons the heterogeneous kinetics are explained in terms of isolated ion-pairs rather than spurs. In all solvents the yields of ions observed are lower than the values currently projected from γ-radiolysis. This may be due to an initial rapid decay of the ions which escape detection because of the finite response time of the apparatus. On the other hand, several fates may await the radiation-produced ions, and only a few such as solvation and capture by solute are capable of being observed in the present work.

The force of attraction between the ions in an ion-pair varies as the inverse of the dielectric strength of the liquid. In liquids of low dielectric strength such as C_6H_{12}, the increased force of attraction between the geminate ions leads to recombination, and only a small fraction escapes into the bulk of the liquid. The fraction escaping recombination increases with increasing dielectric strength. This is observed in the present work, the relative importance of the heterogeneous ion kinetics increasing with decreasing dielectric constant.

The heterogeneous e_{aq}^- decay in water may be explained in terms of the above model of geminate ion pairs. However, it is customary to explain the radiolysis of water in terms of the 'spurs' model, which also explains the observed data in water.

References

ANBAR, M., and NETA, P., 1965, *Int. J. Appl. Radiation Isotopes*, **16**, 227.

COOPER, R., and THOMAS, J. K., 1968a, *J. chem. Phys.*, **48**, 5103.

—— 1968b, *J. chem. Phys.*, **48**, 5097.

HART, E. J., GORDON, S., and FIELDEN, E. M., 1966, *J. phys. Chem.*, **70**, 150.

HUNT, J. W., and THOMAS, J. K., 1967, *Radiat. Res.*, **32**, 149.

KUPPERMANN, A., 1961, *The Chemical and Biological Action of Radiation*, Ed. M. Haissinsky (New York: Academic Press), p. 5.

PLATZMAN, R. L., 1966, *Radiation Research*, Ed. Silini Cortina (Amsterdam: North Holland) p. 20.

SAMUEL, A., and MAGEE, J. L., 1953, *J. chem. Phys.*, **21**, 1080.

SAUER, M. C., and MANI, I., 1968, *J. phys. Chem.*, **72**, 3856.

SKELLY, D. W., and HAMILL, W. H., 1966, *J. chem. Phys.*, **44**, 2891.

THOMAS, J. K., *The Excited State*, Ed. A. Lamola (in press).

THOMAS, J. K., and BENSASSON, R. V., 1967, *J. chem. Phys.*, **46**, 4147.

THOMAS, J. K., JOHNSON, K., KLIPPERT, T., and LOWERS, R., 1968, *J. chem. Phys.*, **48**, 1608.

Molecular oxygen as a primary radiation chemical species in the high LET radiolysis of aqueous solutions

R. P. O. HÜBER

Physics Department, P.O. Box 363, University of Birmingham, Birmingham 15, England

Abstract. The origin of molecular oxygen released in high LET radiolysis from water and aqueous solutions has been investigated with ^3He particles and fast neutrons. The results show that the molecular oxygen is released as a result of reactions in the early stages of the track formation and has, therefore, to be considered a primary radiation chemical species.

1. Introduction

The radiolysis of water has been under investigation for almost as long as ionizing radiation has been known. In spite of this long history and the apparent simplicity of the chemistry of water, its radiolysis and the radiolysis of aqueous solutions are still only partly understood. One particular phenomenon, the decrease of 'oxygen effects' with increasing ionization density of the radiation (or LET), has been studied in various systems but only Alper and Moore (1967a and b) and the work described here, have investigated the phenomenon itself.

As long ago as 1960 Ardashnikov suggested that molecular oxygen released in the track would explain the LET dependence of the oxygen effect. A small amount of oxygen formed in the track could compete effectively with the oxygen dissolved in the bulk of the liquid and thereby reduce the observed oxygen effect. In his paper he suggests a mechanism based on well-known reactions. The track of a densely ionizing particle can be considered as a column of primary species, the hydroxyl radicals concentrated near the axis, and the reducing species spread out around it. This distribution favours the reaction between radicals of the same kind, e.g. the formation of hydrogen peroxide in the centre region. The hydrogen peroxide is then decomposed leading to the release of molecular oxygen in the following reactions

$$H_2O_2 + OH \rightarrow H_2O + HO_2 \qquad (1)$$

$$H_2O_2 + HO_2 \rightarrow H_2O + OH + O_2 \text{ and}$$

$$HO_2 + HO_2 \rightarrow H_2O_2 + O_2$$

The competition between oxygen released in the track and that dissolved in the bulk seems a natural explanation of the phenomenon. The mechanism proposed for the oxygen release has, however, a number of weaknesses, the most obvious being the rate constant of the key reaction (1) which is far too low ($4{\cdot}5 \times 10^7$ M^{-1} s^{-1}) compared with typically, $\sim 10^{10}$ M^{-1} s^{-1} for primary radicals reacting with each other and with solutes. However, this objection is based on rate constants determined with low LET radiation and may, therefore, not apply to reactions occurring in high LET tracks. As the data available from low LET studies are of only very limited use, an experiment had to be designed for the investigation of the hypothesis and the reaction mechanism proposed.

2. Experimental

The experimental approach is based on the dual nature of energetic nuclear radiation to induce radioactivity as well as to radiolyse. A particle, such as an energetic ^{3}He nucleus, may somewhere along its track, interact with an oxygen nucleus in such a way as to form a radioactive oxygen isotope. This radioisotope will be formed in the track region created by the activating particle, secondary particles and its own recoil, all regions of high LET, i.e. the site under investigation. The following two reactions have been used

$$^{16}O(^{3}He, ^{4}He)\ ^{15}O \text{ and}$$

$$^{16}O\ (n, 2n)\ ^{15}O.$$

where the product nucleus ^{15}O is a positron emitter of 2·1 min half-life. The recoiling oxygen isotope loses its energy initially by ionization until its velocity becomes comparable with that of the least tightly bound electron. The remaining kinetic energy is being dissipated predominantly in a number of elastic collisions with other non-radioactive oxygen atoms. The overall result is that the oxygen-15 passes on its kinetic energy, being left in a state comparable with that of an oxygen atom activated somewhere along the track, therefore reacting and labelling all the oxygen-containing radiolytic track species. The main advantage of this method compared with other procedures, lies in the fact that at a known time a specific label is introduced exactly at the site under investigation.

The schematic representation of the experimental arrangement is shown in figure 1.

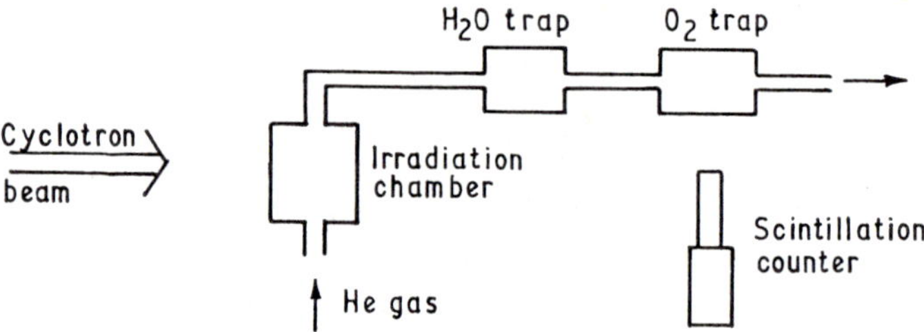

Figure 1. Schematic representation of the apparatus.

The gaseous radiolysis products are extracted from the irradiation chamber by passing a continuous stream of He gas through it. After the removal of the water vapour the oxygen is extracted from the gas in a scintillation counter assembly, by a zeolite molecular sieve at liquid nitrogen temperature. The amount of oxygen-15 in the molecular oxygen is measured by the amount of annihilation radiation detected.*

The ^{3}He experiments were carried out at a particle energy of 11·7 MeV and the fast neutrons were obtained by exposing a Be target to 30 MeV ^{3}He particles.

The conversion of the ^{15}O radioactivity to a yield of molecular oxygen in molecules per 100 eV or $G(O_2)$, is based on the yields obtained from 2 mM $Ce(SO_4)_2$ in 0·4M H_2SO_4 and 2 mM $FeSO_4$/10 mM $CuSO_4$ in 5 mM H_2SO_4 solutions. The $Ce(SO_4)_2$ solution decomposes and releases the hydrogen peroxide-bound oxygen as well as

* A description of the experimental aspects will be the subject of a separate publication.

that bound by perhydroxyl radicals, while the second system decomposes the hydrogen peroxide by the formation of hydroxyl ions, i.e. without the release of molecular oxygen. The difference in labelled oxygen yields corresponds, therefore, to the yield of hydrogen peroxide which has been determined by a number of investigators for various LET's. A value of $G_{H_2O_2}=1{\cdot}06$ has been used for ^3He particles and of $G_{H_2O_2}=0{\cdot}6$ for the fast neutrons.

3. Results

The release of labelled molecular oxygen from pure water as a function of dose is given in figure 2. The yield increases from approximately $G(O_2)=0{\cdot}1$ as the dose increases. The dose dependence of the yield is due to the influence of the hydrogen

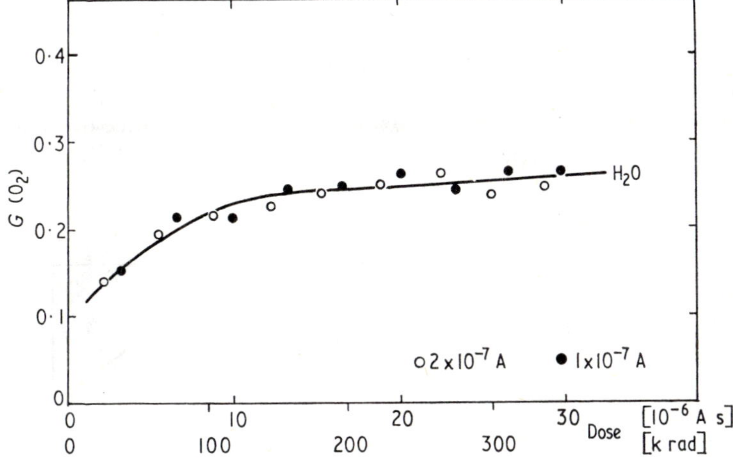

Figure 2. The oxygen yield from H_2O as a function of total dose. Successive irradiations were carried out at different dose-rates.

peroxide, this being the only radiolysis product remaining in the sample. The influence of the hydrogen peroxide can be attributed to its ability to scavenge the reducing species generated in the bulk of the liquid by tracks of low ionization density, thereby diminishing the chance of the oxygen being scavenged. The observation of ^{15}O labelled oxygen in the released molecular oxygen, in itself does not prove that it has been formed in the track, as it may well have been released in the bulk of the water by the decomposition of labelled hydrogen peroxide. The relative importance of track to bulk mechanisms can, however, be estimated without much difficulty by comparing the yield obtained from successive exposures at different dose rate. During the first exposure labelled hydrogen peroxide is formed, so that if the decomposition of this hydrogen peroxide contributes significantly to the oxygen released in a second exposure this will be reflected in an enhanced yield depending on the time allowed for the decay of the label, or what amounts to the same, the yield should be dose-rate dependent. This dependence is largest at low doses and decreases gradually as the level of inactive hydrogen peroxide builds up.

No significant change in yield is observed when the dose rate is changed by a factor of three in water or 0·4M H_2SO_4 solution which suggests that very little oxygen is formed outside the high LET tracks.

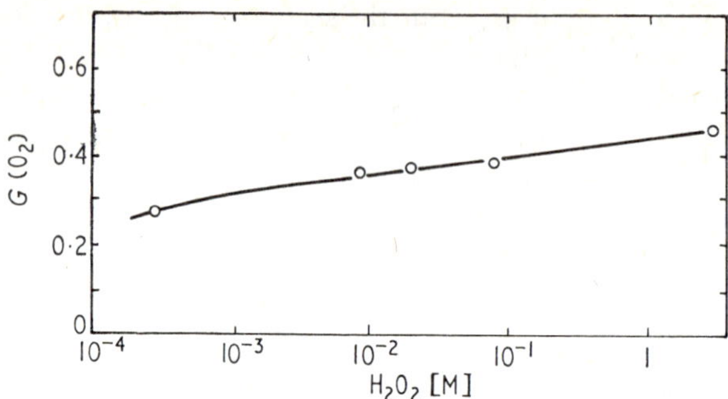

Figure 3. The oxygen yield as a function of 'e^-_{sol}-scavenger' concentration.

Having located the site of oxygen formation, the mechanism can be investigated by examining the influence of scavengers. Figure 3 shows the dependence of the oxygen yield on the concentration of hydrogen peroxide, a scavenger for the reducing species, and figure 4, on the concentration of iodide ions, a scavenger for the hydroxyl radical. The presence of hydrogen peroxide will shift the balance of the reactions by scavenging of the reducing species producing a hydroxyl radical

$$H_2O_2 + e^-_{sol} \rightarrow OH^- + OH.$$

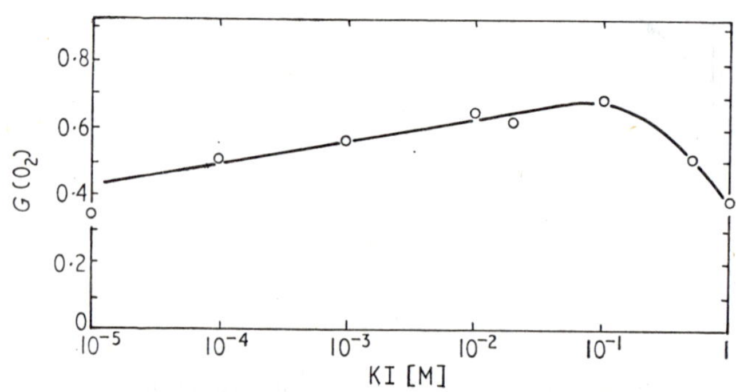

Figure 4. The oxygen yield as a function of 'OH scavenger' concentration.

At low concentrations this will affect the reactions occurring in the bulk only, i.e. radiolytic processes due to the low and intermediate LET fraction of the radiation. As the concentration increases, the interference with track reactions becomes important leading to an increased yield in labelled hydrogen peroxide due to the reduction in the back reaction coupled with an increase in OH radicals which, in turn, enhances the yield of labelled molecular oxygen. Based on the model described earlier, this increase will be reduced to some extent by the competition of inactive H_2O_2 for decomposition.

By adding potassium iodide to the water the OH radicals are converted to iodine atoms $OH + I^- \rightarrow OH^- + I$. This does not affect the decomposition of hydrogen peroxide, as the products formed by the reaction, i.e. iodine atoms and hydroxyl radicals, are the same in this system. However, as a scavenger for the hydroxyl radical it increases the yield of oxidizing species while the yield of hydrogen peroxide is reduced. The increase in yield of oxidizing species enhances the decomposition of hydrogen peroxide, i.e. the release of labelled oxygen. However, this can, only continue until the reduced yield of H_2O_2 formation in the track is matched by its decomposition by the enhanced yield of oxidizing species. If the concentration is increased beyond this point the oxygen yield must fall off as the yield of H_2O_2 is decreased still further. The scavenger concentration at which this occurs is approximately 8×10^{-2}M KI which compares favourably with a concentration of 5×10^{-2}M estimated on the basis of Kuppermann's (1967) calculations for ^{210}Po α-particles. A comparative study with fast neutrons has been carried out only with water and sulphuric acid solutions. No labelled oxygen is released from water or 0·4M H_2SO_4. However, the irradiation of a 10^{-4}M H_2O_2 solution gives a yield almost identical to that obtained from the $FeSO_4/CuSO_4$ in H_2SO_4 solution, which suggests that molecular oxygen is released from the high LET tracks under these conditions also. The overall radiolytic climate is, however, such that the oxygen leaving the track region is scavenged in the bulk of the liquid, so that the oxygen can only be observed when scavenging is prevented by the introduction of a scavenger for the reducing species, or alternatively, the HO_2 is decomposed.

5. Conclusion

From these results it seems evident that molecular oxygen is a radiolysis product formed in the high LET tracks of ionizing radiation. The observation of a net release depends on the LET distribution of the radiation while the formation of oxygen depends only on the ionization density of the track segment concerned.

The changes of oxygen yield with the concentration of scavengers is in agreement with the expectations based on the mechanism proposed by Ardashnikov. If the interpretation of the experimental results is correct, the primary yield of oxygen for an LET of 80 keV μm^{-1} is approximately $G_{O_2}=0{\cdot}2$ of which only about $G(O_2)=0{\cdot}1$ can be detected in the absence of H_2O_2.

Acknowledgements

I would like to express my gratitude to Professor J. H. Fremlin for his help and his continued interest. The help of the Nuffield Cyclotron Technical Staff is gratefully acknowledged.

References

ALPER, T., and MOORE, J. L., 1967a, *Br. J. Radiol.*, **40**, 843–8.

—— 1967b, *Rad. Res.*, **32**, 780.

ARDASHNIKOV, S. N., 1960, *The Role of Peroxides and Oxygen in the First Stages of the Radiobiological Effect*, pp. 146–52, *Izdatel'stvo Akademii Nauk SSSR*, trans. 1966 by J. E. Baker, AERE Harwell.

KUPPERMANN, A., *Radiation Research*, Ed. G. Silini (Amsterdam: North Holland).

Non-ionizing interactions

H. JUNG

Institut für Strahlenbiologie, Kernforschungszentrum Karlsruhe, Germany

Abstract. Experimental studies in dry systems show some non-ionizing interactions such as elastic nuclear collisions and attack by thermal hydrogen atoms to cause biological damage similar to, though naturally differing in quantity and/or quality, from the damage caused by excitations and ionizations. Elastic collisions were investigated using low energy protons. Hydrogen atoms were generated in a gas discharge or, alternatively, by irradiating plastic foils with 2 MeV protons. In either case exponential inactivation curves were obtained with infectious DNA extracted from bacteriophage ΦX174 and with chromatographically purified ribonuclease. The results indicate that, because of processes of diffusion, the local distribution of damaged molecules is necessarily different, to some extent, from the local distribution of excitations and ionizations, i.e. different from the original structure of the track of a charged particle.

Though this conference is mainly concerned with discussing the frequency and the local distribution of excitations and ionizations it may be of interest to recall that not all types of interactions lead to excited and ionized molecules. Part of the energy of fast charged particles traversing matter is dissipated in direct impact with atomic nuclei. This process is generally called elastic nuclear collision. Depending on the momentum and charge of the incident particle, on the charge and the mass of the nucleus encountered, and on the impact parameter, the energy transferred may be sufficient to eject nuclei from their original position in a macromolecule, carrying with them a complete set of electrons. This derangement of atoms represents a chemical change which in biological structures may cause severe and permanent damage (Platzman 1952).

To investigate the effects of elastic nuclear collisions, the use of low energy charged particles appears to be the best of a number of methods available (Jung and Zimmer 1966). Figure 1 shows the cross sections for energy losses by electronic interaction σ_e and by elastic nuclear collisions σ_n as calculated by Neufeld and Snyder (1961) for protons in tissue. The probability for ionization σ_e increases with decreasing proton energy, reaches a maximum at about 50 to 100 keV, the so-called Bragg-maximum, and decreases again in the low energy range. For fast protons the cross section for elastic collisions σ_n is smaller by about three orders of magnitude than the cross section for electronic interaction. It increases steadily with decreasing proton energy. As in the low energy region σ_e decreases while σ_n is still rising, there is an energy limit below which the incident protons lose more energy in nuclear collisions than in ionizations and excitations. According to the calculations by Neufeld and Snyder this limit lies at a proton energy of 1·5 keV.

The technique used in our measurements was to measure the differential inactivation cross section in thin layers of biological material and for various proton energies. If ionization *only* were important for inactivation of the biological material one would expect the inactivation probability to show a similar course as the ionization probability σ_e. If nuclear collisions contribute to the observed effect the inactivation cross section should decrease in the high energy region with decreasing proton energy, but at low energies where the contribution of elastic collisions is predominant over the contribution of ionizations one would expect the inactivation cross section to increase again.

For the macromolecules used in these experiments the condition must be satisfied that a proton, in passing through one of the irradiated entities, must have little chance of inducing more than one inactivating event. If this condition is not fulfilled one cannot anticipate any energy dependence of the inactivation cross-section.

The results of two experiments of the above-mentioned kind are given in figure 2 which shows the cross section for the inactivation of ribonuclease (Jung 1965, 1967) and of infectious DNA of bacteriophage ΦX174 (Kürzinger and Jung 1968) for various proton energies. If ionizations alone were able to cause biological damage the inactivation cross sections should decrease according to the dotted line representing the ionization probability σ_e as given in figure 1. From the deviation of the experi-

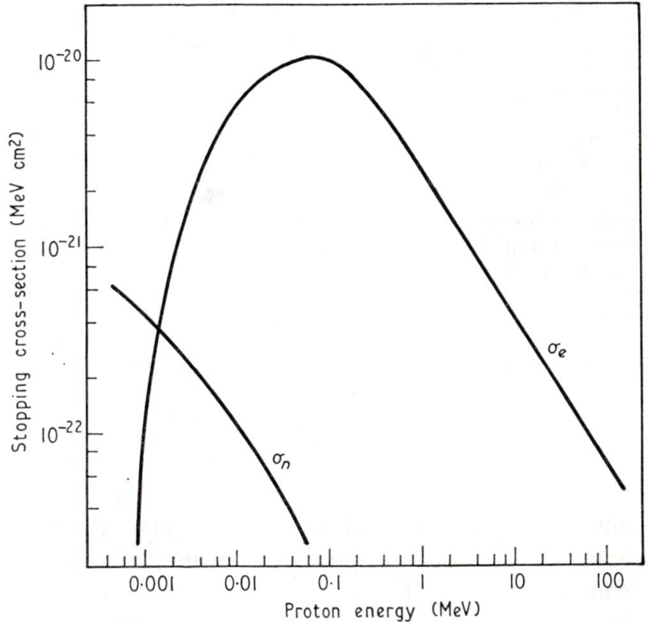

Figure 1. Electronic stopping power (σ_e) and nuclear stopping power (σ_n) for protons in tissue (Neufeld and Snyder 1961).

mental points from the calculated curve in the low energy region it is obvious that elastic nuclear collisions are able to inactivate ribonuclease as well as infectious DNA.

The variation of the inactivation probability with proton energy for ΦX-DNA is much smaller than in the case of ribonuclease. This result is easily explained by the fact that the molecular weight of ΦX-DNA is higher by a factor of about 120 than the molecular weight of ribonuclease. Thus, slow protons passing through a DNA molecule giving the lowest possible energy transfer still produce inactivating events with a probability approaching unity. A more detailed description of this analysis is given by Jung and Kürzinger (1969).

Further differences in the biological action of elastic collisions and that of ionizations appeared when studying the influence of protective agents and low temperature on the radiosensitivity of ribonuclease. When cystamine is present during irradiation dry

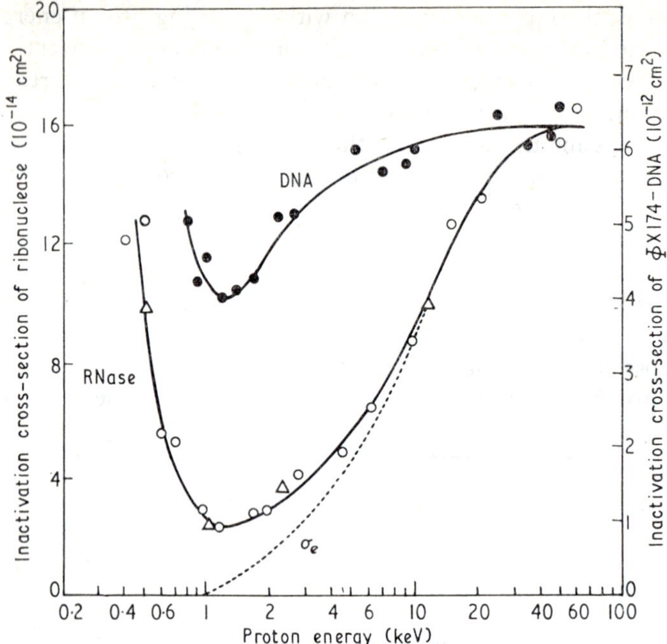

Figure 2. Cross section for the inactivation of ribonuclease and infectious DNA of bacteriophage ΦX174 with protons of different energies. ○ 5 times crystallized RNase; △ Chromatographically purified RNase; ● ΦX174-DNA; ---. Electronic stopping power (σ_e) from figure 1 in relative units.

ribonuclease is protected against the action of 'ionizing' protons of 2 MeV, the dose reduction factor being 1·8. But no protection is observed when inactivation is achieved by elastic nuclear collisions (1·4 keV protons). Similar results were obtained when the irradiations were carried out at different temperatures. With 2 MeV protons the radiosensitivity of ribonuclease was found to be 3 times higher at room temperature than at 120 K, but when using slow protons of 1·4 keV energy the inactivation cross section of RNase turned out to be independent of temperature (Jung 1966).

Besides elastic nuclear collisions there is another type of non-ionizing interaction by which biological structures may be damaged, namely the reaction of diffusible radicals liberated by radiation in dry systems. Several experimental facts indicate that atomic hydrogen may be involved in the damage observed in dry systems.

On irradiating hydrocarbons, amino acids, proteins, and nucleic acid constituents a gas is formed, the major fraction of which is molecular hydrogen. Furthermore, it could be concluded from electron spin resonance experiments that many radicals of irradiated organic molecules are formed by breakage of C–H bonds (Smith 1962).

According to a hypothesis by Braams (1963) a macromolecule, MH, under irradiation dissociates into a macroradical, M·, and a hydrogen radical, H·,

$$MH \rightarrow M\cdot + H\cdot \tag{1}$$

As hydrogen atoms are chemically very reactive, they may interact with undamaged molecules, MH, either by hydrogen abstraction or by addition to the macromolecule:

$$H\cdot + MH \rightarrow M\cdot + H_2 \tag{2}$$

$$H\cdot + MH \rightarrow MH\cdot_2. \tag{3}$$

It is known from the work of Stein and co-workers that in aqueous solution hydrogen atoms destroy the enzymatic activity of trypsin (Mee *et al.* 1965) and ribonuclease (B. E. Holmes *et al.* 1967) as well as the plaque-forming ability of phage T7 (Dewey and Stein 1968). For dry systems it has been shown by ESR-measurements that H atoms react with proteins (Snipes and Schmidt 1966) and with DNA and its constituents (Heller and Cole 1965, Herak and Gordy 1965, 1966a, b, D. E. Holmes *et al.* 1966, 1967, Dertinger 1967a, b), but up to now no experimental proof had been available that this reaction leads to an inactivation of dry enzymes, DNA, and bacteriophage.

To study the action of atomic hydrogen on various specimens of biological importance, two different arrangements were employed (Jung and Kürzinger 1968). System I consists of an electrodeless discharge in a stream of hydrogen gas, maintained at a pressure of 16 torr. The samples to be exposed were freeze-dried on microscope cover slips and attached vertically in such a way as to prevent UV light from reaching the samples. System II is a setup having some similarity to a parallel-plate condenser where a plastic foil is facing the samples to be exposed. Hydrogen atoms are liberated from the surface of the foil by irradiating with 2 MeV protons and react with the specimens on the opposite side of the condenser. Secondary electrons liberated by the protons are kept off the samples by a grid held at a negative potential. This arrangement permitted the separate investigation of the two postulated reactions, namely the generation of hydrogen atoms by irradiation of organic material (equation 1) and the biological effects of the H atoms liberated (equations 2 and 3). Absolute dosimetry was not performed. However, the variation in the intensity of hydrogen atoms produced in the discharge and by the proton beam was so small that the number of hydrogen atoms reaching the samples could be assumed to be linearly proportional to exposure time. Further details concerning the experimental arrangements and the assay methods for ribonuclease and infectious ΦX-DNA have been described by Jung and Kürzinger (1968).

With increasing exposure times the surviving fraction of all specimens inactivated decreases and reaches a constant level (12 to 35 per cent in different experiments) indicating that the thickness of the irradiated samples exceeds the diffusion length of hydrogen atoms. To allow for this indestructible fraction, one has to subtract this constant value determined in each experiment with good statistics by long-time exposure of numerous samples. After doing so the data obtained for the inactivation of all specimens used in these experiments fall on straight lines in a semi-log plot. Figure 3 shows the dose-effect curve for the inactivation of dry ribonuclease by hydrogen atoms generated in the high-frequency gas discharge (system I). The 37% exposure time (t_{37}) amounts to 12·9 min, while in system II we found $t_{37}=8{\cdot}5$ min. The fact that these values obtained in either system happen to be of the same order of magnitude is, of course, accidental.

In figure 4 is plotted the dose-effect curve for the inactivation of infectious DNA of bacteriophage ΦX174 by H atoms using system II. For single-stranded DNA, too, in systems I and II after subtracting the constant fraction, exponential inactivation curves are obtained the slopes of which are much steeper than that of the inactivation curve of ribonuclease. The 37% exposure times amount to 17 and 11·3 s, respectively.

In the search for differences between the action of ionizing radiation and that of hydrogen atoms we found that using H atoms the dose-effect curves for T1-bacteriophages with normal DNA and those with bromouracil-substituted DNA coincide, which means that substitution of thymine by bromouracil does not alter the sensitivity of T1 phage to atomic hydrogen (Jung and Kürzinger 1968). With γ-radiation,

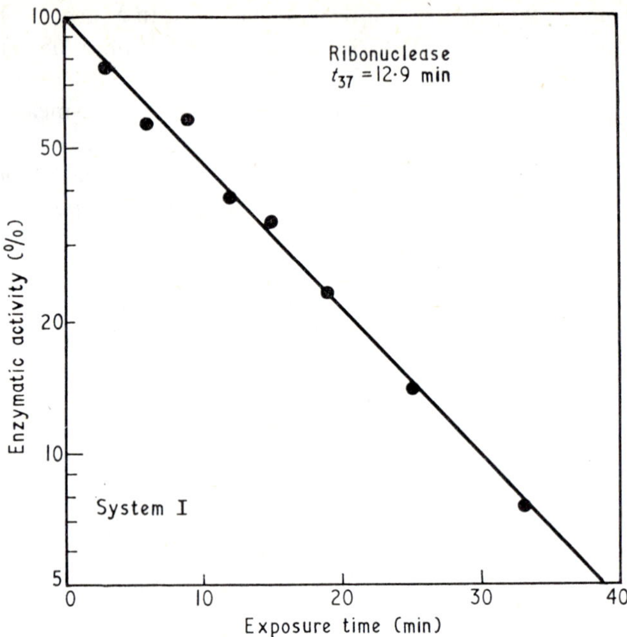

Figure 3. Inactivation of dry ribonuclease by exposure to atomic hydrogen generated in a high frequency gas discharge (system I).

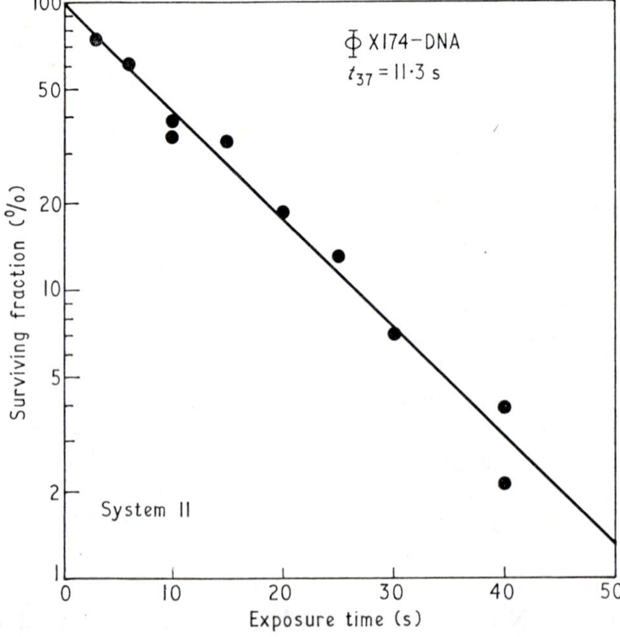

Figure 4. Inactivation of dry infectious ΦX174-DNA by exposure to atomic hydrogen generated by irradiating a foil of polyethylene terephthalate with 2 MeV protons (system II).

however, the sensitivity of BU-T1 phage is higher by a factor of 2·2 as compared to that of normal T1 phages. With UV light in the first decade the two inactivation curves for T1 and BU-T1 differ in slope by a factor of about 3. Our findings are summarized in table 1, where the relative sensitivities of all objects investigated are

Table 1. Relative sensitivities of various specimens to atomic hydrogen in systems I and II and to ^{60}Co γ-radiation compared to that of phage T1

Object	H atoms System I	H atoms System II	Co-γ-radiation
Phage T1	1	1	1
Phage BU-T1	1	1	2·2
ΦX174-DNA	0·67	0·69	1
Ribonuclease	0·0147	0·0154	0·0077

compared. The curves are normalized to a sensitivity ($1/t_{37}$ and $1/D_{37}$, respectively) of phage T1 equalling unity. The good agreement of the results obtained with hydrogen atoms from both experimental arrangements indicates identical mechanisms of inactivation. This conclusion is supported by comparing the action of H atoms with the results of our experiments using γ-radiation. Here the relative radiosensitivities of the various objects are different from the values obtained with hydrogen atoms. Additional information regarding the meaning of our results for molecular radiobiology may be found in a review article by Zimmer (1969) and in a textbook by Dertinger and Jung (1969).

Our experiments have shown that hydrogen atoms are capable of inactivating the biochemical functions of proteins and nucleic acids. As the inactivation kinetics is the same as found in experiments with γ-radiation, the results obtained are in line with the reaction scheme given by the equations 1–3 which ascribes part of the biological effects observed in dry systems to the action of hydrogen atoms. As these hydrogen atoms diffuse away from the place of their liberation, even at low temperatures, they will cause biological changes at some distance from the point of their origin. Thus, the local distribution of damaged molecules is necessarily different, to some extent, from the local distribution of excitations and ionizations, i.e. different from the original structure of the track of a charged particle.

Acknowledgment

The author wishes to express his gratitude to Professor K. G. Zimmer whose continued support and interest made this study possible.

References

BRAAMS, R., 1963, *Nature*, **200**, 752.
DERTINGER, H., 1967a, *Z. Naturforschg.*, **22b**, 1261.
—— 1967b, *Z. Naturforschg.*, **22b**, 1266.
DERTINGER, H., and JUNG, H., 1969, *Molekulare Strahlenbiologie.* (Berlin–Heidelberg–New York: Springer-Verlag).
DEWEY, D. L., and STEIN, G., 1968, *Nature*, **217**, 351.
HELLER, H. C., and COLE, T., 1965, *Proc. Natn. Acad. Sci. U.S.*, **54**, 1486.
HERAK, J. N., and GORDY, W., 1965, *Proc. Natn. Acad. Sci. U.S.*, **54**, 1287.
—— 1966a, *Proc. Natn. Acad. Sci. U.S.*, **55**, 1373.
—— 1966b, *Proc. Natn. Acad. Sci. U.S.*, **56**, 7.
HOLMES, B. E., NAVON, G., and STEIN, G., 1967, *Nature*, **213**, 1087.
HOLMES, D. E., INGALLS, R. B., and MYERS, L. S., 1967, *Int. J. Radiat. Biol.*, **12**, 415.

HOLMES, D. E., MYERS, L. S., and INGALLS, R. B., 1966, *Nature*, **209**, 1017.
JUNG, H., 1965, *Z. Naturforschg.*, **20b**, 764.
—— 1966, *Z. Naturforschg.*, **21b**, 1165.
—— 1967, *Radiat. Res. Suppl.*, **7**, 64.
JUNG, H., and KÜRZINGER, K., 1968, *Radiat. Res.*, **36**, 369.
—— 1969, *Z. Naturforschg.*, **24b**, 328.
JUNG, H., and ZIMMER, K. G., 1966, *Current Topics in Radiation Research*, Vol. II, Eds M. Ebert and A. Howard (Amsterdam: North Holland), p. 69.
KÜRZINGER, K., and JUNG, H., 1968, *Int. J. Radiat. Biol.*, **14**, 493.
MEE, L. K., NAVON, G., and STEIN, G., 1965, *Biochim. biophys. Acta*, **104**, 151.
NEUFELD, J., and SNYDER, W. S., 1961, *Selected Topics in Radiation Dosimetry* (Vienna: International Atomic Energy Agency), p. 35.
PLATZMAN, R. L., 1952, *Symposium on Radiobiology*, Ed. J. J. Nickson (New York: John Wiley & Sons), p. 97.
SMITH, D. E., 1962, *Ann. Rev. nucl. Sci.*, **12**, 577.
SNIPES, W., and SCHMIDT, J., 1966, *Radiat. Res.*, **29**, 194.
ZIMMER, K. G., 1969, *Current Topics in Radiation Research*, Vol. 5, Eds M. Ebert and A. Howard (Amsterdam: North Holland), p. 1.

1970 CH. PART. TR. SOL. LIQ.

Physics of energetic particles moving through condensed matter

J. KISTEMAKER and J. B. SANDERS

FOM-Institute for Atomic and Molecular Physics, Kruislaan 407, Amsterdam/Wgm., The Netherlands

Abstract. Some formulae are derived and some numerical results given for the path lengths and penetration depths of projectiles in target materials. Various combinations of projectile and target species are considered. Also estimates are made of both elastic and inelastic stopping.

1. General aspects

If an energetic heavy ion moves through condensed matter, we can ask questions about the physical and chemical phenomena along its trajectory. We know that the stopping power $S(E)$ for quite fast particles is mainly determined by electronic processes, and for slow particles more and more by quasi-elastic collisions with atoms (Kistemaker *et al.* 1966). For a proton faster than 10 keV, the electronic processes dominate, whereas for a proton slower than 1000 eV the elastic collisions become important. For C, N or O ions the turning point is around 20 keV, and for Cu, Fe or fission products this will even be between 50 and 100 keV. The influence of both fundamental stopping mechanisms is demonstrated in figure 1.

The track length of an energetic particle moving through condensed matter is directly determined by $S(E)$, including both stopping mechanisms. The penetration depth or range has also a close connection with $S(E)$, but it is essentially shorter than the track length, due to scattering processes directly related with elastic collisions (nuclear stopping process).

Typical questions about tracks, like those on the spread in penetration depth, as well as on the spread perpendicular to the axis of primary impact, will be dealt with in section two.

If a fast ion experiences a so-called large angle scattering ($\Theta > 10°$), an appreciable amount of energy is transferred to the recoil particle. Particle trajectories sometimes show such bending points clearly visible to the eye. But much more frequent are bendings of the order of one degree; these are often underestimated in their importance in the energy decay process and in the radiation damage. A recoil particle with a kinetic energy of 500 to 1000 eV can easily throw ten other atoms out of position within a collision cascade volume of (30 Å)^3. A proton of 100 keV only needs to be deflected over an angle of less than one degree to create a recoil proton of 1000 eV. Such small zones of radiation damage, due to displaced atoms, are called collision cascade volumes; they appear like seeds along fast particle trajectories. At the end of a primary particle trajectory this collision cascading begins to exert more and more influence, as shown in figure 1. It will clearly make a difference if we consider radiation damage due to displaced atoms in pure metals, or in mixed crystals like mica, or worst of all in organic material.

Whereas a Cu atom in a hot collision cascade can easily find a place in which it can

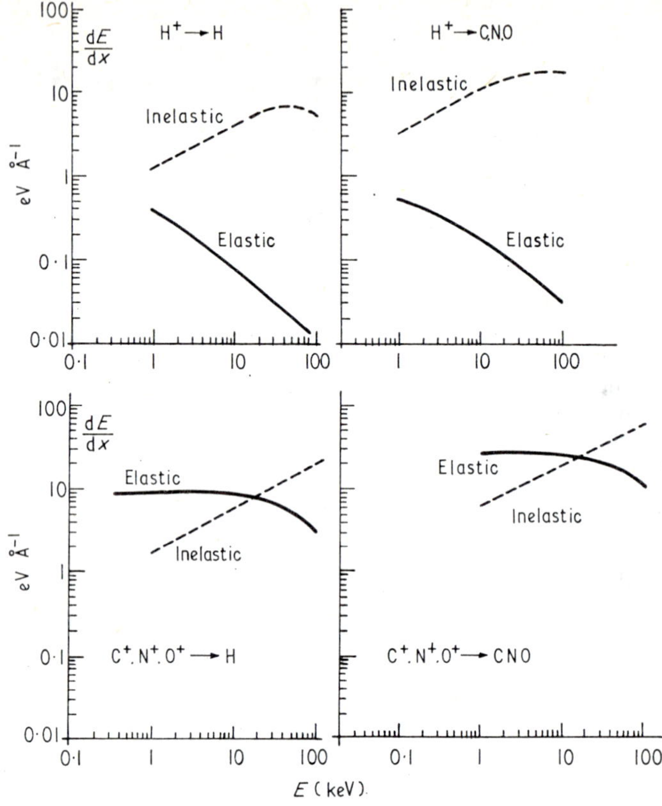

Figure 1. Stopping power (dE/dx) in eV Å^{-1}; target density 10^{23} atoms cm^{-3}. The inelastic curves *above* 9 keV were deduced from (dE/dx) measurements by Phillips, by Ormrod and Duckworth and by Sidenius. Elastic stopping powers calculated by N. Bohr and with Nielsen's approximation were subtracted from the experimental (dE/dx) values. Inelastic stopping powers *below* 9 keV were calculated with $S_e = ke^P$ starting from the deduced inelastic value at 9 keV. To obtain data for the inelastic stopping we always started with experimental values for total stopping, and then subtracted the elastic stopping powers as calculated with N. Bohr's and Nielsen's approximation (see Kistemaker *et al.* 1966).

be frozen in again, in a decent lattice position, this is much more difficult for oxygen atoms in a mica crystal, if pushed into interstitial locations. In complex organic materials, such atomic collision processes are completely irreversible, because of the small chance of restoring an original molecule like a protein or DNA. No wonder Kürzinger and Jung (1968) found extremely large cross sections for the inactivation of DNA under 1 keV proton bombardment (figure 2).

But even in the case of pure single crystal metals, bombarded by heavy ions like Ar^+ or Ne^+, the target can be turned into amorphous material quite easily, if the target temperature is below the annealing temperature (600 K for Ge and Si). In this case, displaced atoms cannot find a regular lattice position fast enough before the hot cascade material freezes again. Thus, local stress zones remain, as can be seen with an electron microscope (Noggle and Oen 1966) in a gold target (figure 3).

It seems most probable that the stress zones are the residues of the very hot recoil energy degradation zones, which have been found at the surface of bombarded metals by Nelson (1968). The lifetime of such thermal spikes is about 5×10^{-12} s for

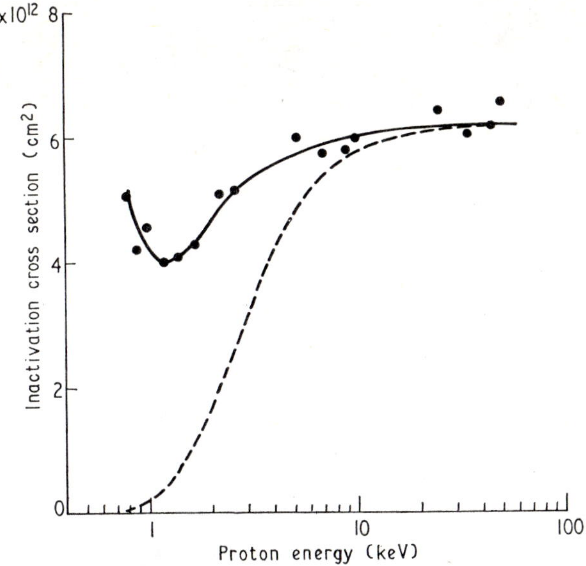

Figure 2. Cross section for the inactivation of dry infectious DNA of bacteriophage ΦX 174 by slow protons of various energies. Broken curve: cross section calculated under the assumption that inactivation is caused by ionization only. Investigation Kürzinger and Jung (1968).

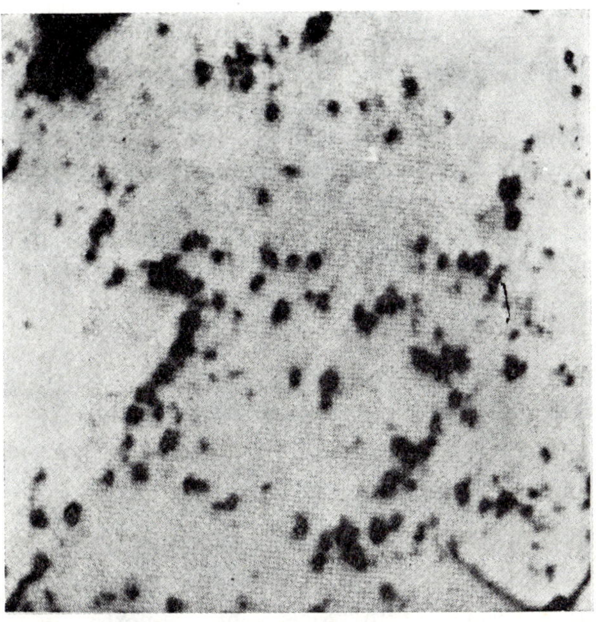

Figure 3(*a*). Black-spot defects produced by 51 MeV ^{127}I ion-bombardment of Au (after Noggle and Oen 1966).

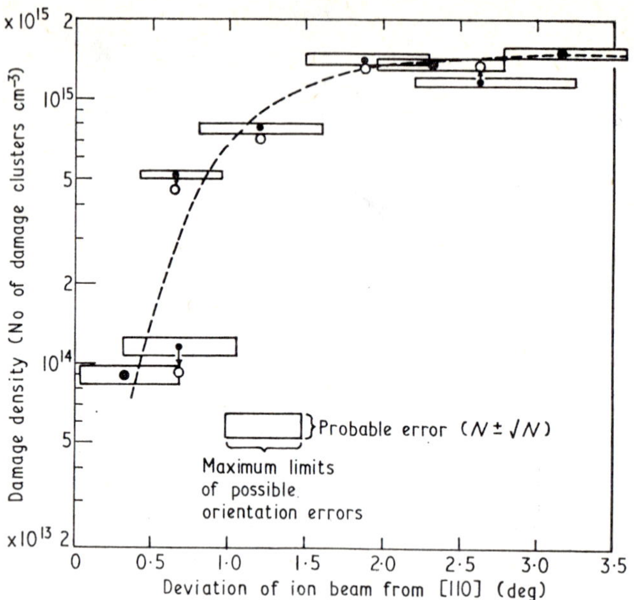

Figure 3(*b*). Variation of black-spot defect density as a function of angle of incidence about a ⟨110⟩ direction (after Noggle and Oen 1966). (●) measured orientation and density, (○) density normalized to $4{\cdot}26 \times 10^{10}$ ions cm^{-2}.

metals. Their sizes are of the order of $(100\ \text{Å})^3$ containing some 10^4 atoms at a temperature of 10^3 to 10^4 K.

It is remarkable that in the work of Noggle and Oen no real tracks are to be seen, although they used fission particles. In mica, however, very 'beautiful' fission tracks were photographed by Silk and Barnes (1959) (figure 4). Their tracks are about 100 Å wide and seem to be stress zones.

To understand this difference, one will have to reconsider the energy degradation process during the passage of the fast fission fragment as will be done in section 3.

Thermal spike zones, as visualized in figure 3, are like seeds along the trajectory at each angular scattering point of about 1° or more. The energy distribution function among the cascade particles determines the various atomic and molecular processes occurring in such a volume. Some calculations have been made by one of us (Sanders 1966) and the result can be seen in figure 5. At a former occasion (Kistemaker *et al.* 1966) we indicated that the radiation damage due to elastic collisions in collision cascades is equal to

$$\text{radiation damage} \sim N_{\text{targ}}\, V_{\text{cascade}}\,(E) \int_{E_{\min}}^{E} \sigma_{\text{reaction}}\,(\epsilon)\, \frac{dn}{d\epsilon}\, d\epsilon$$

It is this type of function that determines the radiation damage due to recoils observed by Jung.

The energy distribution function of the cascade particles is not to be confused with the energy distribution of primary beam particles. The latter is determined by the energy loss fluctuations as derived by Niels Bohr (see next section). Superimposed on such fluctuations is the influence of the regular structure of the target material. This can cause high energy tails in open directions, as demonstrated in figure 6.

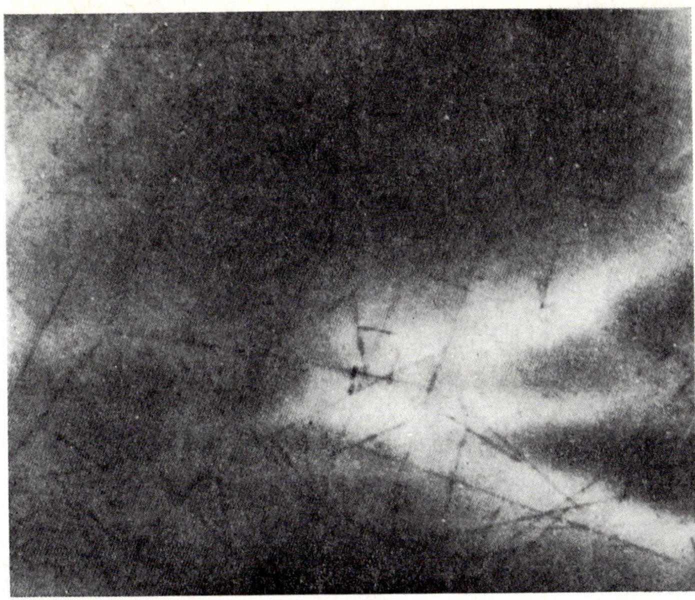

Figure 4. Fission tracks in mica, after Silk and Barnes (1959). This electron microscopic picture shows black lines with a width of 100 Å.

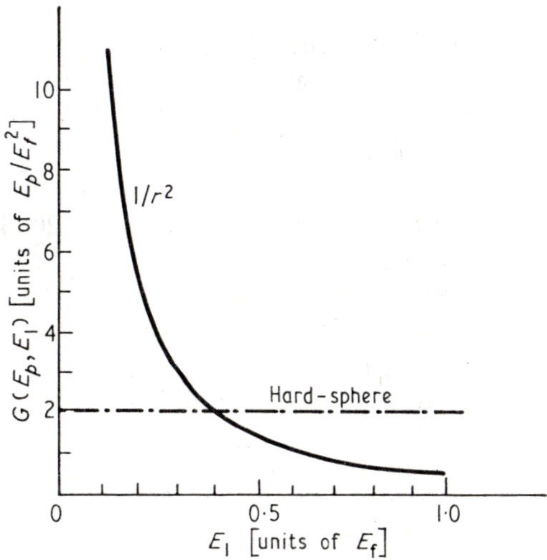

Figure 5. Distribution function in the collision cascade initiated by a primary recoil atom of energy E_p, as calculated by Sanders (1966).

In such open directions, the range can be a factor of 2 to 10 larger than in amorphous materials.

2. Elastic stopping power

An energetic particle which penetrates into matter will lose its kinetic energy in successive collisions with the atoms of this material. Consequently it traverses a path of finite length before being stopped. It is not difficult to calculate the mean length

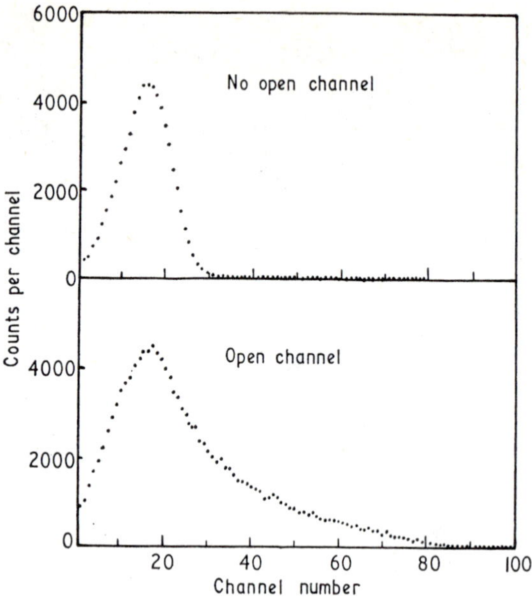

Figure 6. Energy spectra of 2 MeV protons transmitted through a thin Si crystal. Top curve shows the spectra obtained after transmission through a random direction; bottom curve shows the anomalous tail in the energy spectra which occurs when the beam was parallel to a ⟨110⟩ direction (after Dearnaley, 1964).

of this path if we assume a given form for the interaction potential between the projectile and the target atoms. We assume for this potential the form

$$V(r)=\frac{Z_1Z_2e^2}{r^2}\frac{k}{2}a \tag{1}$$

which for particles with kinetic energies in the region 5–200 keV is a fairly good approximation. Here a is a screening length

$$a=0{\cdot}8853a_0(Z_1^{2/3}+Z_2^{2/3})^{-1/2}$$

and k is a constant of the order 1 (we shall take it to be equal to 1 from now on). Lindhard *et al.* (1968) have derived an approximate form for the differential scattering cross section for such a potential. This form is

$$\mathrm{d}\sigma(E,T)=\frac{\pi}{2}\left[\frac{Z_1Z_2e^2}{m_0v^2}a\frac{\pi}{4}\right]T_m{}^{1/2}\frac{\mathrm{d}T}{T^{3/2}}=CE^{-1/2}\frac{\mathrm{d}T}{T^{3/2}} \tag{2}$$

where T is the energy transferred in the collision.

$$C=0{\cdot}8853a_0\left(\frac{\pi}{4}\right)^2 2e^2\left(\frac{Z_1{}^2Z_2{}^2m_1}{m_2(Z_1{}^{2/3}+Z_2{}^{2/3})}\right)^{1/2}$$

Further:

$$m_0=\frac{m_1m_2}{m_1+m_2} \quad \text{and} \quad T_m=\frac{4m_1m_2E}{(m_1+m_2)^2}=\gamma E$$

the maximum possible energy transfer in a collision. The index 1 always refers to the projectile; 2 to the target atoms. We restrict ourselves to two-body interactions only, and we assume only elastic collisions.

Let us consider the mean energy loss of the projectile per unit path length. This mean energy loss, also called the stopping power (figure 1), can be written as

$$\frac{\mathrm{d}E}{\mathrm{d}R}=N\int_0^{\gamma E} T\,\mathrm{d}\sigma(E, T)=NS(E)$$

where N is the number density of the target material.

The average pathlength, which a projectile traverses when it loses the amount of energy $\mathrm{d}E$, is then

$$\mathrm{d}R=\frac{\mathrm{d}E}{NS(E)}$$

and the mean path length traversed when the energy loss equals E_1, is

$$R(E, E_1)=\int_0^{E_1}\frac{\mathrm{d}E}{NS(E)} \tag{3}$$

It is very easy to calculate the stopping power for the r^{-2}-potential as introduced above

$$S(E)=\frac{\pi^2}{8}Z_1Z_2e^2a\left(\frac{m_1}{m_2}\right)^{1/2}E^{-1/2}\int_0^{\gamma E}\frac{\mathrm{d}T}{T^{1/2}}=2C\gamma^{1/2}$$

$$=\frac{\pi^2}{8}Z_1Z_2e^2a\,\frac{m_1}{m_1+m_2}. \tag{4}$$

Hence $R(E, E_1)=\frac{E_1}{2\mathrm{N}C\gamma^{1/2}}$, and the *total mean path length*:

$$\boxed{R(E)=\frac{E}{2NC\gamma^{1/2}}} \quad \text{and} \quad \boxed{\frac{\mathrm{d}E}{\mathrm{d}R}=2NC\gamma^{1/2}} \tag{5 and 6}$$

Hence, there is for this potential a constant mean energy loss along the path.

To give an impression of the magnitude of the quantities involved, we mention here that the stopping power, calculated in this way, of polycrystalline copper for copper projectiles is 3·23 keV Å^2. The number of Cu atoms per Å^3 is 0·086. Hence, the mean length of the track of a copper atom of 100 keV is

$$\frac{100}{3\cdot23\times0\cdot086}=360\text{ Å}$$

If $E=10$ keV the path length equals 36 Å, and so on for every energy.

It is also possible to calculate the fluctuation of the energy loss. N. Bohr has derived on the basis of simple statistical considerations that the mean square deviation in the energy loss over a small distance ΔR is

$$\boxed{\overline{\Delta E^2}-(\overline{\Delta E})^2=N\Delta R\int_0^{\gamma E}T^2\,\mathrm{d}\sigma=N\Delta RC\tfrac{2}{3}\gamma^{3/2}E} \tag{7}$$

Thus, the mean square fluctuation decreases with energy.

Beside the total track length of a projectile it is in principle also possible to calculate the distribution of the vectorial coordinates where the projectiles come to rest (Sanders 1968). Such a distribution is constructed on the bases of moments of the penetration depth of the projectiles and of the transverse extensions. A comparison between a distribution of penetration depths of ^{85}Kr projectiles of various energies in amorphous Al_2O_3 calculated in this way with the experimental results by Domey *et al.* (1964) turns out to be very satisfactory (figure 7).

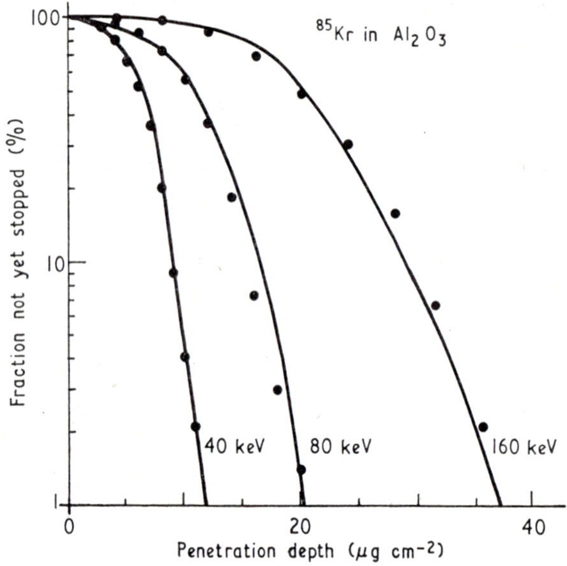

Figure 7. Comparison between a theoretical and an experimental distribution of the penetration depths for the case of ^{85}Kr ions of various energies in amorphous Al_2O_3 as measured by Domey *et al.* 1964. They measure the fraction of radioactive ^{85}Kr ions transmitted through layers of amorphous Al_2O_3 of variable thickness. The lines have been drawn through experimental points and the dots are calculated values for different penetration depths.

With the formalism developed by Sanders (1968) it is possible (Sigmund and Sanders 1967) to calculate for various mass ratios of projectiles and target atoms the following quantities:

(a) The mean depth of penetration $\bar{x}$, divided by E/NC.

(b) The mean square deviation of the penetration depth divided by the square of the average of this depth. The mathematical expression is $(\overline{x^2}-\bar{x}^2)/\bar{x}^2$.

(c) The mean square of the transverse extension (i.e. the distance of the point where the projectile has come to rest to the axis in the incoming direction) divided by the square of the mean penetration depth. This is denoted by $\overline{r^2}/x^2$.

Table

m_2/m_1	$(\bar{x}NC/E)$	$(\overline{x^2}-\bar{x}^2)/\bar{x}^2$	$\overline{r^2}/\bar{x}^2$
0·1	0·842	0·058	0·036
0·25	0·577	0·125	0·088
0·5	0·453	0·195	0·178
1	0·369	0·275	0·352
2	0·297	0·409	0·686
4	0·229	0·710	1·348
10	0·153	1·684	3·342

We see from this that the spread around the average of x and the transverse spread around the initial direction increase with increasing ratio m_2/m_1.

To give a numerical example we consider the following case. Let a Cu atom move through a Cu lattice with an initial kinetic energy $E=100$ keV. Then

$$m_2/m_1=1; \quad N=0{\cdot}086\ \text{Å}^{-3}$$

whereas

$$C=1{\cdot}614\ \text{kev Å}^2.$$

Then

$$\bar{x}=\frac{0{\cdot}369\times 100}{0{\cdot}086\times 1{\cdot}614}=266\ \text{Å}$$

whereas we found already that the mean track length was 360 Å.

$$(x^2-\bar{x}^2)/\bar{x}^2=0{\cdot}275$$

which means that there is a spread of 27·5% in the lengths of various tracks.

$$\overline{r^2}/\bar{x}^2=0{\cdot}352$$

which means that there is a cylindrical column with a radius of 35% of the mean penetration depth which is subject to direct radiation damage.

3. Inelastic stopping processes

We now consider the somewhat more realistic case, when the energetic particle loses its energy not only to target atoms in elastic collisions, but also to a 'free' electron gas such as exists in metals. 'Free' means all electrons of the atomic outer shells which contribute to the polarizability of the target material. We consider these two effects to be separable. Lindhard *et al.* (1968) have found that for projectiles in the keV range the contribution to the stopping power of the electron gas has the form

$$S_{\text{el}}(E)=kE^{1/2}$$

k being a proportionality constant. Recently there has been a good deal of discussion about the value of this constant. Ormrod and Duckworth (1963) have found experimentally that it oscillates with the charge number Z_1 of the projectile (see figure 8).

Due to this energy dependence the total stopping power is therefore not constant along the path of a fast particle. Usually the inelastic part dominates if $E>100$ keV. The method to find the electronic stopping power $S_{\text{el}}(E)$ is then to deduce the calculated elastic stopping power from the measured one. This method has also been used in figure 1.

The stopping power due to electrons has been measured in aluminium and in carbon by Ormrod and Duckworth (1963) for various projectiles. They found a periodic variation in the value of k with the charge number A_1 of the projectile. Theoretical calculations on this subject have been made by Bhalla and Bradford (1968) and Cheshire, Dearnaley and Poate (1968) (figure 8). Explicit values of k can be found in these papers and in the review by Bergström and Domey (1966).

The passage of fission products through dense matter is a subject of special interest (Silk and Barnes 1959, Datz 1965). Datz injected very high energy heavy ions (50 MeV) into open crystal directions and measured energy losses of the order of 1000 eV Å^{-1}, which is a very great energy deposition in the trajectory channel. No doubt, a small part of it goes into elastic collisions creating energetic recoil particles. We have calculated however that this part does not produce energy losses more than 5 eV Å^{-1}.

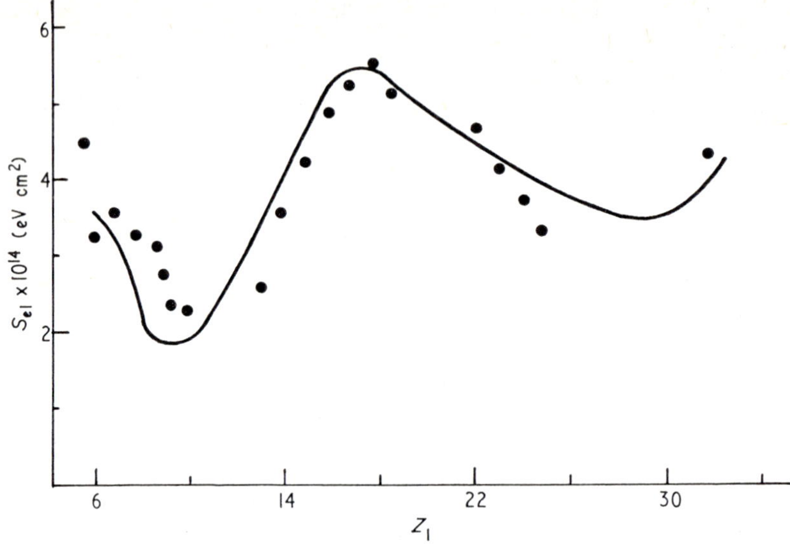

Figure 8. Electronic stopping power, S_{el}, in amorphous carbon as a function of the projectile atomic numbers, calculated by Bhalla and Bradford (1968). The experimental values of Ormrod *et al.* for $Z_1 \leqslant 20$ and of Hvelplund and Fastrup for $Z_1 \geqslant 21$ are also shown as black dots. The jump at $Z_1 = 21$ may be due to normalization difficulties in the two experiments. The velocity of all projectiles was $v = 9 \times 10^7$ cm s^{-1}.

Various other mechanisms of heating can be considered. One of these is the so-called 'shock wave' created in the electron gas due to the passage of a fast particle. Such shock waves cause the creation of so-called *plasmons* and are the origin of the Čerenkov radiation. The energy dissipated into a shock wave front is proportional to the square of the charge:

$$W = \frac{Z^2 e^2}{c^2} \int \omega [1 - c^2 / n^2(\omega)\, v_0^2]\, d\omega$$

integrated over the frequency range where $c < n(\omega)\, v_0$.

Schiff (1955) calculates for a high energy proton ($v_0 \sim c$) moving through water that about 200 photons of 5 eV each can be emitted per cm of track. This corresponds to 10^{-5} eV Å^{-1}, which is extremely small in comparison to the stopping powers shown in figure 1. For fission products this mechanism fails completely, as their speeds are too low ($v_0 < 10^9$ cm s^{-1}), and for a proton with equal speed ($\equiv 300$ keV) the creation of plasmons can be forgotten, because of the same fundamental reason.

This fact shows that the energy dissipation along the tracks goes mainly by discrete electronic excitations in atoms and molecules. From figure 1 we see that the value of $S(E)$ is of the order of 20 eV Å^{-1}, if the projectile only has a kinetic energy of about 100 keV. One has to consider already a complex of ionizations and excitations to understand this high number. But for fission products like Xe, Br or I, with equal speeds, an $S(E)$ of some 1000 eV Å^{-1} requires special consideration. One can only dissipate so much energy, if inner-shell excitations take place in the colliding atoms. For heavy particles this means excitation steps which can go up to the order of some 10 keV, but even for the lighter C, N or O atoms, K- or L-shell excitations require a few times 100 eV.

As the two nuclei have to approach each other at a distance of the order of 10^{-1} Å, the cross section for inner shell excitation processes is of the order of 10^{-18} cm^2 per atom. One has to compare this with normal ionization cross sections of 10^{-15} to 10^{-16} cm^2 per atom. As the energy per event in case of inner-shell excitation is a factor of 100 to 1000 larger than for normal ionizations, one sees that the importance of both excitation processes might be comparable.

If some 1000 eV is stored in a K- or L-shell electron excitation, this energy is released again in 10^{-14} to 10^{-15} s, through an Auger process in the electron system of the atom involved, producing a great number of free electrons. The energy distribution function of such electrons is known now, more or less. There is a very broad maximum between 0 and 10 eV, but also a very important high energetic tail and even discrete maxima corresponding to discrete energy steps in the quantized atomic electron system. These energetic electrons diffuse away from the primary particle trajectory. We know them from ion bombardment of metal surfaces, where they escape from the inside, through the metal surface and where they are known as 'secondary electrons'.

If fast electrons move away from a central track, they create a space charge. But in case of a conducting metal, this space charge is directly annihilated by cold conduction electrons. In an insulating material like mica, this is quite different, however, and perhaps the energetic electrons move apart as far as space charge permits them to do (some 10 eV). There, they come to a standstill and are frozen in immediately.

A simple calculation of the Debye screening length

$$[kT_e/4\pi n_e e^2]^{1/2}$$

taking $kT_e = 10$ eV and $n_e = 10^{20}$ gives about 100 Å diameter for the stress zone. This might be the explanation for the stress zone observed in the experiment of Silk and Barnes (figure 4).

References

Bergström, I., and Domey, B., 1966, *Nucl. Instrum. and Methods*, **43**, 146.

Bhalla, C. P., and Bradford, J. N., 1968, *Phys. Lett.*, **27A**, 318.

Cheshire, I. M., Dearnaley, G., and Ponte, J. M., 1968, *Phys. Lett.*, **27A**, 304.

Datz, S., Noggle, T. S., and Moak, C. D., 1965, *Phys. Rev. Lett.*, **15**, 254.

Domey, B., Brown, F., Davies, J. A., and McCargo, M., 1964, *Can. J. Physics*, **42**, 1624.

Hvelplund, P., and Fastrup, B., 1968, *J. Phys.*, **46**, 489.

Kistemaker, J., de Heer, F. J., Sanders, J., and Snoek, C., 1966, *Radiation Research*, Ed. G. Silini (Amsterdam: North Holland Publishing Company), p. 68.

Kürzinger, K., and Jung, H., 1968, *Int. J. Radiat. Biol.*, **14**, 493.

Lindhard, J., Nielsen, V., and Scharff, M., 1968, *Mat. fys. Med.* Dan. Vid Selsk, **36**, 10.

Lindhard, J., Scharff, M., and Schrøtt, H. E., 1963, *Mat. fys. Med.* Dan. Vid. Selsk, **33**, 14.

Nelson, R. S., 1968, *The Observation of Atomic Collisions in Crystalline Solids* (Amsterdam: North Holland), p. 270.

Noggle, T. S., and Oen, O. S., 1966, *Phys. Rev. Lett.*, **16**, 395.

Ormrod, J. H., and Duckworth, H. E., 1963, *Can. J. Phys.*, **41**, 1424.

Ormrod, J. H., MacDonald, J. R., and Duckworth, H. E., 1965, *Can. J. Phys*, **43**, 275.

Sanders, J. B., 1966, *Physica*, **32**, 2197.

—— 1968, *Can. J. Phys.*, **46**, 455.

Schiff, L. I., 1955, *Quantum Mechanics* (New York: McGraw-Hill), pp. 267–70.

Sigmund, P., and Sanders, J. B., 1967, *Proc. Int. Conf. on Semi-Conductor Technology*, Ed. P. Glotin, Editions Ophyrs, p. 215.

Silk, E. C. H., and Barnes, R. S., 1959, *Phil. Mag.*, **4**, 970.

The aggregation of *F*-centres in LiF*

A. VAN DEN BOSCH

Solid State Physics Department, S.C.K.-C.E.N., Mol (Belgium)

Abstract. In single crystals of lithium fluoride, which are irradiated at room temperature, the radiation damage phenomena can be understood on the basis of a model in which it is assumed that primary defects are created during irradiation. The formation of aggregates of *F*-centres, which are considered as primary defects, depends on the linear energy transfer (LET) of the radiation to the material. It is shown that the high probability for locating an *F*-centre adjacent to another one cannot be directly related to the high probability for energy deposition in tracks.

One of the problems in radiation chemistry, discussed by Drs G. S. Freeman and R. Schiller, is related to the distance at which the so-called primary species, in the tracks of charged particles in condensed media, interact with each other by processes wherein no normal diffusion of atomic or molecular entities is involved. From this point of view, existing data on the aggregation of *F*-centres in irradiated single crystals of lithium fluoride will be discussed here. The model for the *F*-centre, namely an electron trapped at a negative-ion vacancy, is firmly established (Holton Blum, and Slichter 1960). Such a primary defect gives rise to an optical absorption band. When the defect is isolated from the others, the maximum of the absorption is found for light, having a wavelength of about 0·25 μm. When two of these primary defects are located on adjacent lattice sites, their cluster is responsible for the *M*-band (at 0·45 μm) which is well-resolved from the former band. The aggregates of three primary defects, when adjacent to each other, are detected by the *R*-bands (at 0·32 and 0·38 μm). The sequence in the appearance of the bands in the optical absorption spectrum as the dose increases is *F*, *M* and *R*. This sequence can be explained on the basis that during irradiation only primary defects are formed. The aggregates are then built up by the newly-created primary defects which are located adjacent to existing ones. Assuming that, in a random process, every negative ion has the same probability to be substituted by a primary defect, then, in the low dose case (where the concentrations of *R*-centres and higher *F*-centre aggregrates can be neglected) one can say that

$$C_{\mathrm{M}} = K_{\mathrm{MF}} C_{\mathrm{F}}^2/2 \tag{1}$$

In this expression C_{M} represents the number of *M*-centres in the considered sample divided by the number of negative-ions which were originally in the specimen, C_{F} the fraction of the negative-ion lattice sites occupied by an isolated primary defect, and K_{MF} stands for the number of negative-ion sublattice sites per existing *F*-centre in which a newly formed defect gives rise to an *M*-centre.

The values of C_{i} (with i for *F* or *M*) are obtained from the absorption measurements. Discussing the work of Mador *et al.* on X-ray irradiated LiF, Faraday *et al.* (1961) found (K_{MF}) to be equal to about 4400. This value is much higher than 12, the number of nearest neighbour negative-ions about one *F*-centre. Therefore it was

* Work performed under the auspices of the association S.C.K.-R.U.C.A.

concluded that the location of newly-formed primary defects in LiF is not random. When one argues that this high probability of locating an *F*-centre adjacent to another is due to the structure of the track, i.e. when one attributes this probability to the non-homogeneous distribution of energy dissipation in the crystal, one should expect an even larger probability in tracks in which the linear energy transfer (LET) is still higher. This seems however not to be the case. In neutron-irradiated LiF, the primary defects are mostly created by the nuclear reaction

$$^{6}\mathrm{Li}(n, \alpha)^{3}\mathrm{H}+Q \qquad (2)$$

wherein the reaction energy $Q=Q_{\alpha}(2{\cdot}1\ \mathrm{MeV})+Q_{^3\mathrm{H}}(2{\cdot}7\ \mathrm{MeV})$. The range of the α-particle is estimated to be a few micrometers and that of tritium to be a few ten-micrometers. The average LET for the α and the ^{3}H-tracks in neutron-irradiated LiF may be estimated to be respectively 500 and 70 keV $\mu\mathrm{m}^{-1}$. Van den Bosch (1964) pointed out that in neutron-irradiated LiF the $(K_{\mathrm{MF}})_n$ value (in equation (1)) is about 1200. Consequently the probability for locating *F*-centres adjacent to another one is lower for higher LET radiation. The same LET dependence, as the one found for *M*-centre formation, is observed to occur in LiF for the *R*-centre formation

$$(K_{\mathrm{RM}})_{\gamma} > (K_{\mathrm{RM}})_n$$

For much higher radiation doses magnetic susceptibility data on γ and neutron-irradiated LiF single crystals can be understood on the basis of the same model in which it is assumed that the primary defects, which are created during the irradiation, are paramagnetic. The single electrons, associated with these defects, interact when they are sufficiently close to each other. In an oversimplified model, wherein only the negative-ion sublattice of LiF is taken into account, the pairing of electrons has been simulated by Monte-Carlo calculations (Van den Bosch and Schotsmans 1965). The electrons of two defects pair with each other when the newly created defect is designated for one of the *K* neighbouring lattice sites. In this case the value of *K* is somewhat lower than in the low dose case. However, its value is smaller in the neutron-irradiated than in the γ-irradiated one (Van den Bosch 1968). Assuming that a newly-created primary defect does not interact with an existing one when the latter is shielded by u_0 layers of negative ions, one can deduce from the structure of LiF that approximately

$$K_{u_0}=\sum_{u=1}^{u_0}(10u^{2}+2). \qquad (3)$$

At low irradiation doses one finds results $u_{0n}\simeq 7$ and $u_{0\gamma}\simeq 10$; for high doses $u_{0n}\simeq 5$ and $u_{0\gamma}\simeq 6$. In both cases the u_0 value of the neutron-irradiated simple is smaller than the value of the γ-irradiated specimen. In LiF, two negative-ion layers occur in the ⟨100⟩ direction per lattice constant (4 Å). Consequently, in this system without normal diffusion of *F*-centres at room temperature, one has to consider interaction distances between primary species of about 10 to 20 Å, in order to explain the experimental observations on the aggregation of *F*-centres in LiF.

References

Faraday, B. J., Rabin, H., and Compton, W. D., 1961, *Phys. Rev. Lett.*, **7**, 57.

Holton, W. C., Blum, H., and Slichter, C. P., 1960, *Phys. Rev. Lett.*, **5**, 197.

Van den Bosch, A., 1964, *J. Phys. Chem. Solids*, **25**, 1293.

—— 1968, *Abstr. Int. Symp. on Colour Centres in Alkali-halides* (Rome: Tipografia S. PioX), No. 191.

Van den Bosch, A., and Schotsmans, L., 1965, *J. phys. Chem. Solids*, **26**, 1215.

Spur reactions as studied by steady-state radiolysis

B. CERCEK and M. KONGSHAUG*

Paterson Laboratories, Christie Hospital and Holt Radium Institute, Manchester 20, England

Abstract. Hydrated electrons react with p-bromophenol to produce bromide ions. The yields of bromide ions and molecular hydrogen as a function of p-bromophenol concentration in deaerated aqueous solutions give information about spur reactions. These results suggest that four distinct types of spurs exist at times longer than about 10^{-9} s. Contrary to the predictions of the 'diffusion model' there are no spur reactions between 0·5 and 1 ns. The observed spur reactions do not contribute to the yields of the radiolytic transients in the bulk of the solution; the final fate of all the transients in the 'nanosecond-type' spurs is to react with each other. It follows from our results that the radiolytic transients leading to the 'bulk-yields' would be homogeneously distributed at times shorter than 0·5 ns.

1. Introduction

Nanosecond pulse radiolysis has demonstrated the existence of spur reactions at times longer than 1 ns (Thomas and Bensasson 1967). Recent results of a steady-state radiolysis study in which a stepwise increase in the hydrogen ion yields was observed (Cercek and Kongshaug 1969), were consistent with the existence of these spur reactions. This paper presents further evidence on spur reactions obtained from a radiolysis study of deaerated aqueous p-bromophenol solutions in which the bromide ion and molecular hydrogen yields were measured.

2. Experimental section

2.1. *Irradiation facilities*

An underwater ^{60}Co γ-source and a Vickers linear accelerator, giving 5 μs pulses of approximately 8 MeV electrons, were used for irradiations. Details of ^{60}Co irradiation facilities have been described by Keene and Law (1963). Doses in the range from 7000 to 25 000 rad and dose rates from 5 rad s^{-1} to 30 Mrad s^{-1} were used.

2.2. *Materials*

All solutions were prepared using distilled water redistilled from alkaline permanganate. The solutions were deaerated by bubbling with argon for at least 1 h before irradiation. The reagents were used without further purification. p-Bromophenol was Hopkin and Williams Limited 'General Purpose Reagent'; 2-propanol and sodium acetate were of the same origin, but of 'Analar Grade'.

* Fellow of the Norwegian Cancer Society. Permanent address: Norsk Hydro's Institute for Cancer Research, Montebello, Oslo 3, Norway.

2.3. *Analysis*

The yields of bromide ions were determined by potentiometric titrations using 5×10^{-3} or 10^{-2} N silver nitrate solutions. A silver indicator and calomel reference electrode were used and the changes of the electric potentials were measured with a PYE 'Dynacap pH Meter'. Its expandable scale allowed accurate readings of 1 mV. Corrections for the blank (unirradiated solution) were always applied, e.g. for a dose of about 15 000 rad, it was about 20% in 6×10^{-2}M p-bromophenol solution, decreasing proportionally with decreasing p-bromophenol concentration.

Molecular hydrogen was extracted, using the Van Slyke technique, from accelerator-irradiated solutions and its yield was measured by using a 'Perkin–Elmer' gas chromatograph model 452 (molecular sieve No. 5A/2412).

2.4. *Dosimetry*

The Fricke dosemeter (10^{-3} M ferrous sulphate in 0·8 N sulphuric acid solution containing 10^{-3} M NaCl, saturated with air) was used to measure the radiation doses. A value of 15·6 for $G(Fe^{3+})$ was used. In the accelerator irradiations a Faraday cup (abbr. F.C.) was used as a dose monitor in each exposure. For each set of experiments the F.C. was calibrated with a Fricke dosemeter.

3. Results

3.1. *Bromide ion yields*

The yields of bromide ions, $G(Br^-)$ are shown in figure 1A as a function of p-bromophenol concentration. There are four narrowly-defined stepwise increases in the bromide ion yields.

When over 99% of OH radicals and about 70% of the H radicals were scavenged with 2-propanol, the bromide ion yields were reduced by $\Delta G(Br^-) = 0{\cdot}45$ (figure 1B). The ratio of p-bromophenol over 2-propanol, $Q = 0{\cdot}01$, was constant over the whole range of p-bromophenol concentrations. The percentages of H and OH radicals scavenged were calculated using the following rate constants: k(H, OH + p-bromophenol) $= 7 \times 10^9$ M^{-1} s^{-1} (Anbar and Neta 1967, Cercek and Kongshaug 1969) and 2-propanol k(H + 2-propanol) $= 1{\cdot}5 \times 10^8$ M^{-1} s^{-1} (Rabani 1965, Cercek and Kongshaug 1969) and k(HO + 2-propanol) $= 3{\cdot}9 \times 10^9$ M^{-1} s^{-1} (Adams *et al.* 1965). To prevent back reactions between radiolytically produced hydrogen ions and hydrated electrons a (2·5 to 5) $\times 10^{-4}$ M concentration of sodium acetate was used to buffer the solutions. The yields of bromide ions were proportional to doses of radiation over the entire range of doses used, i.e. 7000 to 25 000 rads.

The effect of 2-propanol on the bromide ion yields in a 10^{-3} M p-bromophenol solution is shown in figure 2. After an initial decrease in the bromide yield of $\Delta G(Br^-) = 0{\cdot}45$ in 0·1 M 2-propanol solution, $G(Br^-)$ remains constant with increasing alcohol concentration. The plateau value of $G(Br^-)$ is $2{\cdot}60 \pm 2\%$.

3.2. *Molecular hydrogen yields*

As shown in figure 1C, the relative yields of molecular hydrogen, $G(H_2)/G(H_2)_0$, decrease with increasing p-bromophenol concentrations. $G(H_2)_0$ refers to the molecular hydrogen yield in 10^{-3} M p-bromophenol solution. The results suggest a stepwise decrease in $G(H_2)/G(H_2)_0$ in regions of p-bromophenol concentrations corresponding to those in which a stepwise increase in $G(Br^-)$ was observed in

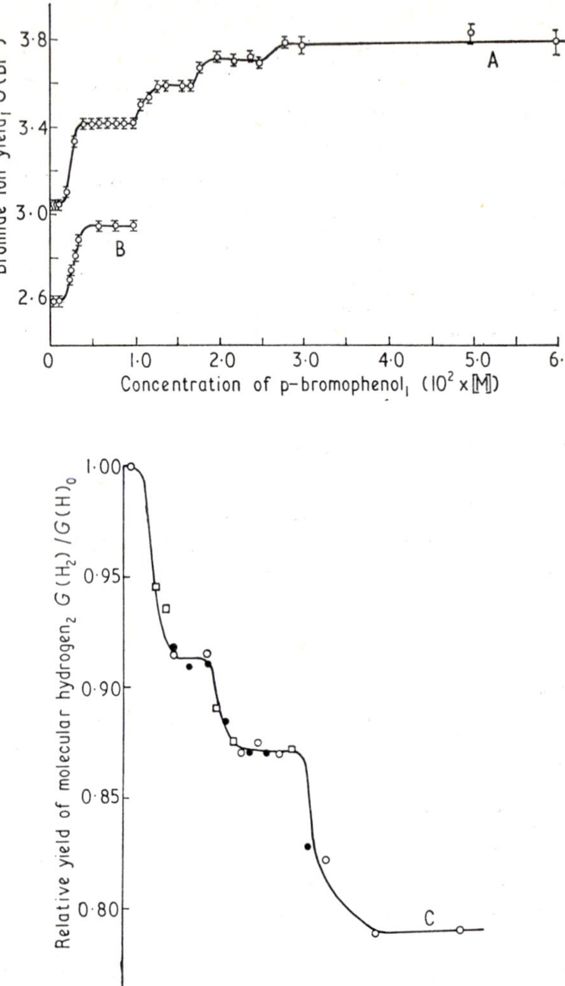

Figure 1. The effect of p-bromophenol on bromide and molecular hydrogen yields: A, p-bromophenol alone; B, In the presence of 2-propanol. p-Bromophenol over 2-propanol, ratio of concentrations, $Q=0{\cdot}01$; C, $G(H_2)_0$ refers to the molecular hydrogen yield in 10^{-3} M p-bromophenol solution. Each point represents the mean value of three or more parallel measurements. Date of experiments: ○—19.3.1969; ●—26.3.1969 and □—1.4.1969.

figure 1A. As in the bromide ion yield determinations, the solutions were also $(2{\cdot}5 \text{ to } 5)\times 10^{-4}$ M in sodium acetate in order to prevent the reaction between hydrated electrons and radiolytically-produced hydrogen ions.

3.3. *Accuracy and reproducibility of results*

One example of the reproducibility and accuracy in the potentiometric determinations of the bromide yields is illustrated in table 1. On the three days different batches of chemicals and solutions were used. It can be seen that the mean values of the bromide yields differ less than 1% and that the maximum deviations from the

Table 1. Reproducibility of bromide ion yields

Experiment No.	$3{\cdot}20\ 10^{-3}$ M PBP, 0·40 M isopropanol, $5{\cdot}10^{-4}$ M NaAC		$2{\cdot}50\ 10^{-2}$ M p-BP, $5{\cdot}10^{-4}$ NaAC		Date of experiment
	Irradiation time (min)	$G(Br^-)$	Irradiation time (min)	$G(Br^-)$	
1	81	2·83	67	3·64	
2	83	2·83	69	3·70	28.3.1969
3	104	2·81	87	3·70	
4	56	2·86	73	3·66	
5	60	2·80	78	3·69	29.3.1969
6	87	2·80	91	3·68	
7	41	2·82	104	3·70	
8	41	2·83	66	3·69	30.3.1969
9	35	2·83	—	—	

mean values are within 2%. The reproducibility of the molecular hydrogen determinations was tested by measuring the yields on different days. Fresh solutions were prepared each day. The results are illustrated in figure 1C. It can be seen that the reproducibility of the measurements was within 0·5 to 1%.

4. Discussion and conclusions

The hydrated electrons, e_{aq}^-, and H radicals react with p-bromophenol producing bromide ions, Br^-, (Anbar Alfassi and Bregman-Rice 1967), whereas OH radicals add to the benzene ring (Cercek and Kongshaug 1969). Thus the total yield of bromide ions produced equals the sum of hydrated electrons, e_{aq}^-, and H radicals, i.e. $G(e_{aq}^-)+G(H)$. The results in figure 1A indicate four definite, but limited increases in the bromide ion yields, $G(Br^-)$. These results are consistent with the scavenging of radiolytic transients which react with p-bromophenol to produce bromide ions, e.g. hydrated electrons, e_{aq}^-, and/or H radicals. Pulse radiolysis has revealed fast spur reactions of the hydrated electrons in deaerated water over a period of about 50 to 100 ns (Thomas and Bensasson 1967, Thomas 1968). The stepwise increases in the bromide ion yields in figure 1A occur over half-lives of the hydrated electron-p-bromophenol reaction from about 50 to 2 ns. We therefore believe our results to be consistent with spur reactions of the hydrated electrons as observed by nanosecond pulse radiolysis. Thomas (1968) reported that about 15% of the hydrated electrons decay in 'nanosecond' spur reactions. In agreement with this observation, we find that about 20% of hydrated electrons decay in these spurs, assuming that the overall increase in the bromide ion yield in figure 1A is due to hydrated electrons and that $G(e_{aq}^-)=2{\cdot}6$. For the first increase in the bromide ion yields, this assumption can be supported by the results in figure 1B. When practically all the OH radicals, about 70% of the H radicals, and none of the hydrated electrons ($k(e_{aq}^-+2\text{-propanol}) < 10^6\ M^{-1}\ s^{-1}$) were scavenged with 2-propanol, the bromide ion yields decreased by a constant amount of $\Delta G(Br^-)=0{\cdot}47$, but the stepwise increase in $G(Br^-)$ remained unaffected.

The decrease in the relative molecular hydrogen yield in figure 1C suggests that about 20% of the molecular hydrogen may be produced by 'nanosecond' spur reactions. It has been argued that less than 30% of the molecular hydrogen could be produced via hydrated electrons as precursors (Anbar and Meyerstein 1966). p-Bromophenol scavenges both hydrated electrons and H radicals; thus, although the observed 20% decrease in molecular hydrogen could be due to scavenging of the hydrated electrons, our results would also be consistent with hydrogen atoms as precursors of the molecular hydrogen.

2-Propanol is an efficient H and OH radical scavenger which is practically inert towards hydrated electrons ($k(e_{aq}^- + \text{2-propanol}) < 10^6\ M^{-1}\ s^{-1}$). The results in figure 2 are in agreement with these properties of 2-propanol. The addition of progressive amounts of 2-propanol to a 10^{-3} M p-bromophenol solution, decreases the yield of bromide ions by an amount not exceeding the yield of hydrogen atoms in the bulk of solution, i.e. $\Delta G(Br^-) = 0{\cdot}45$. It can be seen from figure 2 that a 0·1 M concentration of 2-propanol is enough to reduce the bromide ion yields to the plateau value of $G(Br^-) = 2{\cdot}60$. Contrary to this, it follows from the reactivity of 2-propanol towards H radicals ($k(H + \text{2-propanol}) = 1{\cdot}5 \times 10^8\ M^{-1}\ s^{-1}$), that 0·1 M 2-propanol should reduce the bromide ion yields only by about 70%. To account for the observed effect of 0·1 M 2-propanol, its reactivity towards H atoms in the bulk of the solution would have to be about $10^9\ M^{-1}\ s^{-1}$. This discrepancy in the reactivity of 2-propanol towards H radicals is being investigated. One would expect that the scavenging of OH radicals in the spurs would allow a larger fraction of the hydrated electrons to escape into the bulk of the solution and consequently the bromide ion yields to increase. This was not observed although the half-lives for the reactions of 2-propanol with OH radicals were in the nanosecond region in accordance with the 15 ns half-life for the hydrated electron spur reactions (Thomas and Bensasson 1967). Therefore competition between hydrated electrons and 2-propanol for the OH radicals should have occurred. A possible explanation for no increase in bromide ion yields might be that hydrated electrons which would have reacted with OH radicals react instead with the radiolytically produced hydrogen ions yielding hydrogen atoms which are scavenged by 2-propanol.

The observed steps in the bromide ion yields (figure 1A) and in the molecular hydrogen yields (figure 1C) are in sharp contrast to the continuous changes of the yields with increasing solute concentrations, generally predicted by the diffusion model (Kupperman 1967). This contrast suggests that the postulates of inhomogeneous diffusion controlled kinetics may not be applicable to the 'nanosecond-type' spur reactions. The relatively narrow concentration ranges over which the steps in the yields take place suggest the existence of at least four well characterized spurs with respect to hydrated electron and/or H radical reactivities. A similar sudden stepwise increase in the OH radical yields has been observed by Marketos, Rakintzis and Stein (1968); between pH 2 and 3 the destruction of Safranin T suddenly increased by $\Delta G(-\text{Safranin T}) = 0{\cdot}6$. It has been suggested that it is due to the protonation of a radiolytic transient, e.g. e_{aq}^-, which otherwise would have recombined with an OH radical. An analogous explanation for our steps in the bromide yields is the scavenging of hydrated electrons, e_{aq}^-, in the spurs, which otherwise would have reacted with hydrogen ions, $H_3O_{aq}^+$, or OH radicals and also H radicals reacting with OH radicals.

Generally it is believed that the radiolytic decomposition of water produces 'spurs' containing various numbers of radiolytic species. Our results in figure 1A suggest that four different types of spurs exist at times longer than about 10^{-9} s. It seems reasonable to assume that the first step in $G(Br^-)$ represents spurs with one equivalent of reducing species per spur, such as e_{aq}^- or H radical, and that each further step represents spurs having one unit of reducing equivalent more than the preceding ones. Using the results from figure 1A and assuming: (1) that there are no spur reactions at times less than 10^{-9} s (Hunt and Thomas 1967), (2) spurs with n reducing equivalents are converted into spurs with $n-1$ reducing equivalents; it can be shown that the initial number of spurs per 100 eV which contain 1, 2, 3 and 4 reducing equivalents per spur, $G(\text{SPURS})_n$, are 0·20, 0·05, 0·05 and 0·07, respectively. These G-values were calculated using the relationship

$$G(\text{SPURS})_n = \Delta G(\text{Br}^-)_n - \Delta G(\text{Br}^-)_{n+1}$$

where $\Delta G(\text{Br}^-)_n$ and $\Delta G(\text{Br}^-)_{n+1}$ denote the increase in the bromide yield for step n and $n+1$. If the above assumptions are a reasonable approximation, then the sum

$$\sum_{n=1}^{4} G(\text{SPURS})_n \times n$$

must be equal to the total number of reducing equivalents in this system, i.e.

$$\sum_{n=1}^{4} \Delta G(\text{Br}^-)_n = 0{\cdot}73 \pm 0{\cdot}5$$

In agreement we find

$$\sum_{n=1}^{4} G(\text{SPURS})_n \times n = 0{\cdot}73$$

(see Appendix).

The values of $G(\text{SPURS})_n$ were used to calculate the relative frequency of spurs, F_n, containing various numbers of reducing equivalents, n, by the following relationship

$$F_n = G(\text{SPURS})_n \Big/ \sum_{n=1}^{4} G(\text{SPURS})_n$$

The results are given in table 2. For comparison the distribution of spur sizes in water vapour according to Lea (1955) is included in table 2.

Table 2. Relative frequency of spurs

n	1	2	3	4		This work
Frequency, F_n	0·54	$0{\cdot}13_5$	$0{\cdot}13_5$	0·19		Liquid water
Ion pairs per spur, n	1	2	3	4	>4	Lea's data (1955)
Frequency	0·43	0·22	0·12	0·10	0·13	Water vapour

The above arguments also imply that the final fate of all the transients in these spurs is to recombine, i.e. there is no contribution from the 'nanosecond-type' spurs to the yields of radiolytic transients in the bulk of the solution. The effect of 2-propanol on the bromide yields in figure 2 is consistent with this conclusion. If the 'nanosecond-type' spurs would contribute to the yields of transients in the bulk of solution, then scavenging of H atoms or their precursors, e.g. e_{aq}^-, should decrease the bromide yields. Contrary to this, even in 1 M 2-propanol solutions the yield of bromide ions was the same as in the bulk of the solution, e.g. $G(\text{Br}^-) = 2{\cdot}60$.

The observation by Hunt and Thomas (1967) that no spur-reactions occurred between 0·5 and 1 ns, and also our results, suggest that the radiolytic transients leading to the 'bulk yields' may well be homogeneously distributed at times shorter than

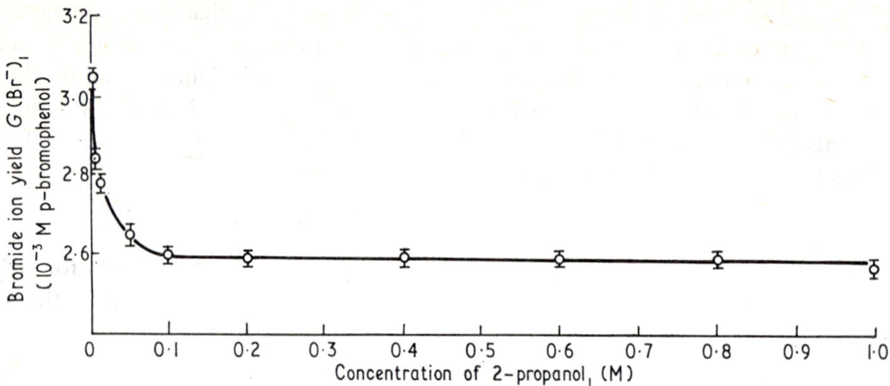

Figure 2. The effect of 2-propanol on bromide ion yields in 10^{-3} M p-bromophenol solution.

0·5 ns. The fast distribution of the radiolytic species may perhaps, at least in part, be accounted for by the mobility of nonhydrated (dry) e^- and positive holes. The evidence for the existence of nonhydrated species was recently discussed by Hamill (1968).

Acknowledgment

We are grateful to Dr M. Ebert for his help during the writing of this paper.

Appendix

$$G(\mathrm{SPURS})_n = \Delta G(\mathrm{Br}^-)_n - \Delta G(\mathrm{Br}^-)_{n+1} \quad (1)$$

The experimental values for $G(\mathrm{Br}^-)$ are

$$\Delta G(\mathrm{Br}^-)_1 = 0{\cdot}37, \quad \Delta G(\mathrm{Br}^-)_2 = 0{\cdot}17, \quad \Delta G(\mathrm{Br}^-)_3 = 0{\cdot}12$$

$$\Delta G(\mathrm{Br}^-)_4 = 0{\cdot}07 \text{ and } \Delta G(\mathrm{Br}^-)_5 = 0$$

From the above values and equation (1) it follows

$$G(\mathrm{SPURS})_1 = 0{\cdot}20,\ G(\mathrm{SPURS})_2 = 0{\cdot}05,\ G(\mathrm{SPURS})_3 = 0{\cdot}05 \text{ and } G(\mathrm{SPURS})_4 = 0{\cdot}07.$$

The sum of the reducing equivalents over all the spurs is

$$\sum_{n=1}^{4} G(\mathrm{SPURS})_n \times n = 0{\cdot}20 \times 1 + 0{\cdot}05 \times 2 + 0{\cdot}05 \times 3 + 0{\cdot}07 \times 4 = 0{\cdot}73$$

To be compared with the experimental value of

$$\sum_{n=1}^{4} \Delta G(\mathrm{Br}^-)_n = 0{\cdot}37 + 0{\cdot}17 + 0{\cdot}12 + 0{\cdot}07 = 0{\cdot}73$$

which we believe to be accurate to within 10%.

References

Adams, G. E., Boag, J. W., Currant, J., and Michael, B. D., 1965, *Pulse Radiolysis* (Eds. M. Ebert, J. P. Keene, A. J. Swallow, and J. H. Baxendale), (London and New York: Academic Press) p. 131.

Anbar, M., Alfassi, F. B., and Bregman-Rice, H., 1967, *J. Am. Chem. Soc.*, **89**, 1263.

Anbar, M., and Meyerstein, D., 1966, *Trans. Farad. Soc.*, **62**, 2121.

ANBAR, M., and NETA, P., 1967, *Int. J. appl. Radiat. Isotopes*, **18**, 493.
CERCEK, B., and KONGSHAUG, M., 1969, *J. Phys. Chem.* **73,** 2056.
HAMILL, W. H., 1968, *J. chem. Phys.*, **49**, 2446.
HUNT, J. W., and THOMAS, J. K., 1967, *Radiat. Res.*, **32**, 149.
KEENE, J. P., and LAW, J., 1963, *Physics Med. Biol.*, **8**, 83.
KUPPERMAN, A., 1967, *Radiation Research*, Ed. G. Silini (North Holland: Amsterdam) p. 214.
LEA, D. E., 1955, *Action of Radiations on Living Cells* (London: Cambridge University Press) p. 27.
MARKETOS, D. G., RAKINTZIS, N. TH., and STEIN, G., 1968, *Z. Phys. Chem.* (Frankfurt an Main), **59**, 177.
RABANI, J., 1965, *Adv. Chem. Ser.*, **50**, 242.
THOMAS, J. K., 1968, *Energetics and Mechanisms in Radiation Biology*, Ed. G. O. Phillips (London and New York: Academic Press), p. 114.
THOMAS, J. K., and BENSASSON, R. V., 1967, *J. chem. Phys.*, **46**, 4147.

Author Index

Subject Index

WITHDRAWN
University of Bristol
UNIVERSITY
OF BRISTOL
PHYSICS